# ANALYSIS OF TRIGLYCERIDES

This cross-and-circle symbol was used by medieval alchemists to designate olive oil as a distinctive chemical species (*302,519*). As such, it represents one of the first chemical nomenclatures for the compounds we now call triglycerides.

# ANALYSIS OF TRIGLYCERIDES

CARTER LITCHFIELD

Department of Biochemistry
Rutgers University
New Brunswick, New Jersey

1972

ACADEMIC PRESS    New York and London

Copyright © 1972, by Academic Press, Inc.
ALL RIGHTS RESERVED.
NO PART OF THIS PUBLICATION MAY BE REPRODUCED OR
TRANSMITTED IN ANY FORM OR BY ANY MEANS, ELECTRONIC
OR MECHANICAL, INCLUDING PHOTOCOPY, RECORDING, OR ANY
INFORMATION STORAGE AND RETRIEVAL SYSTEM, WITHOUT
PERMISSION IN WRITING FROM THE PUBLISHER.

ACADEMIC PRESS, INC.
111 Fifth Avenue, New York, New York 10003

*United Kingdom Edition published by*
ACADEMIC PRESS, INC. (LONDON) LTD.
24/28 Oval Road, London NW1

LIBRARY OF CONGRESS CATALOG CARD NUMBER: 72-77334

PRINTED IN THE UNITED STATES OF AMERICA

To

**Leslie Froomes**
**Tex Isbell**
**Woolsey Motl**
**Raymond Reiser**

my teachers in lipid research

# CONTENTS

*Preface* .................................................... xv
*Acknowledgments* ............................................ xvii

## 1. Introduction

    I. Triglyceride Molecules........................................ 2
        A. Nomenclature ........................................... 2
        B. Complexity of Triglyceride Mixtures ....................... 8
    II. History of Triglyceride Analysis............................. 9
        A. 1815–1955: Crystallization and Oxidation Techniques .......... 9
        B. 1956–1972: Chromatographic and Enzymatic Techniques ........ 14
    III. Applications of Triglyceride Analysis ....................... 15

## 2. Extraction, Isolation, Measurement, and Fatty Acid Analysis

    I. Extraction of Lipids......................................... 17
    II. Isolation of Triglycerides ................................... 19
        A. Column Chromatography on Florisil ........................ 19
        B. Column Chromatography on Silicic Acid .................... 20
        C. Thin-Layer Chromatography on Silicic Acid.................. 22
        D. Other Methods .......................................... 23
    III. Measurement of Total Triglyceride........................... 25
    IV. Fatty Acid Analysis ......................................... 31
        A. Methyl Ester Preparation ................................. 31

|   |   |
|---|---|
| B. Column | 32 |
| C. Identification of Peaks | 33 |
| D. Quantitation | 35 |

## 3. Preparation of Chemical Derivatives before Separation

|   |   |
|---|---|
| I. Reactions at Double Bonds | 38 |
|    A. Hydrogenation | 38 |
|    B. Permanganate Oxidation | 39 |
|    C. Ozonization | 42 |
|    D. Epoxidation | 43 |
|    E. Bromination | 44 |
|    F. Mercuration | 45 |
|    G. Other Reactions at Double Bonds | 46 |
| II. Reactions at Ester Linkages | 46 |
|    A. Estolide Ester Cleavage | 47 |
|    B. Interesterification | 47 |
| III. Reactions of Hydroxy, Epoxy, and Keto Groups | 47 |
|    A. Acetylation | 47 |
|    B. Trimethylsilyl Ethers | 48 |
|    C. Hydrazone Formation | 48 |

## 4. Silver Ion Adsorption Chromatography

|   |   |
|---|---|
| I. Methods | 50 |
|    A. Choice of Method | 50 |
|    B. Thin-Layer Chromatography | 50 |
|       1. Adsorbent | 50 |
|       2. Solvent | 51 |
|       3. Separation Procedure | 52 |
|       4. Quantitation | 53 |
|    C. Column Chromatography | 55 |
|       1. Adsorbent | 55 |
|       2. Solvent | 55 |
|       3. Separation Procedure | 55 |
|       4. Quantitation | 56 |
| II. Applications | 57 |
|    A. Separation by Number of *cis* Double Bonds | 57 |
|       1. Elution Order | 57 |
|       2. Thin-Layer Chromatography | 58 |
|       3. Column Chromatography | 60 |
|    B. Isomer Separations | 61 |
|    C. Other Functional Groups | 64 |
|    D. Oxidized Triglycerides | 64 |
|    E. Derived Diglycerides | 65 |

## 5. Liquid–Liquid Partition Chromatography

I. Methods .................................................... 68
   A. Choice of Method ........................................ 68
   B. Thin-Layer and Paper Chromatography ..................... 68
      1. Preparation of Plates and Paper ....................... 68
      2. Solvent System ....................................... 69
      3. Separation Procedure ................................. 72
      4. Quantitation ......................................... 74
   C. Column Chromatography ................................... 78
      1. Solid Support ........................................ 78
      2. Solvent System ....................................... 79
      3. Separation Procedure ................................. 79
      4. Quantitation ......................................... 80
   D. Countercurrent Distribution ............................. 81
      1. Apparatus ............................................ 81
      2. Solvent System ....................................... 81
      3. Separation Procedure ................................. 85
      4. Quantitation ......................................... 86

II. Applications ............................................... 86
   A. Separation by Partition Number .......................... 86
      1. Elution Order ........................................ 86
      2. Thin-Layer and Paper Chromatography .................. 90
      3. Column Chromatography ................................ 92
      4. Countercurrent Distribution .......................... 93
   B. Separation by Unsaturation .............................. 95
      1. $AgNO_3$-Containing Solvent Systems .................. 96
      2. Brominated Triglycerides ............................. 98
      3. Mercurated Triglycerides ............................. 100
   C. Oxidized Triglycerides .................................. 100
   D. Other Functional Groups and Derivatives ................. 100
      1. *Trans* Unsaturation ................................. 100
      2. Hydroxy Triglycerides ................................ 101
      3. Hydrogenated Triglycerides ........................... 101
   E. Derived Diglycerides .................................... 102

## 6. Gas–Liquid Chromatography

I. Methods .................................................... 105
   A. Apparatus ............................................... 105
   B. Column .................................................. 107
      1. Solid Support ........................................ 107
      2. Liquid Phase ......................................... 108
      3. Column Tubing Material ............................... 109
      4. Column Size .......................................... 111
      5. Single vs. Dual Columns .............................. 112
      6. Column Preparation ................................... 112
   C. Operating Conditions .................................... 114
      1. Sample Injection ..................................... 114

2. Carrier Gas .................................................. 114
3. Column Temperature ...................................... 115
4. Detector .................................................... 117
5. Optimum Operating Conditions ......................... 118
D. Quantitation ................................................. 118
1. Peak Identification ........................................ 118
2. Linearity of Detector Response ......................... 119
3. Calibration ................................................ 120
4. Accuracy .................................................. 124

II. Applications .................................................. 126
A. Separation by Carbon Number ............................ 126
B. Separation by Unsaturation ............................... 128
C. Separation of Isomers ..................................... 129
1. Isomeric Fatty Acids ..................................... 129
2. Triglyceride Positional Isomers .......................... 130
3. Triglyceride Chain-Length Isomers ..................... 131
D. Hydroxy and Epoxy Triglycerides ........................ 131
E. Oxidized Triglycerides .................................... 132
F. Preparative Separations ................................... 133
G. Radioisotope Detection ................................... 135
H. Derived Diglycerides ...................................... 137

## 7. Fractional Crystallization

I. Methods ...................................................... 140
A. Solvent .................................................... 140
B. Procedure ................................................. 141

II. Applications .................................................. 142
A. Separation by Number of Saturated Acyl Groups ....... 142
1. Solubility Considerations ................................ 142
2. Crystallization Sequence ................................ 143
3. Calculation of Triglyceride Composition ............... 145
B. Separation by Number of Double Bonds ................ 146
C. Oxidized Triglycerides .................................... 147
D. Other Derivatives ......................................... 148

## 8. Other Separation Techniques

I. Silicic Acid Adsorption Chromatography ..................... 150
A. Methods ................................................... 150
B. Separation by Molecular Weight ......................... 150
C. Separation by Unsaturation ............................... 153
D. Triglyceride Positional Isomers ........................... 153
E. Oxygenated Triglycerides ................................. 154
F. Brominated Triglycerides ................................. 158
G. Mercurated Triglycerides ................................. 158
H. Derived Diglycerides ...................................... 159

| | |
|---|---:|
| II. Florisil Adsorption Chromatography | 160 |
| III. Aluminum Oxide Adsorption Chromatography | 161 |
| IV. Charcoal Adsorption Chromatography | 162 |
| V. Paper Chromatography | 162 |
| VI. Ion-Exchange Chromatography | 162 |
| VII. Permeation Chromatography | 163 |
| VIII. Thermal Gradient Chromatography | 164 |
| IX. Distillation | 165 |

## 9. Partial Deacylation Reactions

| | |
|---|---:|
| I. Chemical Deacylation Methods | 168 |
|    A. Grignard Reagents | 168 |
|       1. Diglyceride Products | 169 |
|       2. Monoglyceride Products | 172 |
|    B. Other Reagents | 172 |
| II. Enzymatic Deacylation Methods | 173 |
|    A. Pancreatic Lipase | 173 |
|       1. Enzyme | 174 |
|       2. Reaction Conditions | 175 |
|       3. Specificity | 176 |
|       4. Diglyceride Products | 179 |
|       5. Monoglyceride Products | 179 |
|       6. Free Fatty Acid Products | 182 |
|    B. Milk Lipase | 183 |
|       1. Diglyceride Products | 184 |
|       2. Monoglyceride Products | 184 |
|    C. *Rhizopus arrhizus* Lipase | 184 |
|    D. *Geotrichum candidum* Lipase | 185 |
|       1. Diglyceride Products | 186 |
|       2. Monoglyceride Products | 187 |
|    E. Other Lipases | 187 |

## 10. Sterospecific Analysis

| | |
|---|---:|
| I. Methods | 188 |
|    A. *sn*-1,2(2,3)-Diglyceride Method of Brockerhoff | 189 |
|    B. *sn*-1,3-Diglyceride Method of Brockerhoff | 190 |
|    C. Method of Lands | 192 |
|    D. Choice of Method | 194 |
|    E. Deacylation of Triglycerides to Representative Diglycerides | 195 |
|    F. Phosphorylation of Diglycerides | 196 |
|       1. Chemical Synthesis | 196 |
|       2. Enzymatic Synthesis | 197 |
|    G. Hydrolysis of Phospholipid with Phospholipase A | 198 |

|       II. Applications .................................................. | 199 |
|---|---|
|          A. Positional Distribution of Fatty Acids ....................... | 199 |
|          B. Composition of Triglyceride Mixtures ........................ | 200 |
|          C. Composition of Derived Diglycerides ....................... | 205 |

## 11. Physical Properties

|       I. Mass Spectrometry ............................................. | 206 |
|---|---|
|          A. Pure Triglycerides ........................................... | 207 |
|          B. Natural Triglyceride Mixtures .............................. | 209 |
|          C. Derived Diglycerides ....................................... | 211 |
|      II. Melting Point .................................................. | 213 |
|     III. Differential Cooling Curves .................................... | 216 |
|      IV. Infrared Spectroscopy .......................................... | 218 |
|       V. X-Ray Diffraction ............................................. | 221 |
|      VI. Nuclear Magnetic Resonance .................................. | 222 |
|     VII. Rotation of Polarized Light .................................... | 224 |
|          A. Triglycerides ................................................ | 224 |
|          B. Derived Diglycerides ....................................... | 227 |
|    VIII. Piezoelectric Effect ............................................ | 230 |
|      IX. Other Physical Constants ..................................... | 230 |

## 12. Distribution of Fatty Acids in Natural Triglyceride Mixtures

|       I. Positional Distribution Patterns ............................... | 234 |
|---|---|
|          A. Plant Triglycerides ......................................... | 234 |
|             1. Palmitic, Stearic, $C_{20}$, $C_{22}$, and $C_{24}$ Acids ................... | 234 |
|             2. Oleic, Linoleic, and Linolenic Acids ...................... | 236 |
|             3. Other Acids .............................................. | 241 |
|          B. Animal Triglycerides ....................................... | 243 |
|             1. $C_{20}$ and $C_{22}$ Acids .......................................... | 243 |
|             2. Other Acids .............................................. | 246 |
|      II. Triglyceride Composition Patterns .............................. | 248 |
|          A. 1-Random–2-Random–3-Random Hypothesis ................ | 249 |
|          B. 1,3-Random–2-Random Hypothesis ........................ | 250 |
|          C. Other Fatty Acid Distribution Hypothesis .................... | 251 |
|             1. 1,2,3-Random Hypotheses ............................... | 251 |
|             2. Restricted Random Distribution ......................... | 252 |
|             3. Even or Widest Distribution ............................. | 252 |
|             4. Minor Fatty Acid Distribution Hypotheses ................ | 253 |
|          D. Validity of Distribution Hypotheses ........................ | 253 |
|             1. Nonhomogeneous Origin of Many Natural Fats ........... | 253 |
|             2. Comparison of Experimental and Predicted Triglyceride Compositions ............................................. | 255 |
|     III. Biosynthesis of Triglycerides .................................... | 261 |

## 13. Combining Methods for Detailed Analysis of Complex Triglyceride Mixtures

I. Combining Triglyceride Analysis Techniques .................. 266
   A. Separation Techniques ................................. 266
   B. Positional Analysis Techniques ......................... 270

II. Use of Derived Diglycerides of Analysis for Unresolvable Triglyceride Mixtures ..................................... 271

III. Maximum Analysis of Complex Triglyceride Mixtures .......... 272

IV. Major Unsolved Problems of Triglyceride Analysis ............ 280

## References

*Author Index* ................................................ 311
*Subject Index* ............................................... 332

# PREFACE

> Perhaps in no field of study have the tools of the lipid chemist been more inadequate for the task presented (than) in the study of glyceride structure.*
>
> H. J. Dutton

Dutton wrote these words in 1955 as he announced the first successful resolution of a natural triglyceride mixture (linseed oil) by countercurrent distribution. How true they were then! From Chevreul's initial characterization of natural fats as glyceryl esters of fatty acids in 1815 up until 1955 only two relatively unsophisticated techniques were available for determining the triglyceride composition of natural fats: fractional crystallization and permanganate oxidation. Because of the limitations of these methods, natural fats were usually thought of as mixtures of trisaturated, disaturated–monounsaturated, monounsaturated–diunsaturated, and triunsaturated types of triglycerides rather than as made up of individual molecular species as is truly the case.

Between 1955 and the present, the powerful new analytical techniques of silver ion adsorption chromatography, liquid–liquid partition chromatography, gas–liquid chromatography, pancreatic lipase hydrolysis, and stereospecific analysis were introduced and have completely revolutionized the methodology of the field. Using these new tools, it is now possible to distinguish individual molecular species of triglycerides such as $sn$-1-palmito-2-oleo-3-stearin and $sn$-1,2-dipalmito-3-olein in cocoa butter.

---

* H. J. Dutton, *J. Amer. Oil Chem. Soc.* **32**, 652 (1955).

This rapid growth of a new and complex methodology brings with it the need to decide which techniques can best accomplish a given type of analysis for a specific research problem. This monograph was written to provide a comprehensive reference source for those seeking more information on the subject. All reported methods are discussed, and the relative merits and limitations of each are evaluated. Numerous illustrations of practical examples are provided, and applications of both individual techniques and appropriate combinations are described.

Carter Litchfield

# ACKNOWLEDGMENTS

No one writes a first book without realizing that there is more work to bookwriting than he originally thought. It has been my good fortune to have numerous friends and co-workers who have made the task much easier.

I am especially grateful to Frank Gunstone and to Carol Litchfield who read the entire first draft of the manuscript and made many helpful suggestions for its improvement. I also extend my personal thanks to Bob Ackman, Hans Brockerhoff, Bill Christie, Mike Coleman, Earl Hammond, Bob Harlow, Bob Jensen, Fred Padley, Madhu Sahasrabudhe, A. G. Vereshchagin, and Herbert Wessels who reviewed individual chapters in their own areas of specialization. Their advice and comments proved invaluable.

Linda Fisher aided immensely with the editorial work in assembling the final copy; and Diane Cranfield, Anne Greenberg, Susan West, Karen Whitworth, and Dolores Young helped in typing and proofreading the manuscript at various stages in its progress.

To all of these helpful friends and to the many others who answered my innumerable questions along the way, I extend my hearty thanks.

# 1

## INTRODUCTION

The vital role of triglycerides in human life and activities is familiar to almost all who read these lines. Triglycerides are a major form of energy storage for both plants and animals. Man draws upon these sources to provide fatty foods for himself and to obtain fats and oils as industrial raw materials. To better understand these biosynthetic, metabolic, and technological processes involving triglycerides, chemists have developed numerous analytical techniques for characterizing complex triglyceride mixtures.

Two factors make the analytical chemistry of natural fat triglycerides exceptionally difficult: (i) the extremely large number of possible molecular species (Section I,B), and (ii) the very similar chemical and physical properties of most of these molecules. Using the classical techniques of fractional crystallization and permanganate oxidation, only simple separations of groups of triglycerides were possible, and most analyses were semiquantitative in nature. Between 1956 and 1965, however, a series of new chromatographic and enzymatic techniques revolutionized the field, and many of the earlier difficulties have now been overcome. With this proliferation of analytical methods, the former question, "Can I analyze for XYZ triglyceride content?" has now changed to, "Which method should I use to analyze for XYZ triglyceride content?"

The purpose of this monograph is to provide a comprehensive and critical review of the entire field of triglyceride analysis so that the reader can select the best technique or techniques for solving his own specific problem. By devoting an entire book to the subject at a time when the field has reached considerable maturity, triglyceride analysis can now be viewed with a broader perspective than was possible in earlier review papers *(186,240,365,494,550,585,700,898,909)*. It is assumed that the reader is already familiar with the fundamental chemistry of fatty acids and the basic techniques of organic analysis. Therefore, discussions in this book will center on the types of analyses possible and the specific operating conditions necessary when dealing with certain types of triglyceride molecules and their derived diglycerides. Particular emphasis is placed on the experimental details of such work.

Analytical techniques for triglyceride analysis are conveniently subdivided into those for sample preparation (Chapters 2 and 3), molecular fractionation (Chapters 4–8), and positional analysis (Chapters 9–11). Since the analysis of derived diglycerides is an integral part of positional analysis, diglyceride characterization procedures are also covered in Chapters 4–11. Chapter 12 describes the various fatty acid distribution theories for estimating the composition of natural triglyceride mixtures. Finally, Chapter 13 outlines useful combinations of analytical techniques for obtaining maximum compositional information.

## I. TRIGLYCERIDE MOLECULES

### A. Nomenclature

A proper nomenclature for triglycerides must accurately describe the myriad ways in which three fatty acids, either alike or different, can be esterified to glycerol. A number of different systems have been proposed to meet this need, and several are in current use.

Any discussion of triglyceride nomenclature must begin with an understanding of the distinctive stereochemical nature of glycerol. By itself, glycerol is a completely symmetrical molecule.* However, if only one of

$$\begin{array}{c} \text{H} \\ | \\ \text{H}-\text{C}-\text{OH} \\ | \\ \text{HO}-\text{C}-\text{H} \quad \text{----- } plane\ of\ symmetry \\ | \\ \text{H}-\text{C}-\text{OH} \\ | \\ \text{H} \end{array}$$

the primary hydroxyl groups is esterified or if the two primary hydroxyls

---

* The two ends of the glycerol molecule are not stereochemically identical in many enzymatic reactions, however. The $sn$-1- and $sn$-3-hydroxyls are easily distinguished when the molecule forms a three-point attachment to any surface *(754)*. This stereochemical nonidentity of the two —$CH_2OH$ groups is demonstrated by glycerol kinase, which only esterifies phosphate at the $sn$-3-position.

## I. TRIGLYCERIDE MOLECULES

are esterified to different acids, then the plane of symmetry is destroyed, and the central carbon atom acquires chirality. Therefore, an unambiguous

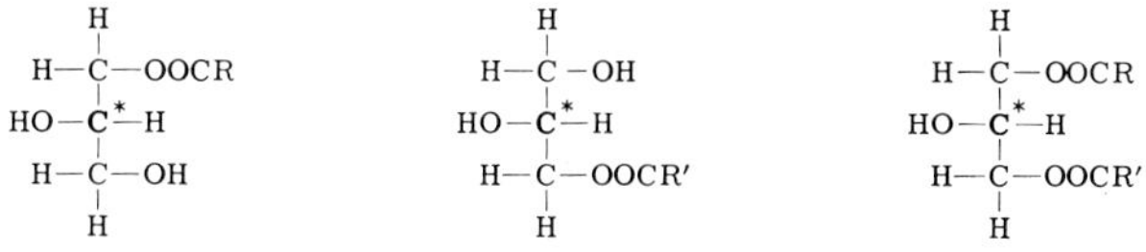

$$
\begin{array}{ccc}
\text{H} & \text{H} & \text{H} \\
| & | & | \\
\text{H}-\text{C}-\text{OOCR} & \text{H}-\text{C}-\text{OH} & \text{H}-\text{C}-\text{OOCR} \\
| & | & | \\
\text{HO}-\text{C}^*-\text{H} & \text{HO}-\text{C}^*-\text{H} & \text{HO}-\text{C}^*-\text{H} \\
| & | & | \\
\text{H}-\text{C}-\text{OH} & \text{H}-\text{C}-\text{OOCR}' & \text{H}-\text{C}-\text{OOCR}' \\
| & | & | \\
\text{H} & \text{H} & \text{H}
\end{array}
$$

C* = *an asymmetric carbon atom*

convention is needed for numbering the three hydroxyl groups so that the attachment of specific acids at specific hydroxyls can be clearly designated. Since both ends of the molecule are —CH$_2$OH groups, any numbering convention is essentially arbitrary.

The convention of Hirschmann (*381*) has now been universally adopted for numbering the three hydroxyl groups of glycerol. If the central carbon atom of the glycerol molecule is viewed with the C—H bond pointing away from the viewer, then each of the three remaining bonds leads to an hydroxyl group (Fig. 1-1). Hirschmann has proposed that the three hydroxyl groups viewed in this manner be numbered in clockwise order, with the 2-position already defined as the hydroxyl attached directly to the central carbon atom. This is equivalent to a standard Fischer projection in which the middle hydroxyl group is located on the left side of the glycerol carbon chain (Fig. 1-2). A more simple view of the same concept is to state that all triglycerides are named as derivatives of L-glycerol and that

FIG. 1-1. Schematic diagram illustrating the Hirschmann stereospecific numbering convention (*381*) for the three hydroxyl groups of glycerol. The central carbon atom of the glycerol molecule is viewed with the C—H bond pointing away from the viewer. The three remaining bonds then lead to the hydroxyl groups, which are numbered in clockwise order with the *sn*-2-position already defined as the hydroxyl attached directly to the central carbon atom.

$$
\begin{array}{ll}
\text{CH}_2\text{OH} & sn\text{-1-position} \\
\text{HO}-\text{C}-\text{H} & sn\text{-2-position} \\
\text{CH}_2\text{OH} & sn\text{-3-position}
\end{array}
$$

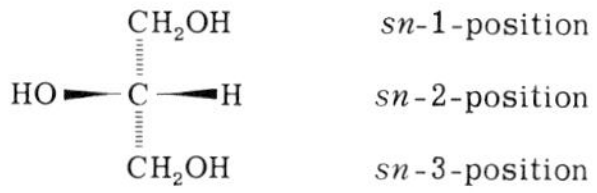

FIG. 1-2. Stereospecific numbering convention applied to the usual Fischer planar projection of glycerol. When the middle hydroxyl group is located on the left side of the glycerol carbon chain, then the carbon atoms are numbered 1 to 3 in the conventional top-to-bottom sequence.

the carbon atoms are numbered in the conventional top-to-bottom sequence. The prefix "*sn-*" (for stereospecifically numbered) is included in the names of all glycerol compounds in which the Hirschmann numbering convention is used (*406*). This *sn-* nomenclature is preferred over the conventional D and L or *R* and *S* notations, since it can describe the stereochemistry of glycerolipid reactions in the most simple and unambiguous manner (*406*).

A number of other prefixes are also commonly used to designate the positioning of substituents in glycerides: "*α-*" refers to the two primary hydroxyl groups, the *sn*-1- and *sn*-3-positions; "*β-*" designates the secondary hydroxyl group, the *sn*-2-position; "*rac-*" (for racemic) precedes the names of glycerides which are equal mixtures of two enantiomers. When no prefix or "*X*" is used, then the positioning of substituents is either unknown or unspecified.

The various systems of triglyceride nomenclature in current use are listed and illustrated in Table 1-1. The alcohol–acid, simplified, and abbreviated systems have received the widest usage and have been adopted throughout this book.

The abbreviated system of triglyceride nomenclature merits a detailed explanation, since it is extensively employed in this book to avoid the use of lengthy systematic names. The abbreviated system is based first of all on the standard letter and number fatty acid abbreviations listed in Table 1-2. Triglyceride abbreviations are then formed by combining the appropriate fatty acid abbreviations in groups of three (Table 1-3). The positioning of the fatty acids within the triglycerides is indicated by the presence or lack of a prefix. An "*sn-*" prefix specifies that the *sn*-1-, *sn*-2-, and *sn*-3-positions are listed in order, thus identifying a single molecular species. A "*rac-*" prefix indicates that the middle fatty acid in the abbreviation is attached to the *sn*-2-position, but the remaining two acids are equally divided between the *sn*-1- and *sn*-3-positions, producing a racemic mixture of two enantiomers. A "*β-*" prefix designates that the middle fatty acid in the abbreviation is esterified at the *β-* or *sn*-2-position and that the positioning of the other two acids on the *sn*-1- and *sn*-3-positions is unknown. Thus "*β-*" specifies a mixture of the two enantiomers in any proportion. The lack of a prefix indicates that all positional isomers that may exist are being referred to.

The complexity of natural triglyceride mixtures has prompted classification of the many molecular species into simple groups according to the kinds of fatty acids they contain. Widely used terms of this nature include:

*Monoacid.* Triglyceride molecules containing only one fatty acid (triolein, trioctanoin, etc.).

## TABLE 1-1
### NOMENCLATURE OF TRIGLYCERIDES

| System of nomenclature | Monoacid | Diacid | Triacid |
|---|---|---|---|
| | $\begin{array}{c} H \\ | \\ H-C-OOC(CH_2)_{12}CH_3 \\ | \\ CH_3(CH_2)_{12}COO-C-H \\ | \\ H-C-OOC(CH_2)_{12}CH_3 \\ | \\ H \end{array}$ | $\begin{array}{c} H \\ | \\ H-C-OOC(CH_2)_7CH=CH(CH_2)_5CH_3 \\ | \\ CH_3(CH_2)_8COO-C-H \\ | \\ H-C-OOC(CH_2)_7CH=CH(CH_2)_5CH_3 \\ | \\ H \end{array}$ | $\begin{array}{c} H \\ | \\ H-C-OOC(CH_2)_{14}CH_3 \\ | \\ CH_3(CH_2)_7CH=CH(CH_2)_7COO-C-H \\ | \\ H-C-OOC(CH_2)_{16}CH_3 \\ | \\ H \end{array}$ |
| Alcohol-acid | Glycerol trimyristate | $sn$-Glycerol-2-decanoate-1,3-dipalmitoleate | $sn$-Glycerol-1-palmitate-2-oleate-3-stearate |
| Acid-alcohol | Trimyristoylglycerol | 2-Decanoyl-1,3-dipalmitoleoyl-$sn$-glycerol | 1-Palmitoyl-2-oleoyl-3-stearoyl-$sn$-glycerol |
| Acid-alcohol with $O$-designation | Tri-$O$-myristoylglycerol | 2-$O$-Decanoyl-1,3-$O$-dipalmitoleoyl-$sn$-glycerol | 1-$O$-Palmitoyl-2-$O$-oleoyl-3-$O$-stearoyl-$sn$-glycerol |
| Simplified | Trimyristin | $sn$-2-Decano-1,3-dipalmitolein | $sn$-1-Palmito-2-oleo-3-stearin |
| Abbreviated | MMM | $\beta$-PoDPo | $sn$-POSt |

## TABLE 1-2
### Standard Abbreviations for Fatty Acids

**Letters**

| | | | |
|---|---|---|---|
| A | Azelaic | N | Nervonic |
| Ac | Acetic | O | Oleic |
| Ad | Arachidic | Oc | Octanoic |
| An | Arachidonic | P | Palmitic |
| B | Butyric | Pe | Petroselinic |
| Be | Behenic | Po | Palmitoleic |
| D | Decanoic | R | Ricinoleic |
| E | Erucic | S | Saturated |
| El | Elaidic | St | Stearic |
| G | Eicosenoic | U | Unsaturated |
| H | Hexanoic | V | Vaccinic |
| L | Linoleic | Ve | Vernolic |
| La | Lauric | X | Unknown long-chain fatty acid |
| Lg | Lignoceric | – | Free hydroxy group |
| Ln | Linolenic | | |
| M | Myristic | | |

**Numbers**

| | | | |
|---|---|---|---|
| 0 | Saturated | 4 | Tetraene |
| 1 | Monoene | 5 | Pentaene |
| 2 | Diene | 6 | Hexaene |
| 3 | Triene | | |

**Carbon Number: Double Bonds**

*cis*-9,*cis*-12-Octadecadienoic acid

*Diacid.* Triglyceride molecules containing two different fatty acids (oleodistearin, laurodierucin, etc.). Diacid triglycerides can take three isomeric forms:

```
   ┌─A         ┌─B         ┌─B
B ─┤         A─┤         B─┤
   └─B         └─B         └─A
```

## TABLE 1-3
### Standard Abbreviations for Triglycerides

**Basic Concept**
Fatty acid abbreviations are combined in groups of three. The presence or lack of a prefix indicates the possible positional distribution of fatty acids within the triglyceride.

**Prefixes**

*sn-*     The fatty acids at the *sn*-1-, *sn*-2-, and *sn*-3-positions are listed in that order, identifying a single molecular species.

*rac-*     A racemic mixture in which two fatty acids are equally distributed between the *sn*-1- and *sn*-3-positions.

*β-*     The fatty acid esterified at the *β*- or *sn*-2-position is known, but the positioning of the other two acids is not. The triglyceride could be a pure enantiomer, a racemate, or an unequal mixture of enantiomers.

No prefix     Positional distribution of the fatty acids is unspecified or unknown, and any mixture of isomers is possible. Prefix is also omitted with monoacid triglycerides which have no isomeric forms.

**Examples**

*sn*-POSt = *sn*-1-Palmito-2-oleo-3-stearin
*rac*-POSt = *sn*-POSt + *sn*-StOP in equal amounts
*β*-POSt = *sn*-POSt + *sn*-StOP in any proportion
POSt = *sn*-POSt + *sn*-StOP + *sn*-PStO + *sn*-OStP + *sn*-OPSt + *sn*-StPO in any proportion
POP = *sn*-OPP + *sn*-POP + *sn*-PPO in any proportion
LLL = trilinolein
111 = trimonoene triglyceride

*Symmetrical.* Diacid triglyceride molecules having the same fatty acid esterified at the *sn*-1- and *sn*-3-positions (*sn*-2-palmitoleo-1,3-dilinolein; *sn*-2-aceto-1,3-dibutyrin; etc.).

*Unsymmetrical.* Diacid triglyceride molecules having different fatty acids esterified at the *sn*-1- and *sn*-3-positions (*sn*-1-palmitoleo-2,3-dilinolein; *sn*-1,2-dibutyro-3-acetin; etc.).

*Triacid.* Triglyceride molecules containing three different fatty acids (oleopalmitostearin, lauromyristopalmitin, etc.). Triacid triglycerides can have six distinct isomeric forms:

*Mixed-Acid.* All di- and triacid triglycerides.

*Glyceride Type.* Kartha (*451,452*) has proposed the term "glyceride type" for the four groups of triglycerides classified according to their content of saturated and unsaturated acids: SSS, SSU, SUU, and

UUU. Unfortunately, "glyceride type" has also been used with different meanings by other workers, and the term has never acquired a standard definition.

*Estolide.* Triglycerides of hydroxy acids in which additional fatty acids are esterified to the hydroxyls on the fatty acid chain.

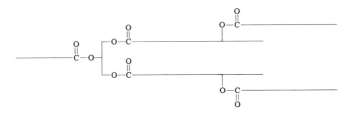

Only the triglycerides of aliphatic fatty acids are discussed in this monograph.

## B. Complexity of Triglyceride Mixtures

When a mixture of $n$ different fatty acids is esterified with glycerol, the number of possible triglycerides that can be formed may be calculated using the formulas in Table 1-4. Regardless of whether isomeric forms are considered, it is obvious that the number of possible triglycerides in-

TABLE 1-4

FORMULAS FOR CALCULATING THE MAXIMUM NUMBER OF POSSIBLE TRIGLYCERIDES THAT CAN BE FORMED FROM $n$ DIFFERENT FATTY ACIDS[a]

| Isomers distinguished | Formula | Example ($n = 2$) |
|---|---|---|
| All isomers distinguished | $n^3$ | SSS<br>*sn*-SSU, *sn*-SUS, *sn*-USS<br>*sn*-SUU, *sn*-USU, *sn*-UUS<br>UUU |
| Optical isomers not distinguished | $\dfrac{n^3 + n^2}{2}$ | SSS<br>β-SSU, β-SUS<br>β-SUU, β-USU<br>UUU |
| No isomers distinguished | $\dfrac{n^3 + 3n^2 + 2n}{6}$ | SSS<br>SSU<br>SUU<br>UUU |

[a] From Daubert (207).

TABLE 1-5

MAXIMUM NUMBER OF TRIGLYCERIDE MOLECULAR SPECIES
IN A SAMPLE CONTAINING $n$ DIFFERENT FATTY ACIDS

| Number of fatty acids $n$ | Maximum number of triglyceride molecular species | | |
|---|---|---|---|
| | All isomers distinguished | Optical isomers not distinguished | No isomers distinguished |
| 1 | 1 | 1 | 1 |
| 2 | 8 | 6 | 4 |
| 3 | 27 | 18 | 10 |
| 4 | 64 | 40 | 20 |
| 5 | 125 | 75 | 35 |
| 10 | 1,000 | 550 | 220 |
| 15 | 3,375 | 1,800 | 680 |
| 20 | 8,000 | 4,200 | 1,540 |
| 30 | 27,000 | 13,950 | 4,960 |
| 40 | 64,000 | 32,800 | 11,480 |

creases very rapidly as additional fatty acids become available for esterification (Table 1-5). The triglyceride mixtures found in many plant seeds contain only 5–10 different fatty acids, which can generate 125 to 1000 possible molecular species of triglycerides. Triglyceride mixtures from animal sources are usually more complex and may contain 10 to 40 different acids, which could form a possible 1,000 to 64,000 different triglycerides. Butterfat, one of the most complex natural fats, is known to contain at least 142 different fatty acids (*419*), which could generate a staggering 2,863,288 possible triglyceride species. Available evidence indicates that most natural fat triglyceride mixtures contain between 50 and 80% of the possible molecular species. This complexity, together with the very similar chemical and physical properties of the various molecules, makes the complete analysis of natural triglyceride mixtures extremely difficult.

## II. HISTORY OF TRIGLYCERIDE ANALYSIS

### A. 1815–1955: Crystallization and Oxidation Techniques

The development of analytical methods to determine the triglyceride composition of natural fats (summarized in Table 1-6) began, as so much of lipid chemistry did, when Chevreul (*160*) first identified fats as the glyceryl esters of fatty acids. In his experiments leading to this discovery, Chevreul (*159*) in 1815 crystallized lard from boiling ethanol, obtain-

TABLE 1-6

HISTORICAL DEVELOPMENT OF MAJOR TECHNIQUES FOR THE ANALYSIS OF MOLECULAR SPECIES OF TRIGLYCERIDES IN NATURAL FATS

| Technique | Year introduced | Authors[a] |
|---|---|---|
| Fractional crystallization | 1815 | Chevreul (*159*) |
| Permanganate oxidation | 1927 | Hilditch and Lea (*355*) |
| Liquid–liquid partition chromatography | 1956 | Dutton and Cannon (*238*) |
|  |  | Priori (*730*) |
| Pancreatic lipase hydrolysis | 1956 | Mattson and Beck (*616*) |
|  |  | Savary and Desnuelle (*796*) |
| Gas–liquid chromatography | 1961 | Huebner (*402*) |
| Silver ion adsorption chromatography | 1962 | de Vries (*218*) |
|  |  | Barrett *et al.* (*55*) |
| Stereospecific analysis | 1965 | Brockerhoff (*111*) |

[a] First workers to demonstrate the usefulness of the technique for semiquantitative analysis of natural fat triglyceride mixtures.

ing a solid precipitate and a soluble oil to which he gave the names "margarin" and "olein." Since only two types of fatty acids had been recognized by 1815 ("margaric," the solid saturated acids; and "oleic," the liquid unsaturated acids), Chevreul believed that he had separated lard into its two basic components: "glyceryl margarate" and "glyceryl oleate." This simple crystallization technique was widely adopted by other workers, who subsequently improved it by using other solvents and multiple crystallizations.

In the same year of 1815, Henri Braconnot (*107*) published an entirely different analytical technique for separating the solid and liquid portions of natural fats. Independently of Chevreul, Braconnot also proposed that all fats are composed of the same solid and liquid components (called "absolute tallow" and "absolute oil" by Braconnot) in varying proportions. He described the separation of these solid and liquid components by placing a sample of fat between pieces of blotting paper; the liquid oil soaked into the blotting paper leaving the solid fat crystals behind.

These early separations of fats into impure solid and liquid glycerides by crystallization and blotting paper techniques were, of course, quite crude by modern standards, yet in 1815 they represented a major step forward. Fats were recognized as mixtures of glycerides rather than a single pure substance. Consequently, the infinite variations in the consistency of natural fats could logically be explained by their containing different proportions of the two basic solid and liquid components; and measurement of these solid and liquid fractions gave a quantitative means of differentiating one fat from another.

## II. HISTORY OF TRIGLYCERIDE ANALYSIS

FIG. 1-3. M. E. Chevreul (1786–1889) crystallized lard from boiling ethanol in 1815 to determine its glyceride composition (*159*). He obtained a solid precipitate and a soluble oil to which he gave the names "margarin" and "olein." Since glycerol was assumed to be a monohydroxy alcohol and since only two types of fatty acids had been recognized by 1815 ("margaric," the solid saturated acids, and "oleic," the liquid unsaturated acids), Chevreul believed that he had separated lard into its two basic components: "glyceryl margarate" and "glyceryl oleate."

During the first half of the nineteenth century, the trihydroxy nature of glycerol was unknown. Fats were assumed to contain glycerol and fatty acids in a 1:1 molar ratio, giving rise to only monoacid glycerides. As early as 1838, Pelouze and Boudet (*707*) isolated a solid glyceride from cocoa butter and crystallized it to a constant melting point of 29°. Saponification of this apparently homogeneous glyceride (probably impure POSt) yielded *both* saturated and oleic acids, indicating that some unknown form of mixed-acid glyceride must exist in nature. It was not until 1854, however, when Berthelot (*75*) synthesized mono-, di-, and triglycerides that the triol character of glycerol was finally recognized and the theoretical possibility of mixed-acid triglycerides in natural fats was pointed out.

Gradually, evidence accumulated to establish the widespread existence of naturally occurring mixed-acid triglycerides. In 1876, Bell (*63*) showed that tributyrin could be isolated from an 8% BBB/92% beef tallow mixture by crystallization from ethanol, but no tributyrin could be isolated from butterfat using the same procedure. He correctly concluded that the 4:0 in butterfat must be present as a mixed-acid triglyceride, possibly as POB (*64*). Further recrystallization experiments on POB in butterfat were reported in 1889 by Blyth and Robertson (*87*). The existence of mixed-acid triglycerides in plant fats became quite clear in 1896 when Heise (*351,352*) found oleodistearin to be the major triglyceride in *Allanblackia stuhlmannii* and kokum butter seed fats.

The realization that natural fats were not simple mixtures of monoacid triglycerides but complex mixtures of mixed-acid triglycerides revealed the true difficulties involved in determining the triglyceride composition of natural fats. The number of molecular species to be analyzed was not $n$ (where $n$ = number of fatty acids) as previously expected, but $n^3$ (Table 1-4). Many individual molecular species would undoubtedly be so similar in their solubilities that they could not possibly be separated by fractional crystallization. To meet this problem, T. P. Hilditch and his associates (*365*) proposed that natural triglyceride species be classified into four simple groups according to their saturated and unsaturated acid content (i.e., SSS, SSU, SUU, and UUU); and during the 1927–1950 era they developed two useful new techniques for analysis of these groups.

The first approach by Hilditch and Lea (*355*) was $KMnO_4$ oxidation of the double bonds in natural fats to produce acidic derivatives. These acidic derivatives were then removed by aqueous extraction, and the remaining SSS, unoxidized by $KMnO_4$, was measured. The second procedure from Hilditch's laboratory (*365*) was a systematic fractional crystallization for segregating natural fats into fractions of widely different iodine value. With semisolid fats, these fractions were sufficiently simple so that their SSS, SSU, SUU, and UUU content could be estimated from their fatty

FIG. 1-4. T. P. Hilditch (1886–1965) and his co-workers introduced two useful procedures that gave the first real understanding of the complex triglyceride composition of natural fats. In 1927 Hilditch and Lea (355) showed that permanganate oxidation of unsaturated fats permitted the unoxidized, fully saturated triglycerides to be recovered and measured. During the 1930's, Hilditch and his students (365) perfected a systematic fractional crystallization procedure for separating fats into fractions of widely different iodine value; with semisolid fats, these fractions were sufficiently simple that their SSS, SSU, SUU, and UUU content could be estimated from their fatty acid composition.

acid composition and from their SSS content determined by permanganate oxidation. Hilditch and his co-workers used these two techniques extensively to give the first real understanding of the complex triglyceride composition of natural fats.

## B. 1956–1972: Chromatographic and Enzymatic Techniques

Triglyceride analysis by the crystallization and oxidation methods developed in the 1815–1955 era suffered from three major disadvantages: (i) large samples were required, about 10–300 g, (ii) procedures were very time consuming, 1–3 months per sample, and (iii) results were at best only semiquantitative. Between 1956 and 1965, a series of new chromatographic and enzymatic techniques revolutionized the field, making it possible to perform very detailed triglyceride analyses on as little as 10–100 mg of sample. The key to this revolution was the introduction of gas–liquid chromatography of fatty acid methyl esters by James and Martin (*411*) in 1952. This technique answered the need for a rapid, accurate micromethod and allowed routine fatty acid analysis of individual triglyceride fractions in a matter of a few hours.

Liquid–liquid partition chromatography was the first chromatographic separation technique successfully applied to natural fat triglycerides. In 1956 Dutton and Cannon (*238*) published the first in a series of studies on the use of countercurrent distribution to fractionate seed fat triglycerides by partition number. Independently of these American workers, Priori (*730*) in 1956 described a paper chromatography technique for liquid–liquid partition separations of natural fat triglycerides on a semimicro scale. In 1961, Huebner (*402*) demonstrated that gas–liquid chromatography could be useful in defining natural fat triglyceride mixtures. This technique separated triglycerides by molecular weight, clearly resolving differences of as little as two carbon atoms. The next year, in 1962, de Vries (*218*) and Barrett *et al.* (*55*) introduced the important silver ion adsorption chromatography of triglycerides, making it possible to fractionate the species in natural fats on the basis of unsaturation. These three rapid, semimicro separation techniques opened up many new fields for triglyceride composition studies, particularly in lipid metabolism.

Concurrent with the development of chromatographic separation methods, new enzymatic techniques were developed to characterize triglyceride positional isomers. In 1956, Mattson and Beck (*616*) and Savary and Desnuelle (*796*) introduced analytical methods for treating triglyceride mixtures with pancreatic lipase, which specifically hydrolyzes the primary ester linkages. Isolation and analysis of the resultant 2-monoglycerides identified the fatty acids attached at the *sn*-2-position, and the composition of the acids at the combined $sn$-1- and $sn$-3-positions could then be readily calculated by difference. Complete positional analysis of triglycerides became possible in 1965 when Brockerhoff (*111*) published a stereospecific analysis technique based on the stereospecific hydrolysis of derived diacyl phosphatidylphenol by phospholipase A. This method permitted the fatty acid compositions of the $sn$-1- and $sn$-3-positions to be distinguished for the first time.

Using various combinations of these new chromatographic and enzymatic characterization methods, one can now perform in hours and with greater quantitative accuracy what had formerly taken several months to accomplish. In addition, this new methodology has finally made it possible, as Hammond (*340*) has pointed out, to perform a complete analysis of all the molecular species of triglycerides in most plant fats and many animal fats. However, no one has yet achieved this goal at the time this book is published. The future will undoubtedly bring new and improved techniques to the field of triglyceride analysis. Recent studies by Hites (*382*) on the mass spectrometry of natural fat triglyceride mixtures indicate one such possibility. However, the methodology developed in the productive 1956–1965 decade will certainly remain standard for many years to come.

## III. APPLICATIONS OF TRIGLYCERIDE ANALYSIS

The analytical chemistry of triglycerides has found numerous applications in both biochemical and technological fields. Some of the more important problems that have been solved using these techniques are listed in Table 1-7.

Careful consideration, however, reveals that there are many disciplines which regularly deal with triglyceride mixtures but rarely concern themselves with individual molecular species. Application of the new triglyceride methodology in these fields might well produce additional useful information. Serum triglyceride level, for example, is a standard analytical procedure in clinical laboratories; but little work has been done to investigate whether certain molecular species of serum triglycerides might possibly reflect specific metabolic disorders. In the technological field, the consistency characteristics of the fats and oils used in shortenings and margarines have usually been correlated with fatty acid composition or with the liquid/solid ratios at various temperatures. Now that rapid triglyceride analytical procedures are available, it might prove more useful to define the consistency characteristics of commercial fats and oils in terms of their SSS, SSU, SUU, and UUU composition. Surely the next decade will see the new triglyceride analysis techniques being widely applied in fields such as these.

TABLE 1-7

SOME IMPORTANT APPLICATIONS OF TRIGLYCERIDE ANALYSIS PROCEDURES TO MAJOR BIOCHEMICAL AND TECHNOLOGICAL RESEARCH PROBLEMS

| Problem | References |
| --- | --- |
| Defining pathways for triglyceride biosynthesis | *12, 15, 124, 314, 370, 428, 432* |
| Defining mechanisms of intestinal fat absorption | *103, 183, 313, 621, 624, 756* |
| Increased utilization of dietary fats in infant feeding | *264, 889* |
| Identification of specific triglycerides as active substances producing aggregation by the beetle *Tribolium confusum* | *879, 879a* |
| Diagnosis of Refsum's disease by characterization of distinctive phytanate triglycerides found in plasma | *448, 557a, 559* |
| Identification of specific triglycerides giving cocoa butter its unique suitability for confectionary chocolate | *154, 443, 782, 863* |
| Detection of adulterated olive oil | *132, 295, 496, 497, 631, 844* |
| Identification of specific lard triglycerides causing graininess in shortenings and margarines | *154, 443, 863* |

The ultimate goal of triglyceride analysis, however, is to define the patterns by which nature assembles glycerol and fatty acids into complex mixtures of triglycerides. Empirical correlations of analytical data show distinct relationships between the fatty acid and triglyceride compositions of natural fats. Enzymatic studies reveal that there are only two or three pathways synthesizing the many triglyceride species found in living organisms. Further research using both these approaches should eventually decipher the fundamental fatty acid distribution patterns found in natural triglycerides. When this goal is finally reached, it may well be possible to estimate the composition of many natural triglyceride mixtures directly from their fatty acid composition.

# 2

# EXTRACTION, ISOLATION, MEASUREMENT, AND FATTY ACID ANALYSIS

The first step in triglyceride analysis is to isolate all the triglycerides from the original sample and to characterize their fatty acid composition. Only then can the appropriate procedures be selected for the analysis of specific triglyceride molecular species. This chapter outlines such preliminary procedures for extraction, isolation, and measurement of sample triglycerides, as well as the analysis of their fatty acid composition by gas–liquid chromatography.

## I. EXTRACTION OF LIPIDS

The complete extraction of triglycerides from tissues, biological fluids, reaction mixtures, or other samples is easily accomplished using standard extraction procedures.

Chloroform/methanol extraction is the preferred method for quantitative recovery of lipids. The various procedures and discussions available in the literature (*86,248,249,276,744*) can be summarized as follows:

> A weighed sample is placed in 10–30 volumes of chloroform/methanol 2/1 (v/v) and homogenized in a suitable high-shear blender or mixer or in a tissue-grinding tube of the Ten Broeck type. The suspension is then filtered, and the homogenizer and funnel are washed with additional chloroform/metha-

nol. For effective extraction, the chloroform/methanol solution must be monophasic at this point; and with samples of high water content, it may be necessary to perform a first extraction with chloroform/methanol 1/2 to achieve this. For complete lipid recovery (>94%), the insoluble residue is extracted a second time using the same procedure.

Nonlipid components in the extract are removed by shaking with 0.2 volumes of distilled water. Separation of the bottom $CHCl_3$ layer containing the lipids may require centrifugation or standing overnight if considerable surfactive material is present. The $CHCl_3$ layer is washed once with methanol/water 1/1 and then evaporated on a rotary flash evaporator to recover the extracted lipids.

This procedure yields quantitative recovery of all neutral lipids (including triglycerides) and most polar lipids except gangliosides and polyphosphoinositides. However, preliminary cell rupture procedures are sometimes necessary for microorganisms having polysaccharide cell walls (*892*).

Many other extraction solvents will give quantitative recovery of triglycerides but are not as effective as chloroform/methanol for the very polar lipids. Bloor's early lipid extraction solvent of ethanol/diethyl ether 3/1 gives quantitative recovery of all neutral lipids (*249*). Petroleum ether will remove neutral lipids from fairly dry (<10% water) materials, such as ground seeds, using a standard Soxhlet extraction apparatus (*24*) or by homogenizing with the solvent. Triglycerides are frequently extracted

TABLE 2-1
GENERAL PRECAUTIONS FOR PREVENTING CONTAMINATION AND
CHEMICAL ALTERATION OF LIPID SAMPLES DURING ANALYSIS

**To Avoid Contamination**
1. Distill all solvents before use.
2. Rinse all glassware with 2/1 chloroform/methanol immediately before use.
3. Avoid sample contact with any rubber or plastic except Teflon.
4. Store samples and solvents in glass containers with glass or Teflon-lined caps.

**To Prevent Oxidation of Polyunsaturated Acids**
1. Perform all laboratory manipulations under nitrogen.
2. Bubble nitrogen through solvents to purge dissolved oxygen.
3. Evaporate solvents below 40° under an inert atmosphere.
4. Add 0.005% antioxidant (w/v) to solutions of stored samples.[a]
5. Store samples in solution under nitrogen at 0° or lower.

**To Avoid Chemical Alteration**
1. Extract lipids *immediately* upon tissue dissection.
2. Avoid prolonged sample contact with any alcohol, particularly under acidic or alkaline conditions.

[a] Care should be taken to select an antioxidant that does not interfere with subsequent analyses. BHT (butylated hydroxytoluene, or 2,6-di-*tert*-butyl-*p*-cresol) is often employed because it elutes in or near the solvent front during GLC of long-chain fatty acid methyl esters

from human blood serum by shaking the sample with solvents such as isopropanol (*198,277,522,677,888*), chloroform (*305,410*), diisopropyl ether (*296*), nonane/isopropanol (*775*), or petroleum ether/ethanol (*294*). In some cases, it is possible to combine the extraction and triglyceride isolation steps by applying a serum sample (*954*) or a small piece of frozen tissue (*157*) directly on a preparative thin-layer chromatography plate and allowing lipid extraction to occur during solvent development. Additional extraction procedures are described by Entenman (*248*).

Special care should be taken to prevent contamination or chemical alteration of samples before, during, or after extraction. Recommended precautions are listed in Table 2-1.

## II. ISOLATION OF TRIGLYCERIDES

The most widely used methods for isolating triglycerides from lipid mixtures are column chromatography on Florisil or column or thin-layer chromatography (TLC) on silicic acid. These procedures are relatively rapid, require only simple equipment, yield good resolution of triglycerides from other lipid classes, and (when properly used) produce no chemical alteration of the sample. Column chromatography handles larger size samples, but TLC gives better resolution of lipid classes.

### A. Column Chromatography on Florisil

Column chromatography on Florisil (nominally magnesium silicate) is probably the most useful method for separating large amounts of triglycerides from lipid mixtures. Florisil is preferable to silicic acid for this purpose, since free fatty acids are more strongly adsorbed on Florisil; consequently there is no possibility of the triglyceride and fatty acid peaks overlapping as sometimes occurs on columns of silicic acid (Section II,B).

A number of procedures for isolating triglycerides on Florisil columns have been described in the literature (*136,137,162,407,505,745*). The following method, based mainly on that of Carroll (*136*), is typical:

> Florisil (100/200 mesh) of suitable activity for column chromatography of lipids is prepared by heating overnight at 260° or alternatively by heating at 650° and then equilibrating with 7% (w/w) water. Twelve grams of this material is slurried with hexane and uniformly packed in a 12 mm i.d. chromatography column after covering the glass filter plate with filter paper disk. The column of Florisil (about 150 mm high) is then covered with a second filter paper disk. Solvent flow is adjusted to 1.5–3.0 ml/minute, and the eluate is collected in 5 ml fractions.

The hexane level is lowered to the top of the packed column; then 100–300 mg of the lipids to be separated is dissolved in a minimum volume of hexane and applied evenly to the top of the column. Hydrocarbons, wax esters, and cholesterol esters are eluted first with ~75 ml hexane/diethyl ether 95/5, followed by ~75 ml hexane/diethyl ether 85/15 to elute the triglycerides. The remaining neutral lipids can be stripped from the column with 100 ml of diethyl ether/acetic acid 96/4 if desired. The fractions containing the triglycerides are identified by analytical TLC, combined, and the solvent evaporated. Total separation time is 2–3 hours.

This procedure can be scaled up to handle larger size samples when necessary. A model separation of tripalmitin from other lipid classes is shown in Fig. 2-1.

The adsorption of triglyceride molecules by Florisil varies slightly with molecular weight and unsaturation, and this effect usually causes the triglyceride peak to be nonhomogeneous. By analogy with silicic acid (Chapter 8, Section I,E), one would expect triglycerides containing oxygenated fatty acids to be adsorbed more strongly than normal triglycerides by Florisil, resulting in the separation of triglyceride molecules by oxygen content.

## B. Column Chromatography on Silicic Acid

The isolation of triglycerides from lipid mixtures using column chromatography on silicic acid is a routine procedure that has been described in numerous variations (*137,223,265,380,395,564,743,777,779*). Silicic acid of uniform size (100/200 mesh is typical) is either purchased or prepared

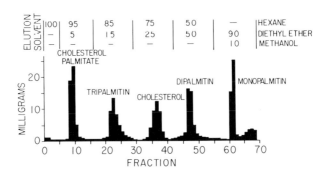

FIG. 2-1. Separation of tripalmitin from a model mixture of lipid classes by column chromatography on Florisil. *Operating conditions:* 20 × 170 mm column containing 30 g of 60/100 mesh Florisil activated at 650°; sample mixture containing 40 mg each cholesterol palmitate, tripalmitin, cholesterol, dipalmitin, and monopalmitin; 1.5 ml/minute stepwise elution with solvent mixtures indicated; separation monitored gravimetrically by collecting and evaporating 10 ml fractions. From Carroll (*136*).

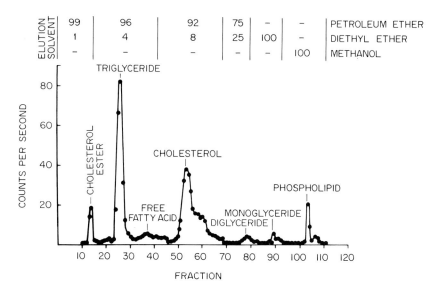

FIG. 2-2. Isolation of triglycerides from rabbit peritoneal macrophage lipids by column chromatography on silicic acid. *Operating conditions:* approximately 18 g of activated silicic acid; 590 cps of $^{14}$C-labeled rabbit peritoneal macrophage lipids together with approximately 10 mg each of unlabeled cholesterol ester, tripalmitin, palmitic acid, cholesterol, and phosphatidylcholine; stepwise elution with solvent mixtures indicated; 20 ml fractions monitored by measuring radioactivity. From Day and Fidge (208).

in the laboratory (380,395). Reproducible activity for column chromatography of lipids is obtained by heating the silicic acid overnight at 200° and then equilibrating with approximately 10% (w/w) water.

Operating procedures for isolating triglycerides on columns of silicic acid are very similar to those described above for Florisil. Figure 2-2 illustrates the separation of rabbit peritoneal macrophage lipids using typical operating conditions. The elution pattern of neutral lipids on silicic acid columns is similar to that with Florisil, except for the free fatty acids. With silicic acid, the fatty acids may (Fig. 2-2) or may not (200) elute at the back of the triglyceride peak, depending on sample composition, activity of the silicic acid, and the elution solvents used. If the triglycerides are contaminated with free fatty acids, the latter are usually removed by passing the sample through a second column of Florisil (Section II,A), Sephadex (9), aluminum oxide (Table 2-2), or ion-exchange resin (396,503) or by extracting with aqueous $Na_2CO_3$ (224).

As in the case of Florisil, differences in triglyceride unsaturation and chain length slightly alter elution volumes, causing nonhomogeneous triglyceride peaks (Fig. 8-4 and Chapter 8, Sections I,B and I,C). Triglycerides containing oxygenated fatty acids are adsorbed more strongly than other triglycerides, making possible the separation of triglycerides according to oxygen content as discussed in Chapter 8.

## C. Thin-Layer Chromatography on Silicic Acid

Preparative TLC on silicic acid is the preferred method for isolating triglycerides from lipid mixtures when sample size is small (<50 mg) or when resolution from other compounds is difficult. By using a large number of preparative TLC plates, however, sample sizes approaching column chromatography methods can be achieved.

General procedures for thin-layer chromatography have been discussed in recent reviews (*88,842,853*) and need not be repeated here. A typical method for isolating triglycerides by preparative TLC on silicic acid proceeds as follows:

> A standard silicic acid absorbent for TLC is mixed with the recommended amount of water and spread 1.0 mm thick on $200 \times 200$ mm glass plates in the usual manner. About 20 g of adsorbent is required for each plate. The plates are dried for 2 hours in air, activated at 110° for 1 hour, and then stored in a desiccator. Before use, each plate is predeveloped in diethyl ether to move any organic contaminants to the top of the plate where they can be removed by scraping off a narrow band of adsorbent.

> The sample, 5–60 mg dissolved in benzene or other suitable solvent, is applied as a narrow streak about 20 mm above the bottom edge of the plate. The sample solution can be applied as a row of overlapping spots using a capillary or a microsyringe, but better band resolution is obtained with a semiautomatic sample streaker which produces a very even sample application in a thin band. A suitable triglyceride standard is also applied at one side of the plate to aid in identifying the triglyceride band.

> Ascending development with petroleum ether/diethyl ether/acetic acid 87/12/1 is carried out in a standard chromatography tank or in a "sandwich" apparatus. As soon as the solvent reaches the top of the adsorbent, the plate is removed and dried in a stream of nitrogen. The developed plate is sprayed lightly with 2′,7′-dichlorofluorescein solution (0.1% in ethanol/water 96/4) and viewed under ultraviolet light to locate the various components. The triglyceride band is marked with a microspatula, scraped from the plate with a razor blade, and placed in a small chromatography column. The triglycerides are eluted with three column volumes of diethyl ether, which is evaporated to recover the pure triglycerides.

Figure 2-3 illustrates a preparative separation of triglycerides from yeast lipids by TLC on silicic acid using slightly different operating conditions from those described above.

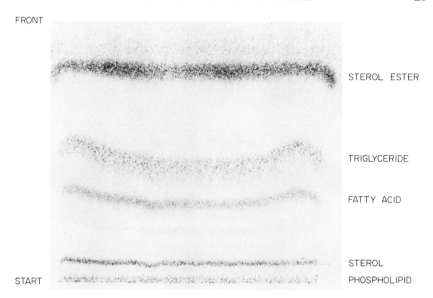

FIG. 2-3. Separation of triglycerides from baker's yeast lipids by preparative thin-layer chromatography on silicic acid. *Operating conditions:* 200 × 200 mm TLC plate coated with 0.25 mm layer of Silica Gel G; sample size, ~18 mg; single development with petroleum ether/diethyl ether/acetic acid 87/12/1; bands visualized by charring with $H_2SO_4/K_2Cr_2O_7$.

If a sample contains either short-chain ($C_2$–$C_8$) or oxygenated (hydroxy, epoxy, keto, acyloxy) fatty acids, which are adsorbed more strongly than normal long-chain acids, then it is possible that several triglyceride bands may occur. Such separations are discussed in detail in Chapter 8, Section I,E.

Care should be taken that the triglyceride band does not become so large that it engulfs any minor components that migrate either slightly ahead of or slightly behind the triglycerides. Diacyl glyceryl ethers, for example, have an $R_f$ only slightly greater than that of triglyceride of corresponding chain length (*805*) and can often be overlooked if very small amounts are present. Diester compounds such as 1,2-ethanediol dioleate migrate so close to the triglyceride band that the two lipid classes cannot usually be separated from one another on silicic acid (*59,68*).

## D. Other Methods

Other methods can be used to isolate triglycerides from lipid mixtures but they have not proven as useful as the Florisil and silicic acid chromatographic procedures described above.

TABLE 2-2

SINGLE-SOLVENT FRACTIONATION PROCEDURES FOR SEPARATING TRIGLYCERIDES FROM MORE POLAR LIPIDS USING COLUMN OR SLURRY TECHNIQUES

| Adsorbent | Solvent | Application | Lipids remaining in solution | Lipids adsorbed | References |
|---|---|---|---|---|---|
| Florisil | Benzene | Natural fats | Triglycerides, hydrocarbons, sterol esters | Diglycerides, monoglycerides, fatty acids, sterols, phospholipids | *576* |
| Aluminum oxide | 90/10 Petroleum ether/ diethyl ether | Reaction mixtures from triglyceride synthesis | Triglycerides, methyl esters | Diglycerides, monoglycerides, fatty acids | *417* |
| Silicic acid | 96/4 Petroleum ether/ diethyl ether *or* 60/40 benzene/hexane | Plasma lipids | Triglycerides, hydrocarbons, cholesterol esters, (fatty acids) | Fatty acids, sterols, phospholipids | *380, 395* |
| Silicic acid *or* Zeolite | Chloroform *or* diethyl ether *or* isopropanol | Plasma lipids | Triglycerides, hydrocarbons, cholesterol esters, fatty acids, cholesterol | Phospholipids | *133, 198, 273, 522, 607, 917* |

Single-solvent fractionation with various adsorbents has often been employed to separate triglycerides from more polar lipids. In this technique, the least polar solvent mixture that will remove all triglycerides from the adsorbent is determined by experiment. The lipid mixture is then applied to a column of the adsorbent, and all triglycerides and any less polar lipids are rapidly eluted with the designated solvent. Alternatively, a slurry of sample, solvent, and adsorbent is mixed and then filtered, followed by additional washing with fresh solvent. In either case, the triglycerides and less polar lipids are found in the eluate or filtrate, while the more polar lipids remain bound to the adsorbent. Table 2-2 lists a number of solvent/adsorbent combinations that have been used for single-solvent fractionation of lipid mixtures into triglyceride and nontriglyceride fractions. The technique has been widely used to separate plasma lipids into neutral lipid and phospholipid fractions prior to further analysis.

The advantage of single-solvent fractionation is its speed; triglycerides are separated from the more polar lipids in just a few minutes, rather than the 1–3 hours required by column chromatography. The disadvantage of single-solvent fractionation is its inability to separate triglycerides from less polar lipids (hydrocarbons, sterol esters, wax esters, and diacyl glyceryl ethers) which remain with the triglycerides and can interfere with further analyses. For example, fatty acid methyl esters prepared from such "triglyceride" fractions might be contaminated by sterol ester acyl groups; or extraneous nontriglyceride bands could appear during subsequent separations on $AgNO_3$-impregnated TLC plates. If, however, lipids less polar than triglycerides are present only in negligible amounts, then single-solvent fractionation can be used with no problem.

Triglycerides have also been isolated from lipid mixtures using chromatography on glass fiber paper impregnated with silicic acid (*723*), TLC on aluminum oxide (*563*), column chromatography on $Al_2O_3$ (*417*), countercurrent distribution (*885*), and permeation chromatography (*9,232*). However, none of these techniques is widely used today.

## III. MEASUREMENT OF TOTAL TRIGLYCERIDE

Once the triglycerides have been isolated from a lipid mixture, the next step is to measure the total amount of triglyceride present.

Table 2-3 lists the many available micromethods for quantitative measurement of total triglyceride. Although each procedure has specific advantages and disadvantages, the method of choice is usually gas–liquid chromatography (GLC) of methyl esters with an internal standard. This GLC method is the fastest if fatty acid composition must also be determined,

TABLE 2-3

MICROMETHODS FOR MEASUREMENT OF TOTAL TRIGLYCERIDE

| Method | Approximate sensitivity for PPP (μg) | Comments | References |
|---|---|---|---|
| **Gas–liquid chromatography** | | | |
| Methyl esters | | | |
| Add internal standard such as triheptadecanoin or methyl pentadecanoate. Prepare fatty acid methyl esters in usual manner. | 0.1 | Fastest method if fatty acid analyses are run by GLC. | *85, 176, 330, 977* |
| Triglycerides | | | |
| Add internal standard such as tridecanoin or trilaurin. | 0.1 | Fastest method if triglyceride carbon number distributions are determined by GLC. | *538, 567* |
| Glycerol | | | |
| React with KOH or LiAlH₄. Recover glycerol and add internal standard such as butane-1,4-diol. Analyze free alcohols or their acetate or trimethylsilylether derivatives. | 0.5 | Slightly less accurate than GLC of methyl esters or triglycerides. | *33, 387, 413, 746, 771* |
| **Spectrophotometry** | | | |
| Hydroxamic acid | | | |
| React with HONH₂HCl and ferric perchlorate to form ferric hydroxamates. Measure absorption at 520 nm. | 20 | Specific for ester linkages. Semiautomated procedure described in *30a*. | *611, 841, 933, 936* |
| Chromotropic acid | | | |
| Saponify. Acidulate. Convert glycerol to formaldehyde with NaIO₄; then destroy excess NaIO₄ with NaAsO₂ or NaHSO₃. Add chromotropic acid and heat for color reaction with formaldehyde. Read at 570 nm. | 5 | Specific for 1,2-diols. Semiautomated procedures described in *521a, 583a*. | *46, 133, 161, 282, 350, 917* |

## III. MEASUREMENT OF TOTAL TRIGLYCERIDE

Phenylhydrazine
 Saponify, acidulate, and remove fatty acids. Convert glycerol to formaldehyde with NaIO₄. Add phenylhydrazine, then potassium ferricyanide and HCl. Read color at 530 nm. — 5 — Specific for 1,2-diols. — *296, 437, 583, 750*

Acetylacetone
 Saponify. Acidulate. Convert glycerol to formaldehyde with NaIO₄; then add NaAsO₂ and acetylacetone reagent. Heat. Measure absorption or fluorescence at 405 nm. — 5 — Specific for 1,2-diols. Semiautomated procedures described in *86a, 198, 634a, 677, 775*. — *273, 305, 791, 848a*

Glycerol kinase/glycerol dehydrogenase
 Saponify, acidulate, and remove fatty acids. Add buffer containing glycerol kinase, ATP, NAD, and MgCl₂. Incubate 10 minutes at 30°, then read absorption or fluorescence at 340 nm. Add glycerol dehydrogenase, wait 30 minutes, read again at 340 nm. Difference represents NADH from reduction of glycerol phosphate. — 5 — Specific for glycerol. Semiautomated procedures described in *343a, 518a*. — *62, 158, 703, 850*

Glycerol kinase/pyruvate kinase/lactate dehydrogenase
 Saponify, acidulate, and remove fatty acids. Add buffer containing pyruvate kinase, lactate dehydrogenase, ATP, phosphoenolpyruvate, NADH, and MgCl₂. Read absorption or fluorescence at 340 nm. Add glycerol kinase, wait 10 minutes, read again at 340 nm. Difference represents NAD produced by sequence of enzymatic reactions. — 5 — Specific for glycerol. Semiautomated procedures described in *64a, 205a, 602*. — *244, 722, 809, 888*

Infrared
 Measure ester band absorption at 1745 cm⁻¹ (5.73 μm). — 500 — Specific for ester linkages. Nondestructive. See Fig. 11-9. — *285, 523, 682*

3-Methylbenzothiazoline-2-one — 2 — Specific for 1,2-diols. — *599, 704*
o-Aminophenol — 5 — Detects glycerol. — *643*
p-Phenazobenzoyl chloride — 2 — Specific for hydrophobic acids, aldehydes, and alcohols. — *473*

Copper soaps — 5 — Specific for hydrophobic acids produced by saponification of triglycerides. See also *470a, 559a, 752a*. — *836*

TABLE 2-3 (continued)

| Method | Approximate sensitivity for PPP (μg) | Comments | References |
|---|---|---|---|
| Dichromate/H$_2$SO$_4$ oxidation | 5 | Detects all nonvolatile organic material. | 22, 609, 612, 702 |
| Sulfophosphovanillin reaction | 5 | Positive reaction with most lipids, but color intensity varies with fatty acid composition. | 978 |
| **Titration** | | | |
| Periodate | | | |
| Saponify, acidulate, and extract fatty acids. Add excess NaIO$_4$ to glycerol and back-titrate with NaAsO$_2$. | 500 | Specific for 1,2-diols. | 441 |
| Cerium perchlorate | | | |
| Saponify, acidulate, and extract fatty acids. Titrate glycerol with cerium perchlorate. | 500 | Detects all alcohols. | 647, 786 |
| KOH *or* NaOH | | | |
| Saponify; then titrate excess KOH with HCl. Alternatively, saponify, acidulate, and extract fatty acids; titrate fatty acids with NaOH. | 500 | First method detects all acids. Second method specific for hydrophobic acids. | 16, 312, 412a |
| **Thin-layer chromatography** | | | |
| Densitometry | | | |
| Spray plate with 50% H$_2$SO$_4$, 70% H$_2$SO$_4$ saturated with K$_2$Cr$_2$O$_7$, or other charring agent. Heat at 200°. Quantitate spots with densitometer. | 1 | Sample hydrogenation before analysis increases accuracy (688). | 78, 84, 233, 688 |

## III. MEASUREMENT OF TOTAL TRIGLYCERIDE

| Method | Description | | Refs. |
|---|---|---|---|
| Fluorimetry | Spray plate with rhodamine 6G solution. Quantitate spots with scanning fluorimeter. | 1 | Sample hydrogenation necessary for highest accuracy. | 674a |
| GLC Detector | Measure pyrolysis or combustion products with GLC detector | 1 | Commercial instruments not yet available. | 196, 335, 489, 701, 873 |
| Spot area | | 1 | Less accurate than densitometry. | 742, 804, 912 |
| **Automatic column chromatography monitor** | | | | |
| Refractive index | Measure refractive index of eluate or of residue after eluate evaporation on moving plastic tape. | 20–100 | Changing elution solvent alters R.I. of unevaporated eluate and changes baseline on chromatogram. Nondestructive. | 89, 379, 948 |
| GLC Detector | Deposit eluate on moving wire or chain and evaporate solvent. Measure nonvolatile material directly or as its pyrolysis or combustion products in GLC detector. | <1 | Several commercial instruments available. | 138, 195, 449, 827, 969 |
| Weight | | 500 | Mostly for column chromatography work. Seldom used with TLC because of insensitivity and difficulties with impurities. Nondestructive. | 220, 490, 521, 863 |
| Radioactivity | Transesterify with $^{3}$H- or $^{14}$C-methanol. Alternatively, saponify, acidulate, isolate fatty acids, and react with $^{14}$C-diazomethane, $^{63}$Ni(NO$_3$)$_2$, or $^{60}$Co(NO$_3$)$_2$. Isolate methyl esters and measure radioactivity. | 0.1 | Most sensitive procedure available. Inorganic salts not suitable for <C$_{14}$ acids. | 270, 383, 384, 958 |

is applicable to almost any sample size, is as accurate as most other procedures, and (assuming proper isolation of triglycerides) is not usually troubled by interfering substances. The internal standard should be a glyceryl, methyl, or other ester of a $C_{12}$–$C_{24}$ acid that will undergo the same chemical reactions as the sample during methyl ester preparation. The methyl ester derived from the standard must not overlap any GLC peak from the original sample; hence proper choice requires prior knowledge of the fatty acids present. Triglycerides or methyl esters of odd-carbon fatty acids such as 13:0, 15:0, 17:0, and 21:0 are frequently selected as convenient standards which are commercially available in 99% purity. An aliquot of internal standard solution is added to an aliquot of the triglyceride sample, methyl esters are prepared by any standard method, and fatty acid composition is determined as usual by GLC (Section IV). A typical calculation procedure for quantitating total triglyceride by GLC of methyl esters with an internal standard is presented in Table 2-4.

If fatty acid analyses are not being run and only total triglyceride con-

TABLE 2-4
Typical Calculation for Measuring Total Triglyceride by GLC of Derived Methyl Esters with an Internal Standard

Approximate weight of triglyceride sample = 1–3 mg
Weight of methyl pentadecanoate[a] standard added = 0.450 mg = 1.76 μmoles
Fatty acid composition by GLC

|        | Weight % | Mole % |
|--------|----------|--------|
| 15:0   | 20.0     | 22.3   |
| 16:0   | 5.0      | 5.3    |
| 18:0   | 1.0      | 1.0    |
| 18:1   | 40.0     | 38.4   |
| 18:2   | 34.0     | 33.0   |

Weight of methyl esters derived from sample = $\dfrac{80.0}{20.0} \times 0.450 = 1.80$ mg

Weight of triglycerides in sample[b] = $1.80 \times \dfrac{885.5}{889.5} = 1.79$ mg

μmoles of triglyceride in sample = $\dfrac{77.7}{22.3} \times \dfrac{1.76}{3} = 2.04$ μmoles

[a] If tripentadecanoin is used, the weight of an equivalent amount of methyl pentadecanoate should be calculated so that both the weight of the internal standard and the GLC weight percent data refer to methyl esters.

[b] $\dfrac{885.5}{889.5} = \dfrac{\text{molecular weight of triolein}}{3 \text{ (molecular weight of methyl oleate)}}$. This correction factor is very small and is based on an average molecular weight of the fatty acids found in the sample.

tent is desired, then various non-GLC techniques are preferable. The actual choice of method depends on the sample size available, the possible presence of any interfering compounds, the laboratory equipment available, and the manner in which the triglycerides were isolated. For the clinical determination of serum triglyceride levels, six spectrophotometric procedures have been widely tested and can be recommended: hydroxamic acid, chromotropic acid, phenylhydrazine, acetylacetone, glycerol kinase/glycerol dehydrogenase, and glycerol kinase/pyruvate kinase/lactate dehydrogenase. Semiautomated procedures utilizing a Technicon AutoAnalyzer have been developed for five of these methods, as noted in Table 2-3. Separation of lipid mixtures by TLC followed by charring and densitometry to determine triglyceride content has proven satisfactory for general research applications. Quantitation of triglycerides in individual column chromatography fractions is always a tedious process using any manual procedure—hence the recent development of automatic column chromatography monitoring systems which, although still in need of further improvement, provide a rapid method for triglyceride quantitation.

The accuracy of the various techniques listed in Table 2-3 has been extensively investigated with varying results. The reproducibility and precision of most of the procedures listed generally lie within a range of 1 to 5 relative percent, except near the lower limits of sensitivity. An extensive survey of the literature indicates that variations within this range are apparently due more to operator skill and to the care with which variables are controlled rather than to any greater inherent accuracy of one procedure over another. For example, semiautomated procedures have usually proven slightly more accurate than the corresponding manual methods.

## IV. FATTY ACID ANALYSIS

The determination of fatty acid composition is an essential part of triglyceride analysis. The method of choice for fatty acid analysis is clearly gas–liquid chromatography (GLC), and its introduction has made possible many of the microanalytical techniques for triglycerides described in the following chapters. A complete discussion of GLC procedures would require a book in itself and is definitely beyond the scope of this monograph. However, the specific technique for fatty acid methyl esters will be briefly described here; and for further details the reader should consult one of the excellent recent reviews on the subject (*5,6,412,532,855*).

### A. Methyl Ester Preparation

The conversion of triglycerides to fatty acid methyl esters suitable for GLC analysis can be accomplished by a number of satisfactory procedures. Methanolysis using HCl (*85,472*), $H_2SO_4$ (*12,705*), KOH (*110*), $NaOCH_3$ (*26,747*), or $BF_3$ (*668*), as a catalyst is commonly employed. One of the simplest procedures is the KOH-catalyzed methanolysis method of Brockerhoff (*110*):

> Dissolve 1–40 mg of triglyceride in 0.5 ml of diethyl ether and add 1.0 ml of 0.5 N methanolic KOH solution. Shake. After 10 minutes at room temperature, add 1.0 ml of 1.0 N HCl. Shake. Extract the methyl esters with $3 \times 1.0$ ml of petroleum ether. Combine the extracts in a screwcap test tube and evaporate with nitrogen. Wash down the sides of the tube with sufficient carbon disulfide to redissolve the methyl esters for GLC analysis.

Alternatively, the triglycerides can be saponified and then acidulated to yield free fatty acids, which are converted into methyl esters with $CH_3OH/HCl$ (*472*), $CH_3OH/H_2SO_4$ (*717*), $CH_3OH/BF_3$ (*644*), or diazomethane (*204,799*).

Fatty acids with ten or less carbon atoms present a special problem since their methyl esters are volatile and slightly water soluble, causing losses to occur during extraction and solvent evaporation. Two ways to avoid this problem with short-chain acids are (i) to analyze short-chain acids as esters of higher molecular weight alcohols such as butanol (*197,784*) or 2-chloroethanol (*693*), and (ii) to perform no evaporation during methyl ester preparation but inject the entire extract or reaction mixture into the chromatograph (*177,586*).

Special procedures are required for preparing chromatographable methyl esters of fatty acids containing reactive functional groups such as epoxides (*49*) or cyclopropene rings (*747*).

For a more comprehensive review of methyl ester preparation procedures, the recent review by Christie (*169*) should be consulted.

## B. Column

For methyl ester analyses, 1.5–3.0 m glass or stainless steel columns containing 3–20% (w/w) of liquid phase coated on an inert GLC support are generally used. Two types of liquid phases are commonly employed:

> (a) Polyesters (diethyleneglycol succinate, butanediol succinate, etc.) which elute unsaturated esters *after* saturated esters of the same chain length.
> (b) Long-chain hydrocarbons (Apiezon L) which elute unsaturated esters *before* saturated esters of the same chain length.

In general, polyesters of high polarity are the most useful since they give the best resolution of component fatty acid peaks. However, comparison of elution patterns on several liquid phases of different polarity greatly

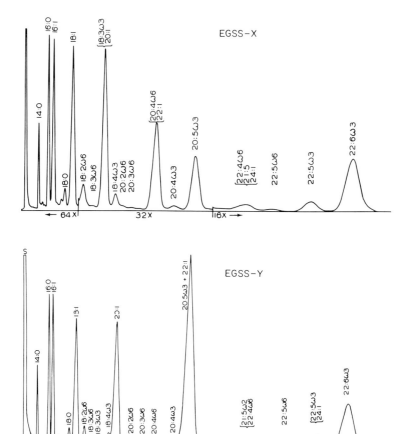

FIG. 2-4. Gas chromatograms of cod liver oil fatty acid methyl esters on EGSS-X and EGSS-Y packed columns. Note changes in elution patterns with change in polarity of the liquid phase. *Operating conditions:* 1.83 m × 3 mm i.d. glass column packed with 14% EGSS-X or 15% EGSS-Y polyester-silicone coated on 100/120 mesh Gas Chrom P; 200°; argon carrier gas; flame ionization detector. From Ackman (5).

aids peak identification (see below). Figure 2-4 shows typical gas chromatograms for isothermal analyses of cod liver oil fatty acid methyl esters on two different polyester liquid phases.

## C. Identification of Peaks

The multiplicity of peaks in most methyl ester gas chromatograms necessitates a systematic approach to peak identification. Seven major techniques

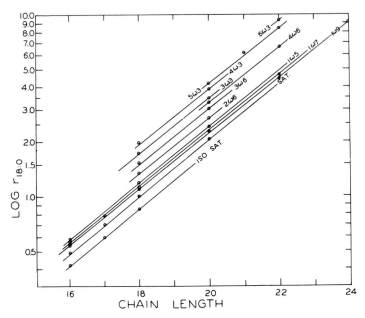

Fig. 2-5. Semilog plot of retention time vs. chain length for cod liver phospholipid fatty acid methyl esters on a capillary column coated with butanediol succinate polyester. Parallel linear relationships exist for each homologous series. Thus if the retention time of one member of an homologous series is known, the retention time of any other member can be accurately predicted. Data taken from chromatogram in Fig. 2-6. From Ackman (5).

have proven extremely useful for this purpose:

>*Known Standards.* Co-chromatography or comparison of retention times between the sample and known standards indicates which peaks have identical retention times.
>
>*Predicting Retention Times.* For an homologous series of long-chain methyl esters, there is a linear relationship between the log of the retention time and the number of carbon atoms (Fig. 2-5). Since the lines for the various homologous series are parallel, the retention time of any member of a series can be predicted if the retention time of any one member of that series is known. Systematic interrelationships between the retention times of different unsaturated homologous series (i.e., acids with the same number of double bonds or the same $\omega$ classification) have also been worked out, allowing the retention times of one series to be predicted from the retention times of a related series. A detailed description of retention time prediction techniques can be found in articles by Ackman (5) and Jamieson (412).

*Hydrogenation.* Hydrogenation of methyl esters before GLC analysis reveals the exact distribution of chain lengths within the sample. All peaks not falling in the *n*-saturated homologous series after hydrogenation are probably branched-chain or oxygenated acids.

*TLC Fractionation.* Fractionation of methyl esters by various TLC techniques prior to GLC analysis aids identification of distinct groups of peaks. TLC on silicic acid will separate oxygenated from nonoxygenated methyl esters (*665,717*). TLC on $AgNO_3$/silicic acid will fractionate methyl esters according to the number of double bonds they contain (*663,734*). TLC using liquid–liquid partition chromatography will separate methyl esters on the basis of integral partition number (Chapter 5, Section II,A,1) into groups of predictable composition (*342,600*).

*Different Liquid Phase.* A second GLC analysis on a liquid phase of different polarity provides a different peak elution pattern. Overlapping peaks are often separated (compare $18:3\omega3$ and $20:1$ in the two chromatograms in Fig. 2-4), and an additional opportunity for comparing predicted and actual retention times is provided.

*Capillary Column.* The greatly increased resolution of capillary columns (Fig. 2-6) makes overlapping peaks less likely and identification by retention time more accurate. Many double-bond positional isomers can also be distinguished.

*Isolation–Oxidation.* Isolation of an unknown peak by preparative GLC permits reductive ozonolysis or permanganate oxidation (*731*) reactions to be employed. Identification of the degradation products by GLC yields positive identification of double-bond locations within the fatty acid chain.

GLC on packed, polyester columns has proven adequate for peak identification in over 95% of the fatty acid analyses involved in triglyceride work. The extent to which the above identification methods are utilized depends, of course, on the complexity of the sample and the type of information desired. Other chemical and physical identification techniques are only resorted to when GLC procedures indicate unusual and unexpected retention times.

## D. Quantitation

Accurate analysis of fatty acid composition by GLC requires the use of calibration mixtures resembling the composition of the unknown sample. Fortunately, quantitative mixtures of methyl esters similar to those derived from various natural fats are now available from many lipid supply houses.

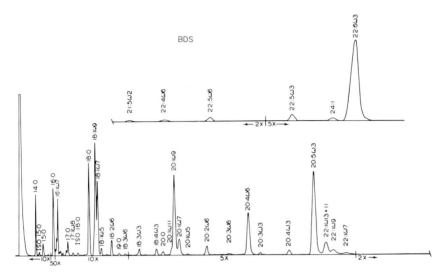

FIG. 2-6. Gas chromatogram of cod liver phospholipid fatty acid methyl esters on a capillary column coated with butanediol succinate polyester (BDS). Note the greatly increased resolution compared with packed polyester columns (Fig. 2-4), particularly with double-bond positional isomers. *Operating conditions:* 46 m × 0.25 mm i.d. capillary column coated with BDS; 170°; helium carrier gas; flame ionization detector. From Ackman (5).

These should be used to calibrate a given column and instrument under typical operating conditions to establish (i) that detector response is linear with sample size and (ii) the relative detector response to various methyl esters. With a proper flame ionization detector, peak area should be directly proportional to component weight within a narrow range of methyl ester chain lengths and for molecules with 0–3 double bonds (5). Below $C_{10}$, calibration factors are definitely required because of the wide variation of oxygen content of the molecules; and 5- and 6-double-bond acids are often subject to systematic losses during analysis (5).

After proper identification and calibration procedures have been completed, the unknown sample is analyzed under identical conditions. The area of each peak is measured and multiplied by any necessary calibration factors. Corrected peak areas are then summed and the respective weight percentages calculated. Fatty acid compositions for triglyceride analysis are usually converted to a mole percent basis, so that fatty acid and triglyceride composition data are readily interconvertible. Under optimum operating conditions, the component fatty acids in any triglyceride mixture can be determined by GLC with an accuracy exceeding 5 relative percent for major acids [>10% of total mixture] and 0.5 absolute percent for minor acids [<10% of total mixture] (353,392).

# 3

# PREPARATION OF CHEMICAL DERIVATIVES BEFORE SEPARATION

As a general rule, it is best to perform triglyceride separations on the unaltered molecules found in the original sample. However, the separation of different species of triglycerides and of diglycerides derived from triglycerides can sometimes be improved by making chemical derivatives of the original compounds. The major problem in preparing such derivatives for analytical work is to obtain quantitative conversion of the original material to the expected end product without significant by-product formation. When conversion approaches 99%, as in hydrogenation and acetylation reactions, then derivative formation sometimes offers decided analytical advantages. However, when yields fall below 95%, as in mercaptan addition and ozonization, then analysis of the original sample is definitely preferable to using the derivatized form.

Glycerides contain three functional groups which are potential sites for derivative formation: the double bonds, the ester linkages, and (sometimes) hydroxy, epoxy, or keto groups. This chapter reviews the chemical reactions available for preparing derivatives at each of these sites and evaluates the suitability of the various reactions for quantitative analysis. Specific analytical applications of the various derivatives are discussed in Chapters 4–8.

# 3. PREPARATION OF CHEMICAL DERIVATIVES

## I. REACTIONS AT DOUBLE BONDS

### A. Hydrogenation

Samples of unsaturated triglycerides can be readily converted into fully saturated molecules by hydrogenation. The reaction is essentially quantitative and can be carried out at room temperature in the presence of a highly active Pt catalyst. The following simple procedure has proven effective for routine hydrogenation of triglyceride samples in our laboratory:

> The sample (1–50 mg) is dissolved in 3–5 ml of freshly distilled dioxane and placed in a 25 ml flask with a ground glass fitting. Ten to 20 mg of platinum oxide (Adams' catalyst) and a small Teflon-coated magnetic stirring bar are added, and the flask is closed with a double inlet stopper as shown in Fig. 3-1. After thorough flushing with hydrogen, the stopcock is closed, and a positive pressure (1–3 psi) of hydrogen is maintained throughout the reaction. The magnetic stirrer is then switched on and run for 30 minutes. Proper reaction is indicated by the catalyst powder turning from dark brown to black and eventually forming long strands. After reaction, 10 ml of $CHCl_3$ is added, and the spent catalyst is removed by passing the solution through a bed of filter aid followed by 30 ml of $CHCl_3$. The filtrate is then washed with water, dried over $Na_2SO_4$, filtered, and evaporated to recover the hydrogenated triglycerides.

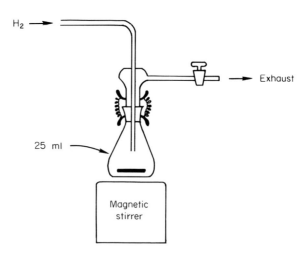

FIG. 3-1. Simple semimicro hydrogenation apparatus suitable for triglyceride samples. Operating procedure given in text.

Sample concentration during hydrogenation must be kept low enough so that triglycerides do not precipitate out of solution before becoming fully saturated. Alcoholic solvents can sometimes cause transesterification (727) and should be avoided unless transesterification is shown to be absent (7). The complete hydrogenation of all double bonds is checked by GLC analysis of the derived methyl esters. Some natural fats, particularly marine oils and Cruciferae seed oils, contain catalyst poisons which markedly retard the rate of hydrogenation. This can be overcome by using more catalyst or by repeating the hydrogenation procedure several times. When hydrogen uptake is to be used as a measure of sample unsaturation, more complex hydrogenation equipment is necessary (698,828,905).

Miwa et al. (654) and Brown et al. (123) have described a novel type of hydrogenation procedure for measuring unsaturation by titrating with $NaBH_4$. Their method utilizes a Pt catalyst prepared by in situ reduction of chloroplatinic acid by $NaBH_4$, in situ generation of hydrogen from $NaBH_4$, and an automatic burette valve that introduces standardized $NaBH_4$ solution into the reaction mixture only as long as hydrogenation is proceeding. This method was designed for the routine determination of unsaturation in a large number of fat samples, and the hydrogenated triglycerides were not recovered. However, the procedure could easily be adapted for the rapid hydrogenation of analytical samples, provided that reaction conditions were chosen so that no transesterification occurred.

It is also possible to hydrogenate triglycerides directly on paper chromatograms or thin-layer chromatography plates. This is an effective procedure when used in combination with two-dimensional development; after an initial separation of the unsaturated triglycerides, they can be hydrogenated and further resolved by a second development perpendicular to the first. Kaufmann et al. (487) have hydrogenated fatty acids and glycerides directly on undecane-impregnated TLC plates by spraying the sample area with a 1–2% solution of colloidal palladium. After drying, each plate was placed in a vacuum desiccator which was evacuated and filled with hydrogen several times. After 1 hour in a hydrogen atmosphere at room temperature, the sample was completely hydrogenated, and the plate was ready for the second development.

## B. Permanganate Oxidation

The oxidation of natural fats with potassium permanganate has been used to determine the SSS, SSU, SUU, and UUU content of natural fats, although the quantitative nature of the reaction has been the subject of considerable controversy. In theory, $KMnO_4$ will cleave each double bond in a triglyceride to form two carboxyl groups:

$$\begin{array}{c} \text{H} \\ \text{H}-\overset{|}{\text{C}}-\text{OOC}(\text{CH}_2)_7\text{CH}=\text{CH}(\text{CH}_2)_7\text{CH}_3 \\ \text{CH}_3(\text{CH}_2)_{14}\text{COO}-\overset{|}{\text{C}}-\text{H} \\ \text{H}-\overset{|}{\text{C}}-\text{OOC}(\text{CH}_2)_{16}\text{CH}_3 \\ \text{H} \end{array}$$

$$\downarrow \text{KMnO}_4$$

$$\begin{array}{c} \text{H} \\ \text{H}-\overset{|}{\text{C}}-\text{OOC}(\text{CH}_2)_7\text{COOH} \\ \text{CH}_3(\text{CH}_2)_{14}\text{COO}-\overset{|}{\text{C}}-\text{H} \quad\quad + \quad \text{HOOC}(\text{CH}_2)_7\text{CH}_3 \\ \text{H}-\overset{|}{\text{C}}-\text{OOC}(\text{CH}_2)_{16}\text{CH}_3 \\ \text{H} \end{array}$$

Triglycerides containing unsaturated fatty acids are oxidized to glycerides of dicarboxylic acids. Thus a triglyceride mixture containing 16:0, 18:0, 18:1, 18:2, and 18:3 would be converted into a mixture of SSS, SSA, SAA, and AAA, which could then be separated by various chromatographic procedures. In practice, however, permanganate oxidations are only semiquantitative because of the following difficulties:

(a) Incomplete reaction leaving some of the double bonds unoxidized
(b) Hydrolysis of some of the ester linkages
(c) By-product formation including additional ester groups and unidentified neutral material
(d) Secondary degradation of the carbon chain ("overoxidation") resulting in some dicarboxylic acids containing one or two carbon atoms less than expected

Three main procedures for the permanganate oxidation of unsaturated triglycerides have been described: (i) the Hilditch method using $KMnO_4$ in acetone solution; (ii) the Kartha method using $KMnO_4$ in acetone/acetic acid solution; and (iii) the von Rudloff method using $KMnO_4/NaIO_4$ in *tert*-butanol solution.

Hilditch and Lea (*355*) first described the oxidation of natural triglyceride mixtures in acetone solution by refluxing with powdered $KMnO_4$. This procedure with its subsequent modifications (*4,77,166,193,366,562*) is only semiquantitative due to three limitations. The reaction is usually incomplete, and oxidized samples often have an iodine value of 2 to 8 (*193,358,562*). Appreciable hydrolysis occurs during oxidation, probably because of alkaline products formed by permanganate reduction (*163,165,253,451*). Overoxidation also occurs (*42,61,253*) since it is characteristic of any permanganate cleavage of double bonds.

Kartha (*451*) has found that ester hydrolysis can be reduced by maintaining an excess of acetic acid during permanganate oxidation of unsaturated triglycerides in acetone solution. Even with this modification, however, the reaction remains semiquantitative. Several workers (*164,338,552,553,901*) still report an objectionable amount of hydrolysis, although Kartha (*455,463,464*) denies this. With Kartha's original procedure, from 2 to 8% of the double bonds remained unoxidized after exposure to $KMnO_4$ (*465–467*), but this level can be reduced using altered reaction conditions (*460,470*). Overoxidation definitely takes place (*61,253,338*) despite Kartha's claims to the contrary (*468,469*), which are based on a dubious analytical technique (*528*). However, the most serious objection to the use of acetic acid in permanganate oxidation is the apparent formation of new acetate ester groups (*253,338*) and the appearance of unidentified material which contains no free carboxyl groups and which can be chromatographically separated from the unoxidized SSS (*253,552,797*). It has been suggested that ammonium sulfate (*253*), magnesium sulfate (*942*), or carbon dioxide (*968*) might be more effective than acetic acid for pH control during permanganate oxidation; but according to Eshelman and Hammond (*253*), none of these materials gives much improvement over the Hilditch procedure.

Von Rudloff (*937*) has shown that the double bonds in triglycerides can be oxidized in a *tert*-butanol solution of $NaIO_4$ containing only catalytic amounts of $KMnO_4$. The periodate serves to regenerate the permanganate after use so that only a very small amount of $KMnO_4$ is needed for oxidation. Hence some of the side reactions caused by permanganate are considerably reduced. Youngs and Subbaram (*971,974*) have adapted the von Rudloff oxidation procedure for general application to triglyceride mixtures from natural fats:

> Twenty milligrams of triolein, dissolved in 5 ml of *tert*-butanol, is added to a mixture of 5 ml of oxidant solution (21 g $NaIO_4$ + 25 ml 0.1 $M$ $KMnO_4$ per liter), 4 ml of distilled water, 1 ml of $K_2CO_3$ solution (8 mg/ml), and 10 ml of *tert*-butanol. The reaction mixture is stirred for 2.5 hours at 65°, cooled in tap water, and ethylene gas is bubbled in until the pink color disappears. The *tert*-butanol is then removed under reduced pressure at 70–80° in a rotary evaporator. HCl (2–3 drops) is added to make the contents acidic; and after saturating the aqueous solution with NaCl, the oxidized glycerides are extracted with three 25 ml portions of chloroform. The chloroform layer is separated and the solvent removed in a current of air. The short-chain, monocarboxylic acids formed by the oxidation are stripped from the sample by placing it in a vacuum oven overnight at 85°. The proportion of reagents used for the oxidation is adjusted according to the fatty acid composition. Linoleic and linolenic acids require three and five times, respectively, as much as that required for oleic acid.

Of the three main procedures for permanganate oxidation of triglycerides, the von Rudloff method probably comes closest to giving the expected reaction products. However, side reactions are not completely eliminated. Kaimal and Lakshminarayana (*445*) report varying amounts of hydrolysis and residual unsaturation, with the largest losses occurring with higher iodine value samples. Youngs and Subbaram (*971,974*), on the other hand, claim an absence of unreacted double bonds and less than 1% hydrolysis, the latter being confirmed by Takahashi (*874*). TLC has revealed 2–4% unexpected reaction by-products after $KMnO_4/KIO_4$ oxidation (*445,874*), but they remain unidentified. Overoxidation also occurs (*874,900,971*), and 1–5% $C_7$ and $C_8$ dicarboxylic acids appear with the expected $C_9$ azelate.

It is obvious from the above discussion that permanganate oxidation of unsaturated triglycerides is always accompanied by unwanted side reactions and cannot be recommended for general use. Now that unsaturated triglycerides can be more effectively separated by chromatographic techniques, it seems preferable to avoid the problems of oxidized derivative formation wherever possible. When oxidized triglycerides are required, however, then the $KMnO_4/KIO_4$ procedure is recommended for the lowest level of by-product formation.

The free carboxyl groups in oxidized triglycerides are often esterified prior to chromatographic separation. Quantitative conversion to methyl esters is easily achieved by reaction with diazomethane (*974*). Vander Wal (*912,913*) has formed allyl esters by reacting the silver salts with allyl iodide, but he did not report the percent conversion obtained.

## C. Ozonization

Privett and Blank (*733*) have described the addition of ozone to unsaturated fats to produce triglyceride ozonides, which can be separated on silicic acid according to their oxygen content (Chapter 8, Section I,E). Ozonization was carried out at $-65°$ by pouring a pentane solution of the sample into a pentane solution of ozone and evaporating all solvent after 1–2 minutes. The reaction was originally thought to be quantitative

$$O_3 + {-}HC{=}CH{-} \xrightarrow{-65°} {-}HC\underset{O-O}{\overset{O}{\diagup\diagdown}}CH{-}$$

(*733*), but subsequent work by Privett and Nickell (*739*) has shown that pentane ozonization of methyl oleate can yield as little as 72.1% of the expected ozonide. Losses are due to the formation of hydroperoxides, acids, aldehydes, and also to the interchange of alkyl residues between different ozonides (*738,739*). (Note, for example, the "O₃ residue" spot on the chromatoplate shown in Fig. 8-7.) The ozonization of triglycerides apparently is not a quantitative reaction under the given conditions, and losses become greater as unsaturation increases.

Triglyceride ozonides can be quantitatively split to yield the corresponding aldehydes by hydrogenation with a Lindlar catalyst (*732,733*). Ozonization plus hydrogenation converts the original unsaturated triglycerides into "aldehyde cores." Like permanganate oxidation, this reduction

$$\begin{array}{c}
\text{H} \\
| \\
\text{H}-\text{C}-\text{OOC}(\text{CH}_2)_7\text{HC}\overset{\text{O}}{\underset{\text{O}-\text{O}}{\diagdown}}\text{CH}(\text{CH}_2)_7\text{CH}_3 \\
| \\
\text{CH}_3(\text{CH}_2)_{14}\text{COO}-\text{C}-\text{H} \\
| \\
\text{H}-\text{C}-\text{OOC}(\text{CH}_2)_7\text{HC}\overset{\text{O}}{\underset{\text{O}-\text{O}}{\diagdown}}\text{CHCH}_2\text{HC}\overset{\text{O}}{\underset{\text{O}-\text{O}}{\diagdown}}\text{CH}(\text{CH}_2)_4\text{CH}_3 \\
| \\
\text{H}
\end{array}$$

Ozonide

$$\xrightarrow[\text{Pd (Pb poisoned)}]{\text{H}_2}$$

$$\begin{array}{c}
\text{H} \\
| \\
\text{H}-\text{C}-\text{OOC}(\text{CH}_2)_7\text{HC}=\text{O} \\
| \\
\text{CH}_3(\text{CH}_2)_{14}\text{COO}-\text{C}-\text{H} \\
| \\
\text{H}-\text{C}-\text{OOC}(\text{CH}_2)_7\text{HC}=\text{O} \\
| \\
\text{H}
\end{array}$$

"Aldehyde core"

yields the same derivative for several different triglycerides (PPO, PPL, and PPLn all produce the same product, for example). Other ozonide cleavage methods are available (*739,748,854*), but their yields with triglyceride ozonides have not been evaluated.

## D. Epoxidation

Piguelevsky and co-workers (*720,721*) have reacted unsaturated triglycerides with peracetic acid to form higher-melting epoxide derivatives which were then separated by crystallization. Under the reaction conditions de-

$$-\text{HC}=\text{CH}- \xrightarrow{\text{CH}_3\text{COOOH}} -\text{HC}\overset{\text{O}}{\diagup\diagdown}\text{CH}-$$

scribed by Piguelevsky and others (*266,291,596,720*), however, epoxidation is apparently only 65–87% complete. Hence epoxy derivatives cannot be recommended for quantitative work.

Peracetic acid has also been used for the oxidation of unsaturated triglycerides directly on TLC plates during separation by liquid–liquid partition chromatography. Mangold (*603*) has reported that 10% peracetic acid in the developing solvent will quantitatively oxidize the unsaturated triglycerides. The more polar oxidized molecules move with the solvent front, leaving only the saturated compounds in their normal positions on the chromatoplate. Mangold did not characterize the oxidation products of this reaction, but some by-products are no doubt produced along with the expected epoxides.

**E. Bromination**

Bromine can be added to the double bonds of unsaturated triglycerides to appreciably change their relative solubilities and thus improve their separation characteristics. A commonly used bromination procedure has been the dropwise addition of a dilute $CHCl_3$ solution of bromine to a dilute $CHCl_3$ solution* of the sample near $-10°$ and with vigorous agitation until a faint yellow color persists (*259,645,919,930*). However, the quantitative aspects of this reaction have never been rigorously tested with pure triglycerides; and while the yield is undoubtedly high, the reaction may not go entirely to completion. Substitution reactions may also occur simultaneously with addition at the double bonds, and some stereoisomeric addition products are formed (*645*).

Where brominated triglycerides are to be used for analytical purposes, it is advantageous to adopt one of the bromination procedures developed for determining iodine values (*106,288,474,642*). Here, at least, bromine uptake is known to be equivalent to the unsaturation present, although side reactions have not been fully investigated. Vereshchagin (*921*) and Kaufmann and Khoe (*480*) have used this type of procedure in preparing brominated triglycerides for chromatographic separations:

> Weigh the proper amount of triglyceride (100 mg for iodine values above 120, 200 mg for iodine values between 60 and 120, 500 mg for iodine values below 60) into a flask and dissolve in 10 ml of chloroform. Add 25 ml of Kaufmann reagent (5.2 ml of liquid bromine added to 1000 ml of a saturated solution of NaBr in absolute methanol). Some NaBr precipitates at this point. If the triglyceride forms a separate liquid phase, add sufficient $CHCl_3$ to achieve a homogeneous solution (*921*). Allow to stand in a dark place for 30–120 minutes depending on the iodine value. Add 15 ml of KI solution and then sufficient $Na_2S_2O_3$ solution to discharge the blue color from starch indicator. The brominated triglycerides are isolated by adding 50 ml of water and extracting with chloroform (*480*).

* The greater solubility of brominated triglycerides in $CHCl_3$ makes it a better solvent than petroleum ether for achieving complete bromination.

Conjugated double bonds and acetylenic bonds are not fully brominated by the procedure on page 44.

Kaufmann *et al.* (*487*) have also brominated triglyceride samples directly on the TLC plates used for their separation, thereby altering the $R_f$ values of the unsaturated but not the saturated triglycerides. Quantitative bromination is claimed when 0.5% (v/v) bromine is added to the developing solvent, but substitution reactions were not checked. This type of bromination procedure cannot be carried out in solvents such as acetone which react with bromine.

Black and Overley (*81*) and Dasso and Cattaneo (*206*) maintain that brominated triglycerides can be converted back into the original unsaturated compounds by debromination with Zn in acetone or ethanol. This reaction needs further study, however, to determine whether a quantitative conversion is achieved and to establish the absence of interesterification. A small amount (2–5%) of *cis–trans* isomerization of double bonds always accompanies bromination–debromination (*409*) and this may complicate subsequent analyses in some cases. Debromination of brominated triglycerides to produce the corresponding methyl esters is easily accomplished by reaction with zinc dust in methanolic HCl (*409,919*).

## F. Mercuration

Mercuric acetate will react with the double bonds of unsaturated triglycerides to increase the polarity of the molecules in proportion to the amount of unsaturation they contain:

$$\begin{array}{c} | \\ HC \\ \| \\ HC \\ | \end{array} \quad \xrightleftharpoons[\text{HCl}]{\substack{Hg(OOCCH_3)_2 \\ CH_3OH}} \quad \begin{array}{c} | \\ HC-Hg-OOCCH_3 \\ | \\ HC-OCH_3 \\ | \end{array}$$

Such derivatives are readily separated chromatographically according to the number of polar mercury atoms (i.e., number of original double bonds) they contain. Hirayama (*371*) has used the following reaction for quantitative mercuration of triglyceride mixtures:

> The triglyceride mixture (500 mg) is dissolved in 2 ml of benzene, and 5 ml of a solution of mercuric acetate (20% excess) is added. The reaction mixture is heated to 60° for 30 minutes, cooled, and evaporated to dryness. The triglyceride addition products are then taken up in petroleum ether and used directly for analysis.

Other workers (*505,678,769*) have varied these reaction conditions slightly. Noda (*678*) showed that no transesterification occurred during mercuration, but he did find evidence for a trace amount of hydrolysis.

Rheineck et al. (*769*) report the quantitative mercuration of linseed oil using 1-propanol or acetic acid in place of methanol to form the corresponding propoxy or acetoxy derivatives.

The original triglycerides can be quantitatively regenerated from their mercuric acetate addition products by reaction with dilute HCl (*505*). To avoid any possibility of transesterification, this should be done in a nonalcoholic solvent such as tetrahydrofuran. The double bonds regenerated from mercurated compounds retain their original geometric configuration (*952*).

## G. Other Reactions at Double Bonds

Eshelman et al. (*254*) have investigated the addition of mercaptoacetic acid to the double bonds of triglycerides as a means of separating SSS from unsaturated molecules:

$$-HC=CH- \; + \; HS-CH_2COOH \; \longrightarrow \; -H_2C-\underset{\displaystyle |}{CH}- \quad \overset{\displaystyle S-CH_2COOH}{\phantom{-H_2C-CH-}}$$

The reaction was incomplete under the conditions tested. After removal of the mercaptoacetic glycerides by ion-exchange chromatography, the remaining neutral triglycerides still contained appreciable unsaturation, even when the reaction was repeated several times. The procedure is not recommended for quantitative triglyceride analysis.

Gunde and Hilditch (*322*) and Eibner and Schmidinger (*245*) have evaluated catalytic *cis–trans* isomerization as a means of converting natural fats into higher-melting, "elaidinized" triglyceride mixtures which might be more easily separated by crystallization. This reaction is unsatisfactory

$$\underset{cis}{\overset{H\phantom{=}H}{-C=C-}} \quad \underset{200°}{\overset{Se}{\rightleftarrows}} \quad \underset{trans}{\overset{H\phantom{=C}}{-C=C-}\phantom{}_H}$$

for analytical purposes, however, because complete conversion of *cis* bonds to *trans* bonds is impossible (isomerization is an equilibrium reaction) and because polyunsaturated acids are partially destroyed by polymerization and other side reactions.

## II. REACTIONS AT ESTER LINKAGES

Most reactions at the ester linkages of triglycerides produce cleavage of the fatty acid chain from the glycerol moiety and partially destroy the original molecular structure. Hence there is usually no point in using such reactions prior to a triglyceride fractionation procedure.

## A. Estolide Ester Cleavage

One unusual estolide ester cleavage reaction has proven useful in preparing triglyceride derivatives before separation, however. The estolide ester linkage in a triglyceride containing a conjugated dienol such as coriolic (13-hydroxy-*cis*-9-*trans*-11-octadecadienoic) acid is subject to preferential methanolysis in 0.065 $M$ methanolic $H_2SO_4$ at room temperature. Phillips and Smith (*716*) have used this reaction to convert the tetraacid estolide triglycerides of *Monnina emarginata* seed oil into triacid triglycerides for further analysis. The quantitative aspects of the reaction have not been investigated, however.

## B. Interesterification

Norris and Mattil (*681*) have suggested the analysis of natural fat triglyceride composition before and after random interesterification to determine whether the reaction changed the triglyceride composition. Since a 1,2,3-random pattern of fatty acid distribution (Chapter 12, Section II,C,1) is easily calculated from the fatty acid composition and since modern results indicate that no natural fat has a 1,2,3-random distribution pattern, this approach is no longer useful.

## III. REACTIONS OF HYDROXY, EPOXY, AND KETO GROUPS

Derivatization of hydroxy, epoxy, and keto groups in triglycerides is usually unnecessary for their isolation, since triglyceride mixtures are easily fractionated according to the number of oxygenated fatty acids per molecule using adsorption chromatography on silicic acid (Chapter 8, Section I,E). In some cases, however, specific derivatives have proven useful for stabilizing or altering the physical properties of glycerides.

## A. Acetylation

Acetylation of hydroxy groups with acetic anhydride is a well-known quantitative reaction that has been widely used in organic analysis. Privett and Nutter (*740*) describe a typical procedure suitable for hydroxyacyl- and diglycerides:

> A solution containing 1–100 mg of sample is placed in an ampule and the solvent evaporated with a stream of nitrogen. After adding 2.5 ml of acetic anhydride and 0.5 ml of pyridine, the ampule is sealed, heated for 3 hours in a boiling water bath, and then cooled. The ampule is opened, the solvent

evaporated under reduced pressure, and the crude reaction mixture is partitioned in 60 ml of chloroform/methanol/water 32/16/12 (v/v/v). The acetylated glycerides dissolve in the $CHCl_3$ layer, which is separated and dried over $Na_2SO_4$. After filtration, the solvent is evaporated and the product purified by preparative TLC.

## B. Trimethylsilyl Ethers

The free hydroxy groups of hydroxy triglycerides and derived diglycerides are frequently converted to trimethylsilyl ethers prior to GLC to increase their volatility and to prevent dehydration or acyl migration during analysis. Quantitative conversions are readily obtained in 10 minutes or less by merely mixing with a suitable reagent. O'Brien and Klopfenstein (*690*) describe a typical reaction:

> One hundred micrograms of diolein is dissolved in 20 µl of heptane/N-bis(trimethylsilyl)acetamide 10/1. After 10 minutes, ~1 µl of the reaction product is injected directly into the gas chromatograph.

Other silylating reagents such as hexamethyldisilazane/trimethylchlorosilane (*531,965*) can also be used. Pyridine is a better reaction solvent than heptane for dissolving saturated glycerides. Silyl ethers are not stable over long periods of time and should be prepared immediately before use.

## C. Hydrazone Formation

Van der Ven (*907*) has used hydrazone formation to isolate the minute quantity (~0.04%) of butterfat triglycerides containing keto groups. After preparing the hydrazones with Girard-T reagent [$H_2NNHCOCH_2N(CH_3)_3Cl$], the reaction products were dissolved in petroleum ether and the hydrazones were qualitatively (62–67%) extracted with chloroform/water 80/20 (v/v). The derivative hydrazones decomposed in acidic, aqueous solution so that the original keto triglycerides could be recovered (*907*), except when the ketone was alpha to the carboxyl group (*908*).

Timmen et al. (*887*) has employed the 2,6-dinitrophenylhydrazone of pyruvyl chloride to form colored derivatives of the hydroxyglycerides in butterfat. The resultant hydrazones were isolated in 81% yield by consecutive chromatography on acidic alumina and silicic acid.

The nonquantitative nature of both the above procedures makes them undesirable for general application. However, where minute amounts of hydroxy or keto triglycerides must be isolated from large amounts of interfering material, as in butterfat, they constitute useful alternative procedures.

# 4

# SILVER ION ADSORPTION CHROMATOGRAPHY

One of the most useful triglyceride separation techniques introduced in the last decade is silver ion or argentation chromatography. The ability of silver ions to chromatographically separate molecules by unsaturation has been widely used for determining the triglyceride composition of natural fats.

Silver ion adsorption chromatography of triglycerides was introduced in 1962 by de Vries (*218*) and Barrett *et al.* (*55*). The technique is based on the weak interaction between Ag$^+$ and the $\pi$-electrons of double and triple bonds. The Ag$^+$/olefin complex is of sufficiently low energy that it can be made and broken during standard lipid chromatographic procedures. Silver ion adsorption chromatography is accomplished by impregnating AgNO$_3$ into normal lipid adsorbents such as silicic acid or Florisil. After suitable adjustment of the solvent system, a given class of lipids can be fractionated according to the number of double bonds per molecule. Exposure to silver nitrate does not produce any chemical alterations of normal triglycerides; hence the fractions separated can be recovered unaltered from the impregnated adsorbent for further analysis.

This chapter discusses the methods and applications for silver ion *adsorption* chromatography of triglyceride mixtures. The complementary technique of silver ion *partition* chromatography (i.e., using Ag$^+$ in solution) is covered in Chapter 5.

## I. METHODS

### A. Choice of Method

Both thin-layer and column chromatography can be used for the separation of triglycerides with a AgNO$_3$-impregnated adsorbent. The choice of method for fractionating a given sample depends on the type of resolution desired and the amount of material to be separated. Column operation has the advantage of a slightly larger sample size (80–150 mg) with less possibility for atmospheric oxidation; but columns are slower, more difficult to monitor, and generally give poorer resolution than TLC plates. Thin layers of adsorbent give much more rapid separations with better resolution and easier location of the components resolved. However, TLC plates have slightly less capacity and can cause more exposure to the oxygen in the air than columns. Today most workers prefer the speed and resolution of TLC plates for both analytical and preparative separations. Atmospheric oxidation can be minimized by conducting all operations under nitrogen, and the capacity of a 1.0 mm thick TLC layer is sufficient (20–100 mg) so that two to five plates can separate as much sample as a column.

### B. Thin-Layer Chromatography

*1. Adsorbent*

Thin-layer plates impregnated with AgNO$_3$ are usually prepared by mixing a commercial silicic acid suitable for TLC with an aqueous solution of AgNO$_3$ and spreading in the usual manner. According to Morris (*663*), any AgNO$_3$ level above 2% (i.e., AgNO$_3$/silicic acid = 2/98, w/w) will give the same sample resolution, and he recommends 5% AgNO$_3$ as the optimum and most economical level of impregnation for general use. Higher levels are beneficial with certain types of samples, however. Bottino (*100*) reports 8% silver nitrate as the most suitable amount for separating highly unsaturated triglycerides of marine origin. Resolution of double-bond positional isomers is enhanced by using AgNO$_3$ levels as high as 23–30% (*666,951*).

Wood and Snyder (*963*) have recommended that the AgNO$_3$ solution used to slurry the adsorbent should contain 30% ammonium hydroxide. Better methyl ester separations are claimed for plates prepared this way, and it seems reasonable to expect improved separations with triglycerides also. Wood and Snyder attribute the better resolution on such plates to the presence of the Ag(NH$_3$)$_2^+$ ion, which may form a stronger complex with olefins than Ag$^+$.

TLC plates can also be impregnated with AgNO₃ by spraying (*605,658,659*), dipping (*659,710*), or developing (*202*) them with a solution of AgNO₃. These procedures are quite useful for impregnating the ready-made sheets of silicic acid which are now available commercially.

The adsorbent layer on AgNO₃-impregnated TLC plates is usually made 0.25 mm thick for analytical work and 0.5–1.0 mm thick for preparative separations. AgNO₃ solutions will severely corrode conventional metal TLC spreaders; but this problem can be avoided by using plastic, glass, anodized aluminum, silver-plated, or stainless steel spreaders. Any adsorbent containing AgNO₃ should be stored in a dark place so that the AgNO₃ will not be inactivated by light. However, Muldrey (*669*) reports that substituting silver zirconyl phosphate for AgNO₃ gives a light-insensitive adsorbent with the same separation characteristics.

TLC plates impregnated with AgNO₃ should be activated for 2–4 hours at 190°–195° rather than at the usual 100°–120°. According to Åkesson (*12*) and Bottino (*101*), the higher activation temperature results in better resolution of the triglyceride bands. Bottino (*101*) and Renkonen (*763*) attribute this improved resolution to greater dehydration of the silicic acid at 195°, causing decreased triglyceride/silicic acid hydrogen bonding and thus enhancing the effects of $\pi$-complexing during development.

TLC plates used for preparative work should be freed of any interfering lipid contaminants by predeveloping them with a solvent more polar than that used for the subsequent triglyceride separation (*203*). Predevelopment moves the impurities to the top of the plate where they can be removed by scraping off a narrow band of adsorbent.

## 2. Solvent

Two general types of solvent mixtures have been widely used for separating triglycerides on AgNO₃-impregnated TLC plates: (i) chloroform containing 0–6% methanol, ethanol, or acetic acid; and (ii) various benzene/diethyl ether mixtures. Typical solvent systems for separating triglyceride mixtures of varying unsaturation are listed in Table 4-1, and many related combinations can be found in the literature. The exact solvent mixture used for a specific separation varies slightly from one laboratory to another since $R_f$ values for specific triglycerides are influenced by temperature, activity of the silicic acid, relative humidity, and AgNO₃ content of the adsorbent (*205*). Hence any published solvent system should be tested and, if necessary, adjusted before use. Our experience indicates that a chloroform/ethanol solvent system gives better resolution of triglyceride bands than a benzene/diethyl ether system when plates are developed in a large chromatography tank. On the other hand, plates de-

TABLE 4-1
TYPICAL SOLVENT SYSTEMS FOR SEPARATING TRIGLYCERIDES
BY UNSATURATION ON $AgNO_3$-IMPREGNATED TLC PLATES

| Double bonds per triglyceride molecule | Solvent systems (v/v) | | References |
|---|---|---|---|
| | Chloroform[a] / Methanol Ethanol Acetic acid | Benzene/ Diethyl ether | |
| 0–4 | 100/0 to 99/1 | 100/0 to 80/20 | *55, 221, 308, 576* |
| 1–6 | 99.2/0.8 to 98.5/1.5 | 90/10 to 80/20 | *142, 313, 330, 444, 951* |
| 5–9 | 97.5/2.5 | 0/100 | *142, 328* |
| 7–12 | 94/6 | — | *100* |

[a] Chloroform normally contains 0.5–1.0% ethanol as a stabilizer.

veloped by the sandwich procedure show better band resolution with a benzene/diethyl ether solvent system.

The use of cyclic olefins such as cyclohexene should be avoided, since they form stronger $\pi$-complexes with $Ag^+$ than unsaturated triglycerides, resulting in decreased resolution (*127*).

## 3. Separation Procedure

The triglyceride sample is applied in as small a spot or as thin a band as possible in order to achieve maximum resolution. For analytical work, a single spot (3–30 μg of triglyceride dissolved in a suitable solvent) is applied near the bottom edge of the plate using a capillary tube, micropipette, or a microsyringe. For preparative work, 30–100 mg of triglyceride can be separated on 200 × 200 or 200 × 400 mm TLC plates having a 1.0 mm thick layer of adsorbent. The sample solution can be applied as a row of overlapping spots, but better band resolution is obtained with a semiautomatic sample streaker which produces a very even sample application in a thin band. The use of a wedge-shaped layer of adsorbent can give better resolution and improved visibility for minor components which migrate to the narrow portion of the wedge (*307*).

Plates are developed in a standard chromatography tank or in a sandwich apparatus which has been flushed with nitrogen to prevent sample oxidation. De Vries and Jurriens (*222,438*) have suggested that colored compounds also be spotted at the origin so that the positions of the triglyceride bands can be followed during elution. They have found that the dinitrophenylhydrazone of decanal has the same $R_f$ value as tristearin, that the dinitrophenylhydrazone of *cis*-2,*cis*-7-decadienal travels with triolein, and that Sudan II (Fluka, Buchs, Switzerland) contains two pigments which migrate with triglycerides containing one and two *cis* double bonds. Multiple development with the same solvent system improves resolution (*202,576,951*), as does development at −22° (*951*). Separations of triglycerides by Ag⁺ adsorption TLC are not always reproducible from plate to plate in the same laboratory; band resolution on one plate in six may be poorer than expected.

After development, the TLC plate is dried in a stream of nitrogen, and the triglycerides are located by one of several techniques. The most convenient nondestructive method for locating triglyceride spots and bands (*537,544*) is spraying with 2′,7′-dichlorofluorescein solution (0.05% in methanol/water 50/50) and viewing under ultraviolet light where lipids appear as yellow spots against a purple background. The best sensitivity is obtained when the dichlorofluorescein solution contains water (*951*) or when the plate is humidified after spraying (*912*). Rhodamine 6G (*760*), dibromofluorescein (*56*), and sodium fluorescein (*576*) have been similarly used. Destructive methods for locating triglyceride spots include drawing the flame of a glassblower's torch across the plate (*325*), or spraying with 50% orthophosphoric (*56*) or 50% sulfuric (*663*) acid followed by charring in an oven at 200°–400°. Iodine vapor is not suitable for locating triglycerides in the presence of $AgNO_3$ (*663*).

### 4. Quantitation

Thin-layer chromatograms can be quantitated in two ways: recovery of the triglyceride bands from the adsorbent followed by a suitable quantitative analysis procedure, or direct quantitation of the spots on the chromatoplate.

Sample recovery is the more useful technique, since the fatty acid composition of each fraction can then be determined. The usual diethyl ether extraction of the silicic acid is supplemented by treatment with chloride ion to ensure complete breakup of the Ag⁺ complexes with the more unsaturated triglycerides (*369, 403, 740,772*). Hill *et al.* (*369*) recommend the following procedure:

> Triglyceride bands are visualized under ultraviolet light after spraying with 0.1% 2′,7′-dichlorofluorescein solution. Each band is scraped into a test tube, and a solution of 1% NaCl in methanol/water 90/10 is added in portions (approximately 0.5 ml) with vigorous mixing until the characteristic red color of the silver-dichlorofluorescein complex is destroyed. Then 5.0 ml of diethyl ether/methanol 90/10 containing 0.1 mg/liter of BHT is added, and the adsorbent is sedimented by brief centrifugation. The supernatant solvent is decanted, and the silicic acid residue is washed two times with 4 ml of the same solvent. The extracts are combined, and the solvent is removed under a stream of nitrogen.

TABLE 4-2
ACCURACY OF GLC INTERNAL-STANDARD TECHNIQUE FOR QUANTITATIVE
ANALYSIS OF TRIGLYCERIDE MIXTURES ON AgNO$_3$-IMPREGNATED TLC PLATES

| Triglyceride | Analysis (wt %) | | | | Average | Known composition | Absolute error | Range |
|---|---|---|---|---|---|---|---|---|
| | 1 | 2 | 3 | 4 | | | | |
| PPP[a] | 25.6 | 27.0 | 26.1 | — | 26.2 | 25.6 | +0.6 | 1.4 |
| PPO | 20.4 | 22.1 | 23.4 | — | 22.0 | 22.4 | −0.4 | 3.0 |
| POO | 25.6 | 19.6 | 22.6 | — | 22.6 | 23.2 | −0.6 | 6.0 |
| OOO | 28.4 | 31.4 | 27.9 | — | 29.2 | 28.8 | +0.4 | 3.5 |
| PPP[b] | 37.4 | 31.9 | 32.2 | 34.0 | 33.9 | 34.1 | −0.2 | 5.5 |
| OOO | 15.9 | 14.8 | 14.3 | 15.2 | 15.1 | 14.2 | +0.9 | 1.6 |
| LLL | 46.6 | 53.3 | 53.5 | 50.3 | 50.9 | 51.6 | −0.7 | 6.9 |

[a] From Blank et al. (85).
[b] From Gordis (313).

During preparative TLC work, a blank plate should always be run as a control; the extract from an area equal to a large sample band should show no impurities during subsequent analysis.

The recovered triglycerides can be quantitated by most any of the micromethods listed in Table 2-3. The method of choice is clearly GLC of the derived methyl esters with an internal standard (Table 2-4, *85,176,330*), since this permits both the fatty acid composition and the amount of triglyceride present to be determined simultaneously. Blank et al. (*85*) and Gordis (*313*) have evaluated the accuracy of this technique (Table 4-2) and found an error of ±0.9 absolute percent for major components *when triplicate or quadruplicate analyses are averaged;* single analyses showed errors as high as 3–4 absolute percent. Triglyceride fractions recovered from Ag$^+$ TLC plates have also been quantitated by triglyceride GLC with an internal standard (*538*), GLC of derived glycerol with an internal standard (*746*), the chromotropic acid color reaction (*46,496,576*), the hydroxamic acid color reaction (*319,935*), infrared spectrophotometry (*69*), titration of excess KIO$_4$ after oxidation of derived glycerol (*441,442*), and weighing (*19,495*). Gravimetric quantitation is not recommended for accurate results, however, unless the smallest fraction weighs more than 5 mg and the adsorbent has been preextracted to remove any contaminants.

Triglyceride fractions can also be quantitated directly on the TLC plate by photodensitometry of the charred spots (*56,261,488*) or by measuring spot area (*912*).

## C. Column Chromatography

### 1. Adsorbent

The adsorbent for silver ion column chromatography is prepared by soaking silicic acid in a solution of $AgNO_3$. A typical preparation has been described by de Vries (219):

> One hundred grams of uniform size silicic acid is suspended in 200 ml of 50% (w/v) aqueous $AgNO_3$ solution. The mixture is heated at 100° for 30 minutes. After cooling, it is filtered through a Büchner funnel and dried at 120° for 16 hours. The product will contain 0.3–0.4 g of $AgNO_3$ per gram of adsorbent.

Activation at 195° rather than 120° might result in improved resolution as with TLC (Section I,B,1). De Vries originally recommended grinding the adsorbent before use, but other workers (135,325,660) have found this unnecessary. Gunstone et al. (325) report an alternative procedure in which the slurry of silicic acid in $AgNO_3$ solution is evaporated rather than filtered. Silicic acid already impregnated with $AgNO_3$ can now be purchased commercially (Applied Science Laboratories, State College, Pa.) in a wide range of mesh sizes.

### 2. Solvent

The separation of triglyceride mixtures by column chromatography on $AgNO_3$-impregnated silicic acid is accomplished by stepwise elution with petroleum ether/benzene/diethyl ether mixtures. Table 4-3 shows the stepwise elution patterns used by de Vries (220) and Subbaram and Youngs (863) for separating triglycerides containing zero to four double bonds. The exact solvent ratios employed can vary slightly from one batch of adsorbent to another (219). Column fractionation of triglycerides containing more than four double bonds per molecule has been attempted (228,229) but has not proven successful in separating specific groups of triglycerides on the basis of unsaturation.

### 3. Separation Procedure

Columns of $AgNO_3$-impregnated silicic acid are packed in the usual manner using a slurry of the activated adsorbent in petroleum ether. A filter aid (such as Celite 535) is sometimes mixed with the adsorbent before packing to obtain higher solvent flow rates (219) Ten to twenty grams of adsorbent are required to separate 80–150 mg of triglyceride mix-

TABLE 4-3
SOLVENT SYSTEMS USED FOR SEPARATING
TRIGLYCERIDE MIXTURES BY COLUMN
CHROMATOGRAPHY ON $AgNO_3$-IMPREGNATED
SILICIC ACID[a]

| Double bonds per triglyceride molecule | Eluting solvent (v/v) |
|---|---|
| 0 | Petroleum ether/benzene 60/40 |
| 1 | Petroleum ether/benzene 45/55 to 40/60 |
| 2 | Petroleum ether/benzene 20/80 |
| 3 | Petroleum ether/diethyl ether 80/20 or Benzene |
| 4+ | Diethyl ether |

[a] From de Vries (*220*) and Subbaram and Youngs (*863*).

ture on an 11–18 mm diameter column (*220,863*) and the technique can be scaled up to handle larger samples if necessary. Columns are wrapped in black paper or aluminum foil to protect the adsorbent from light. In his original publication, de Vries (*220*) recommended that the column temperature be kept constant at 15°, but other workers (*325,863*) have obtained satisfactory results operating at room temperature.

A typical separation procedure for the fractionation of triglycerides by column chromatography on $AgNO_3$-impregnated silicic acid is outlined in the diagram and caption of Fig. 4-4. Other operating conditions follow standard column chromatography practice. Sample separation is usually monitored by examining an identical amount of each fraction on $AgNO_3$-impregnated TLC plates. Spot positions show how effectively the original mixture was separated according to unsaturation, and spot intensity is a guide to the amount of triglyceride in each fraction.

## 4. Quantitation

Any of the micromethods listed in Table 2-3 can be employed for quantitation of column chromatography separations. The most widely used method is to evaporate each fraction in a tared tube and weigh the amount of triglyceride recovered (*220,228,863*), but this is extremely tedious. Automatic detector systems for liquid chromatography work (Table 2-3) are much less work if they are available. Alternative quantitation procedures used include direct GLC of an aliquot from each fraction (687) or densitometry of aliquots placed on TLC plates and charred without development (737).

## II. APPLICATIONS

### A. Separation by Number of *cis* Double Bonds

#### 1. Elution Order

Gunstone and Padley (328) have studied the separation of triglycerides containing saturated, monoene, linoleic, and linolenic acids and found the following elution order during Ag⁺ adsorption chromatography:

```
top  000  001  011  002  111  012  112
022  003  122  013  222  113  023  123
223  033  133  233  333  bottom
```

It is apparent from this list that elution order is not based solely on the number of *cis* double bonds. The affinity of the silver ions for the double bonds is greater when they are clustered in one fatty acid chain than when they are divided between several chains. Hence a linoleic chain forms a stronger complex with Ag⁺ than two oleic chains. Similarly, the three double bonds in one 18:3 chain produce a stronger $\pi$-complex than the four double bonds in two 18:2 chains. This elution order has generally been confirmed by other workers (55,85,220,951), although two exceptions have been noted. Under certain experimental conditions (possibly different Ag⁺ concentration), **011** and **002** have identical $R_f$ values (Fig. 4-1), and **003** can be adsorbed more strongly than **122** (951).

Gunstone and Padley (328) noted that the elution order of triglycerides in Ag⁺ adsorption chromatography can be predicted by assigning arbitrary values to the $\pi$-complexing power of each fatty acid chain: saturated $= 0$, monoene $= 1$, diene $= 2 + a$, and triene $= 4 + 4a$, where $a < 1$. Thus the triglyceride **033** with six double bonds has a complexing power of $8 + 8a$, which is greater than the value $8 + 6a$ for **322** with seven double bonds. Similar predictable relationships exist for the entire elution series.

Although major separation characteristics are determined by the Ag⁺/olefin complex, three minor factors also influence the exact $R_f$ of a triglyceride. Larger molecules travel ahead of smaller molecules (29, 56,221,331) due to their poorer adsorption by the silicic acid (Chapter 8, Section I,B). This is particularly noticeable in samples containing acids shorter than $C_{14}$ or longer than $C_{20}$. The position of a double bond in the fatty acid chain also has some influence on where its triglyceride will elute (Section II,B). Triglyceride positional isomers such as β-**001** and β-**010**, β-**002** and β-**020**, and β-**011** and β-**101** can even be separated to some extent (Section II,B). These three factors often broaden band width and decrease resolution.

## 2. Thin-Layer Chromatography

A maximum of five to ten triglyceride bands can be resolved on a single TLC plate. For more complex mixtures, two separations are necessary using solvent systems of different polarity (Table 4-1). Figure 4-1 illustrates a typical separation of triglycerides from **000** to **222**.

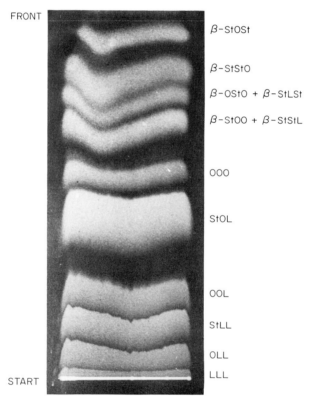

FIG. 4-1. Preparative separation of a mixture of synthetic triglycerides by $Ag^+$ adsorption thin-layer chromatography. *Operating conditions:* 200 × 400 mm TLC plate coated with 1.0 mm layer of $AgNO_3$/Kieselgel G; sample size, 60 mg; horizontal development with benzene/diisopropyl ether 85/15; bands visualized under ultraviolet light after spraying with 2′,7′-dichlorofluorescein solution. From den Boer (*213*).

## II. APPLICATIONS

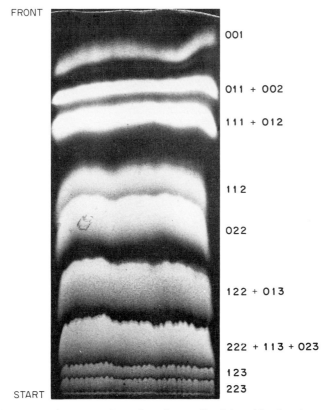

FIG. 4-2. Preparative separation of soybean oil triglycerides by Ag⁺ adsorption thin-layer chromatography. Major triglycerides in each band are tentatively identified using the data of Jurriens and Kroesen (438,443) and Gunstone and Padley (328). *Operating conditions:* 200 × 400 mm TLC plate coated with 1.0 mm layer of $AgNO_3$/silicic acid 3/7; sample size, about 60 mg; single horizontal development with benzene/diethyl ether 85/15; bands visualized under ultraviolet light after spraying with 2′,7′-dichlorofluorescein solution. From Jurriens (438).

Preparative separations of soybean and cod liver oil triglycerides are shown in Figs. 4-2 and 4-3.

Band resolution decreases as triglycerides become more unsaturated. Triglycerides from **000** to **222** can be separated as single bands if no linolenic acid is present (213,495). When a fat contains 18:3, triglycerides from **222** to **333** are frequently only partially resolved, and preparative TLC will often recover ternary or quaternary mixtures (328). Natural fat triglycerides containing four-, five-, and six-double-bond fatty acids are relatively poorly resolved, and only a few distinct bands can be obtained (100).

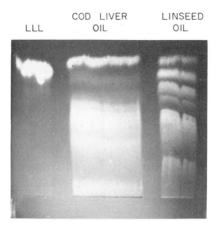

FIG. 4-3. Preparative separation of cod liver oil triglycerides by Ag⁺ adsorption thin-layer chromatography. Trilinolein and linseed oil standards run on same plate for comparison. *Operating conditions:* 200 × 200 mm TLC plate coated with 0.5 mm layer of AgNO$_3$/Adsorbosil-1 8/92; sample size, 20–25 mg; single vertical development with chloroform/ethanol 94/6; bands visualized under ultraviolet light after spraying with 2′,7′-dichlorofluorescein solution. From Bottino (*100*).

Silver ion TLC is a useful method for determining the SSS content of margarines, shortenings, or the component fats used in their formulation. Suitable techniques for isolating and quantitating SSS for this purpose have been described by Persmark and Töregård (*714*) and Bandyopadhyay (*46*). Triglycerides containing oxygenated fatty acids have also been successfully separated on the basis of unsaturation using Ag⁺ TLC (*328,332*).

After preparative TLC, the component triglycerides in a recovered fraction can usually be calculated from: (i) its molar composition of fatty acids, (ii) its relative position on the TLC plate, and (iii) how cleanly the band was resolved. Table 4-4 outlines two methods for calculating the triglycerides present in fractions isolated from natural fats.

### 3. Column Chromatography

Current techniques for column chromatography of triglycerides on AgNO$_3$/silicic acid can resolve only six triglyceride fractions: **000, 001, 002, 011, 111 + 012**, and more unsaturated molecules. Attempts to resolve more unsaturated triglyceride mixtures (*228,229,767*) have so far been unsuccessful. Figure 4-4 shows an efficient separation of palm oil triglycerides by column chromatography. The components in each recovered fraction are calculated in the same manner as for TLC separations (Table 4-4).

II. APPLICATIONS 61

TABLE 4-4
METHODS FOR CALCULATING THE TRIGLYCERIDE COMPOSITION OF FRACTIONS
RECOVERED FROM Ag$^+$ ADSORPTION CHROMATOGRAPHY
OF NATURAL FAT TRIGLYCERIDES

| Fatty acid composition (mole %) | | | Calculations |
|---|---|---|---|
| \multicolumn{4}{c}{Band Recovered from Olive Oil Triglycerides (497)} |

| 16:0 | 57.1 | ⎱ 64.6 | (a) Fatty acid composition and position on the TLC plate indicate this fraction is mostly **001**. |
| 18:0 | 7.5 | ⎰ | |
| 16:1 | 1.6 | ⎱ 35.4 | (b) **001** is well separated from **000** and **011** on TLC plate. Thus deviations from theoretical composition (66.7% saturated, 33.3% monoene) are due to experimental error. |
| 18:1 | 33.8 | ⎰ | |
|  |  |  | (c) % SSPo = (33.3/35.4)(1.6)(3) = 4.5% |
|  |  |  | % SSO = (33.3/35.4)(33.8)(3) = 95.5% |

Band Recovered from Wild Rose Seed Oil Triglycerides (328)

| 16:0 | ⎱ 11 | (a) Fatty acid composition and position on TLC plate indicate this fraction is mainly **033** and **133**. |
| 18:0 | ⎰ | |
| 18:1 | 18 | (b) Since band is not cleanly separated on the TLC plate, **223** and **233** may also be present based on theoretical elution order. |
| 18:2 | 8 | |
| 18:3 | 63 | |
|  |  | (c) % **033** = 3 × 11 = 33% |
|  |  | % **133** = 3 × 18 = 54% |
|  |  | (d) **223** + **233** = 100 − 33 − 54 = 13 |
|  |  | $\tfrac{2}{3}$(**223**) + $\tfrac{1}{3}$(**233**) = 8 |
|  |  | Solving these two simultaneous equations gives |
|  |  | % **223** = 11% and % **233** = 2% |

## B. Isomer Separations

Four types of isomeric triglycerides can be resolved by silver ion adsorption chromatography:

*Isomeric Positioning of Double Bonds Anywhere within the Triglyceride Molecule.* Excellent separations of OOO and StOL, **022** and **013**, and **020** and **011** are shown in Figs. 4-1, 4-2, and 4-4.

*Isomeric Positioning of Double Bonds within a Single Fatty Acid Chain.* Wessels and Rajagopal (951) were able to resolve isomeric oleoyl and petroselinoyl triglycerides by triple development at −22° (Fig. 4-5). Related studies on the double-bond positional isomers of fatty acid methyl esters (167,327,666) indicate that this kind of separation is possible only with a few specific isomer pairs. Certain combi-

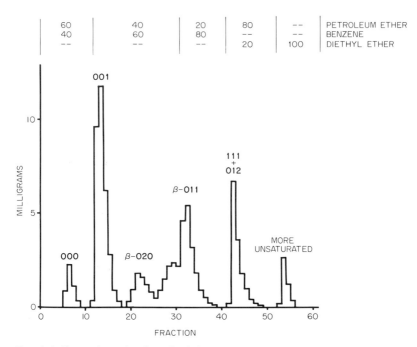

FIG. 4-4. Separation of palm oil triglycerides on a column of AgNO$_3$-impregnated silicic acid. Apparent anomoly of **002** eluting before **011** is due to presence of β-**020** and β-**011** isomers (compare with Fig. 4-1 and *442*). *Operating conditions:* 10 g of AgNO$_3$/silicic acid 1/3 mixed with 5 g of Celite 535 filter aid in an 11 × 400 mm column; sample size, 97 mg; 30 ml/hour stepwise elution with solvent mixtures indicated; separation monitored gravimetrically by collecting and evaporating 10 ml fractions. From de Vries (*220*).

nations of double-bond position and fatty acid chain length can sometimes yield isomerlike separations. For example, Gunstone and Qureshi observed subfractionation of 18:1ω9* and 20:1ω11* triglycerides during Ag⁺ TLC of *Gmelina asiatica* seed fat. Similar distinctions between 16:1ω9* and 18:1ω11* glycerides have also been observed (*331,916*).

*Geometric Isomers of Double Bonds.* Trans olefins form weaker π-complexes than *cis* olefins; hence *cis–trans* isomers can be separated by silver ion adsorption chromatography. De Vries and Jurriens (*220,221*) and Wessels and Rajagopal (*951*) have demonstrated the resolution of isomeric oleoyl and elaidoyl triglycerides by this means

---

* Assuming normal positioning of the double bond in these monoene acids.

## II. APPLICATIONS

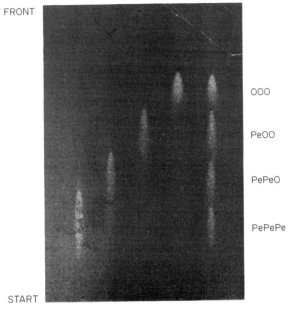

FRONT

OOO

PeOO

PePeO

PePePe

START

FIG. 4-5. Analytical separation of isomeric triglycerides of oleic and petroselinic acids by Ag⁺ adsorption thin-layer chromatography. *Operating conditions:* 200 × 200 mm TLC plate coated with 0.5 mm layer of AgNO₃/Kieselgel G 23/77; sample size, 15 μg each component; triple development with toluene/diethyl ether 75/25 at −22°; spots visualized under ultraviolet light after spraying with 2′,7′-dichlorofluorescein solution. From Wessels and Rajagopal (*951*).

(Fig. 4-6). Such separations of isomeric *cis* and *trans* triglycerides have proven useful in the analysis of hydrogenated fats (*185*).

*Isomeric Esterification of Fatty Acids to Glycerol.* Separations of β-**001** and β-**010**, β-**011** and β-**101**, and β-**002** and β-**020** by Ag⁺ TLC have been reported (Fig. 4-1, *55, 768, 951*). The symmetrical isomer has the higher $R_f$ value in each case.

Isomer separations in silver ion adsorption chromatography are most prominent when the triglyceride molecule has relatively few double bonds. This is particularly true for isomeric positioning of a double bond within a single fatty acid chain and for isomeric esterification of fatty acids to the glycerol. Isomeric forms of more-unsaturated triglycerides must also be adsorbed differently, but these differences are apparently too small a part of the total π-complexing to effect a separation.

Chain length isomers (i.e., PPP and MPSt) and optical isomers (i.e., *sn*-POL and *sn*-OPL) of triglycerides cannot be separated by silver ion adsorption chromatography.

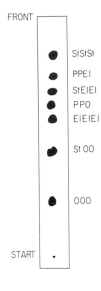

Fig. 4-6. Analytical separation of isomeric triglycerides of oleic and elaidic acids by $Ag^+$ adsorption thin-layer chromatography. *Operating conditions:* 200 × 200 mm TLC plate coated with 0.3 mm layer of $AgNO_3$/Kieselgel G 30/70; sample size, ~25 µg each compound; single development with benzene; spots visualized under ultraviolet light after spraying with 2′,7′-dichlorofluorescein solution. From de Vries and Jurriens (*221*).

## C. Other Functional Groups

Other functional groups having π-electrons form different strength complexes with $Ag^+$, producing useful but more complex triglyceride separations than is encountered with *cis* double bonds. Padley (*699*) has used $Ag^+$ TLC to fractionate seed triglycerides containing conjugated trienoic acids. The unique separations of natural fat triglycerides containing cyclopentene fatty acids (chaulmoogric, gorlic, etc.) have been characterized by Mangold and his co-workers (*45,202*). No doubt triglyceride mixtures containing allenic or acetylenic unsaturation could also be usefully separated on $AgNO_3$-impregnated silicic acid, judging from results with the methyl esters of these acids (*336*).

Triglycerides containing cyclopropene fatty acids (malvalic and sterculic) cannot be separated by silver ion adsorption chromatography because the cyclopropene ring reacts with $AgNO_3$ to produce a complex mixture of products (*429,506*).

## D. Oxidized Triglycerides

During the permanganate oxidation of triglycerides, all unsaturated acids are converted to saturated dicarboxylic acids having one carboxyl group free and the other attached to glycerol. If these free carboxyl groups are then esterified with allyl alcohol, new triglyceride molecules are produced in which each unsaturated acid in the original mixture has been replaced

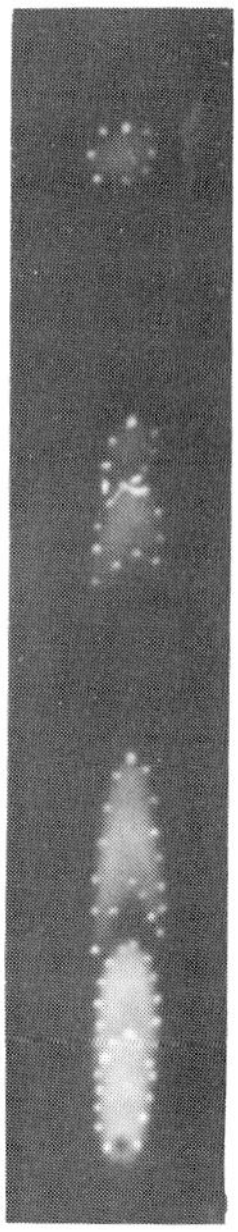

II. APPLICATIONS  65

FIG. 4-7. Analytical separation of the allyl esters of oxidized lard (interesterified) triglycerides by Ag⁺ adsorption thin-layer chromatography. *Operating conditions:* 50 × 150 mm TLC plate coated with 0.09 mm layer of AgNO$_3$/Kieselgur D5/powdered cellulose/sucrose 5/20/2/1; sample size, ∼100 μg; developed with CHCl$_3$/CH$_3$COOH 99.5/0.5; spots visualized under ultraviolet light after spraying with 2′,7′-dichlorofluorescein solution; spots outlined with small holes to aid in quantitating spot area. From Vander Wal (*912–914*).

by a monoenoic residue. Silver ion adsorption chromatography can now separate these compounds according to the various combinations of S and U in the original sample. Vander Wal (*912,913*) has developed such a technique for determining the SSS, β-SUS, β-SSU, β-USU, β-SUU, and UUU content of natural fats; a typical separation is shown in Fig. 4-7.

## E. Derived Diglycerides

The diglycerides produced by selective deacylation of triglycerides (Chapter 9) or by dephosphorylation of phospholipids can also be effectively separated by unsaturation using silver ion adsorption chromatography. Although free diglycerides can be used for this purpose, it is preferable to acetylate the sample first to avoid any isomerization caused by acyl migration. The resultant diglyceride acetates are, of course, triglycerides, which can then be separated by the procedures described above (Fig. 4-8). The best resolution is obtained when the original diglycerides are all the *sn*-1,2(2,3)- or all the *sn*-1,3-isomer, since these two isomers are adsorbed differently by silicic acid (Chapter 8, Section I,H).

Based on the experiments of Kuksis and his co-workers (*390,543*) and Renkonen (*761–763,765,766*) with diglyceride acetates and the triglyceride data of Gunstone and Padley (*328*), the known elution order for di-

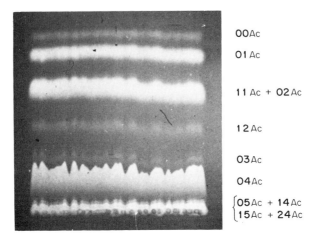

FIG. 4-8. Preparative separation of diglyceride acetates derived from rat liver lecithins using Ag$^+$ adsorption thin-layer chromatography. *Operating conditions:* 200 × 200 mm TLC plate coated with 0.5 mm layer of AgNO$_3$/Silica Gel G 2/8; sample size approximately 20 mg; developed with chloroform/methanol 99.2/0.8; bands visualized under ultraviolet light after spraying with 2′,7′-dichlorofluorescein solution. From Kuksis *et al.* (*537*).

glyceride acetates is

| top | 00Ac | 01Ac | 11Ac | 02Ac | 12Ac | 22Ac | 03Ac |
|---|---|---|---|---|---|---|---|
| 04Ac | 14Ac | 05Ac | 06Ac | $\begin{bmatrix} 24\text{Ac} \\ 15\text{Ac} \end{bmatrix}$ | $\begin{bmatrix} 55\text{Ac} \\ 56\text{Ac} \\ 66\text{Ac} \end{bmatrix}$ | bottom | |

It is surprising to find **06**Ac with a higher $R_f$ value than **24**Ac and **15**Ac (*543*), since Gunstone and Padley (*328*) found Ag$^+$/olefin π-bonding to be stronger when two or three double bonds are clustered together in one chain. Apparently this effect does not always hold for six-double-bond acids.

If handled rapidly enough, underivatized diglycerides can be separated by silver ion adsorption chromatography without appreciable isomerization (*12,15,916*), provided the polarity of the developing solvent is increased slightly.

Useful separations of isomeric diglycerides and diglyceride acetates by Ag$^+$ TLC have also been reported. Renkonen and Rikkinen (*768*) have separated β-StOAc and β-OStAc in this manner. Two isomers of *sn*-**03**–, one containing 20:3-*5c,8c,11c* and the other 20:3-*7c,10c,13c*, were resolved by van Golde and van Deenen (*915*). Even more remarkable is the report by these same authors that *sn*-OPo–, *sn*-OO–, *sn*-PPo–, and *sn*-PO– can be separated into four separate spots (*916*).

# 5

# LIQUID–LIQUID PARTITION CHROMATOGRAPHY

One of the most versatile triglyceride separation methods available to the lipid chemist today is liquid–liquid partition chromatography. This technique separates molecules on the basis of their differential solubilities in two immiscible solvents, one of which is mobile while the other remains stationary. Triglycerides are fractionated into distinct groups according to their *partition number* (see Section II,A,1), which is a predictable function of both molecular weight and unsaturation. This principle can be utilized in four familiar chromatographic procedures: thin-layer chromatography (TLC), paper chromatography (PC), column chromatography, and countercurrent distribution (CCD). The very mild conditions employed make the technique ideal for use with highly unsaturated samples.

Liquid–liquid partition chromatography was first applied to the analysis of natural fat triglycerides in 1956 by Dutton and Cannon (*238*) employing CCD and by Priori (*730*) using PC. The technique was developed further in a number of laboratories in the 1960s, especially by Kaufmann and his co-workers (*475,476,478–487,490–499*) for TLC and PC applications. A major advance was introduced in 1964 when Vereshchagin and Skvortsova (*924,927*) added $AgNO_3$ to the polar phase and produced separations based mainly on unsaturation. Today, partition chromatography is one of the most widely used chromatographic techniques for the analysis of molecular species of triglycerides.

## I. METHODS

### A. Choice of Method

The selection of thin-layer, paper, column, or countercurrent distribution methods for liquid–liquid partition chromatography of triglycerides depends mainly on three factors: sample size, resolution desired, and equipment available. TLC and PC require very simple equipment and can separate 5–25 $\mu$g analytical samples with the highest resolution of 1–40 mg preparative samples with slightly less resolution. Column chromatography fractionates 10–300 mg samples with resolution approaching preparative TLC and necessitates only a modest investment in equipment. On the other hand, CCD handles larger 1–10 g samples with appreciably less resolution but requires expensive apparatus. TLC is the most widely used procedure because it is the most rapid and most easily monitored. However, the recent introduction of automatic monitoring equipment for column chromatography (Table 2-3) will no doubt increase the popularity of column operation.

### B. Thin-Layer and Paper Chromatography

Techniques for separating triglycerides by thin-layer and paper partition chromatography are very similar, and the two techniques are conveniently discussed together. In most laboratories, however, TLC has completely replaced PC because the former is much more rapid (30–90 minutes vs. 4–16 hours for development) and gives somewhat better resolution (*478,484,485,547*).

#### 1. Preparation of Plates and Paper

TLC plates for partition work are prepared in the usual manner using kieselguhr (*143,485,495,547*), silicic acid (*144,646,696,838*), or silanized silicic acid (*30,567*), dried 1 hour at 110°, and stored in a desiccator. The layer thickness is usually 0.25 mm for analytical work and 0.5–1.0 mm for preparative separations. Kieselguhr is the preferred support since it exhibits little or no sample adsorption and makes spot location easier (Section I,B,3). Gypsum layers have also been employed (*480*) when the developed plates are to be dipped in aqueous reagents for spot location. Preparative TLC plates should be cleaned up prior to use by predevelopment with a solvent such as diethyl ether, which will move organic contaminants to the top of the plate where they can be removed by scraping off a narrow band of the coating.

Paper chromatograms are prepared from Whatman No. 1 (*603*), Schleicher & Schüll 2040b or 2043b (*484*), or similar grade chromatographic papers. Before use, the paper is dried 20 minutes at 100°, extracted with benzene, dried again at 100°, and stored in a desiccator (*484*).

Chromatograms are impregnated with the nonpolar stationary phase by:

(a) Immersing them (*very slowly* for TLC plates) (*141,321,483, 646*), developing them (*13,27*), or spraying them (*678*) with a 0.5–15% solution of the stationary phase in a volatile solvent.

(b) Spreading the original thin layer by slurrying the solid support in a solution of the stationary phase in a volatile solvent (*30,567*).

(c) Silanizing the chromatogram by reaction with an alkylchlorosilane (*485,695,696*).

The amount of stationary phase deposited is usually between 6 and 15% of the weight of the solid support (*31,321,567,925*) and can be controlled by varying the concentration of the impregnating solution (a, b) or the moisture content of the solid support (c). After solution impregnation, the residual solvent is evaporated, being careful, if necessary, to minimize loss of a volatile stationary phase. Silanized chromatograms must be developed in methanol before they can be used with $AgNO_3$-containing solvents (*696*) and with dioxane before they can be used for preparative work (*695*).

It should be noted that the *amount* of stationary phase impregnated onto a chromatogram has a marked influence on the $R_f$ values for specific triglycerides (*27,485*). When the amount of impregnated stationary phase is not specified in a publication, it is frequently necessary to adjust the composition of the polar phase to obtain comparable results.

## 2. Solvent System

The selection of a suitable solvent system for separating triglycerides by partition chromatography is largely empirical. However, numerous stationary and mobile phases have been evaluated for thin-layer and paper (Tables 5-1, 5-3, and 5-4), column (Table 5-6), and countercurrent distribution (Table 5-7) work; and suitable solvent systems for separating most triglyceride mixtures have been found. Fortunately, solvent systems are practically interchangeable between these four forms of partition chromatography, so that an effective solvent for TLC can often be adopted for CCD work and vice versa. Note, for example, that linear hydrocarbons are the major stationary phases used in all four methods; and polar phases such as nitroethane (TLC, CCD), acetone/methanol 60/40 to 90/10

TABLE 5-1
NONPOLAR STATIONARY PHASES USED TO SEPARATE TRIGLYCERIDES BY
THIN-LAYER OR PAPER PARTITION CHROMATOGRAPHY

| | Nonpolar stationary phase | References |
|---|---|---|
| | **Nonvolatile liquids** | |
| Hydrocarbons | Hexadecane | *567* |
| | Paraffin oil (bp > 270°) | *857, 951* |
| | Dodecylbenzene | *679* |
| Silicones | Silicone oil (dimethylpolysiloxane, 5–50 cs viscosity) | *485, 603* |
| | Methylsilyl ethers bonded to support | *696* |
| | **Volatile liquids** | |
| Hydrocarbons | Petroleum hydrocarbon (bp 140°–170°) | *678* |
| | Undecane or petroleum hydrocarbon (bp 190°–220°) | *482, 484* |
| | Dodecane | *924, 926* |
| | Tetradecane or petroleum hydrocarbon (bp 240°–250°) | *478, 485* |
| | Tetralin | *678, 679* |

(TLC, PC, column), and acetone/water 95/5 (PC, column) have found application in two or more partition separation methods.

Nonpolar stationary phases for TLC and PC (Table 5-1) are classified as *permanent* or *temporary* depending on whether they can be completely evaporated after development. Volatile liquids permit the use of a wider choice of reagents for spot location (Table 5-5) and are more easily removed from the sample in preparative work, but they do not give the highest resolution. Two types of compounds have proven effective as stationary phases in TLC and PC of triglycerides: hydrocarbons and silicones. The highest resolution to date has been obtained with long-chain linear alkanes (*478*); and resolution improves as the hydrocarbon chain becomes longer (*478,494,569*). Aromatic hydrocarbons have only been tested with mercurated triglycerides. Silicone stationary phases can be commercial silicone oils or alkylsilyl ethers bonded directly to the solid support. The chemically bonded silicone is preferable for preparative work since it is not easily extracted with the sample (*695*), but only methylsilyl ethers have so far been used for triglyceride separations (*696*). It would seem advantageous to combine the high resolution of long-chain hydrocarbons with the non-extractable nature of bonded silicones by using a reagent such as hexadecyltrichlorosilane to produce a bonded, long-chain stationary phase (*2,34,349a,508*).

It is customary to select the nonpolar liquid as the stationary phase in TLC and PC (i.e., so-called "reversed phase" chromatography). Long-chain triglycerides are more soluble in the nonpolar than in the polar phases used in partition chromatography. Therefore, the volume of polar liquid should be larger than the volume of nonpolar liquid for the most effective separation; and this is most easily achieved by choosing the polar liquid as the mobile phase.

Typical polar liquid phases for TLC and PC of various triglycerides are listed in Table 5-3. Most are binary mixtures based on the eluotropic series given in Table 5-2 so that $R_f$ values can be varied by changing the ratio of the two components. The more polar the triglycerides, the greater the polarity of the mobile phase used to separate them. Binary mixtures of acetone/acetonitrile and acetone/methanol have been the most widely used because (i) they give good resolution, (ii) their low viscosity produces faster development, (iii) small changes in solvent composition have no pronounced effect on $R_f$ values, and (iv) they provide considerable latitude for adapting the solvent ratio to specific samples and operating conditions. Nitroethane, propionitrile, and ethanol should also prove useful for the same reasons. Acetic acid has the disadvantage of a high viscosity, which leads to longer development times (484).

Vereshchagin and Skvortsova (924,927) have shown that the addition of $AgNO_3$ to the polar phase causes $Ag^+$/olefin complexing to occur and enhances the solubility of the unsaturated triglycerides in the mobile liquid. This radically changes the partition coefficients of the triglycerides, produc-

TABLE 5-2

Eluotropic Series of Solvents Used for the Polar Phase in Partition Chromatography of Triglycerides[a]

| Polarity | Established eluotropic series | Estimated positions |
|---|---|---|
| Least | Heptane | |
| | Chloroform | |
| | | Nitropropane |
| | | Isopropanol |
| | Acetone | |
| | Acetic acid | |
| | | Ethanol |
| | | Nitroethane |
| | Methanol | |
| | Acetonitrile | |
| | | Furfural |
| Most | Water | |

[a] Based on the experimental results summarized in Tables 5-3, 5-6, and 5-7 and on the relative dielectric constants of the solvents.

## TABLE 5-3
### Typical Polar Mobile Phases Used to Separate Triglycerides by Partition Number in Thin-Layer or Paper Partition Chromatography

| Partition numbers of triglycerides separated | Polar mobile phase[a] | Proportions (v/v) | References |
|---|---|---|---|
| 52–60 | Acetone/acetonitrile | 80/20 | *319, 495, 951* |
|  | Acetone/methanol | 80/20 | *143* |
| 36–52 | Acetone/acetonitrile | 80/20 to 64/36 | *478, 499, 951* |
|  | Acetone/methanol | 90/10 to 64/36 | *143, 321, 499, 857* |
|  | Acetone/acetic acid | 85/15 to 0/100 | *373, 499, 547, 646, 921, 926* |
|  | Acetone/water | 95/5 | *792* |
|  | Nitroethane | 100 | *567* |
|  | Methanol/acetonitrile/propionitrile | 48/38/14 | *485* |
| 24–36 | Acetone/acetonitrile/water | 62/33/5 | *13* |
|  | Acetone/acetonitrile | 80/20 to 70/30 | *483–485* |
|  | Acetic acid/water | 100/0 to 90/10 | *482, 603* |
|  | Methanol/acetonitrile | 56/44 | *485* |
|  | Methanol/chloroform/water | 71/24/5 | *603* |
| 16–22 | Acetic acid/water | 60/40 | *603* |
|  | Acetonitrile/water | 60/40 | *603* |

[a] All polar mobile phases must be 70–100% saturated with stationary phase.

ing a separation based on the number of double bonds per molecule with some subfractionation by partition number (*696;* Section II,B,1). Table 5-4 lists suitable $AgNO_3$-containing polar solvents for TLC and PC of triglycerides.

If the stationary phase is not bonded to the solid support, the polar mobile phase must be saturated with the nonpolar liquid by shaking the two phases together in a separatory funnel before use. If this is not done, the mobile phase will strip the stationary liquid from the solid support. Kaufmann and Makus (*483,484*) have suggested that the mobile phase be only 80–90% saturated with stationary phase (i.e., 80 ml equilibrated mobile phase + 20 ml unequilibrated mobile phase = 80% saturated). This assures continuous equilibrium during development by preventing a slight drop in temperature from precipitating additional stationary phase onto the chromatogram.

### 3. Separation Procedure

A dilute solution of the triglyceride sample is placed in a micropipette, microsyringe, or sample streaker and applied near the bottom edge of the

## I. METHODS

TABLE 5-4

AgNO₃-Containing Polar Mobile Phases Used to Separate Triglycerides by Unsaturation in Thin-Layer and Paper Partition Chromatography

| Triglycerides separated | Polar mobile phase | Proportions (v/v) | References |
|---|---|---|---|
| Cottonseed oil | Methanol/AgNO₃ | 100/sat.[a] | 924, 927 |
| Seed oils of high 18:3 content | Methanol/water/AgNO₃ | 98/2/sat. 95/5/sat. 80/20/sat. | 683, 684, 926 |
| LLL, OLL, StLL, OOO, POO, StOO, StStL | Acetone/ethanol/water/ acetonitrile/AgNO₃ | 83/8/7/2/sat. | 696 |
| OOO, POO, StOO, POP, StOSt | Acetone/ethanol/water/ acetonitrile/AgNO₃ | 72/18/8/2/sat. | 696 |

[a] Saturated with AgNO₃.

chromatogram in as small a spot or as thin a band as possible. The usual sample size is 2–10 μg per spot for analytical work and 1–6 mg per band (180 mm wide) in preparative work. The chromatogram is then developed in the appropriate mobile phase by the ascending technique in a standard chromatography tank or sandwich apparatus or by using horizontal development in a linear or radial manner. Radial development on circular paper chromatograms (Fig. 5-1) is particularly useful in preparative work where large (5–40 mg) samples are used (27,499,679). Multiple development (two to five times) with the same mobile phase gives improved resolution of components (143,478,951). It is often necessary to add antioxidants to the mobile solvent (672) or carry out the development in a nitrogen atmosphere to avoid oxidation of highly unsaturated samples. Constant temperature during development is essential for reproducible results

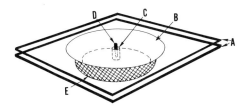

FIG. 5-1. Apparatus for preparative liquid–liquid partition chromatography of triglycerides using chromatograms developed between two glass plates: **A**, 600 × 600 mm glass plate having 8 mm diameter hole at the center; **B**, paper chromatogram prepared from round filter paper; **C**, watchglass holding solvent; **D**, wick; **E**, cork ring. From Kaufmann et al. (499).

(*484,800*). After development, the chromatogram is dried in a stream of nitrogen.

The location of triglyceride spots and bands on the developed chromatogram is complicated by the presence of the stationary phase, which is also a lipid. Three solutions to this problem have been proposed:

> (a) When kieselguhr is used as the solid support in TLC, triglyceride spots can be located by spraying with aqueous fluorescein (*950,955*) or rhodamine B (*140*) solution, even though a paraffin oil stationary phase is present. The reason for this unique phenomenon with kieselguhr is unknown.
> 
> (b) A reagent is chosen that will react with the triglycerides but not with the nonpolar liquid.
> 
> (c) A volatile (bp < 250°) stationary phase is selected so that it can be evaporated after development by heating the chromatogram at 100°–200° for 1–2 hours. Then any lipid-sensitive reagent can be employed.

The various reagents and procedures employed for triglyceride spot location are listed in Table 5-5 along with comments on their usefulness and sensitivity. Fluorescein or rhodamine on kieselguhr TLC plates is the method of choice here, since these reagents are nondestructive, extremely sensitive (<1 μg), and detect both saturated and unsaturated molecules. The widespread use of iodine, iodine/starch, iodine/cyclodextrin, and phosphomolybdic acid reagents attests to their effectiveness in locating unsaturated triglycerides, but these treatments destroy the sample and cannot be used for preparative separations unless only part of the chromatogram is exposed to the reagent (*499,567*). Volatile stationary phases have three limitations: decreased resolution (Section I,B,2), destruction of unsaturated triglycerides during heating, and slightly decreased sensitivity to spray reagents.

## *4. Quantitation*

Thin-layer and paper chromatograms can be quantitated in two ways: by extraction of the separated triglycerides followed by a suitable quantitative analysis procedure, or by direct quantitation of the spots on the chromatogram. Sample extraction is the more useful procedure since the fatty acid composition of each fraction can then be determined.

The recovery of individual triglyceride fractions requires only a simple extraction procedure. TLC bands are usually scraped into a small chromatography column and eluted with diethyl ether (*567*), while bands cut from paper chromatograms are first soaked, then washed in a suitable triglyceride solvent (*921*). Considerable stationary phase is usually recovered with each triglyceride fraction; and this can be removed, if necessary, by TLC on silicic acid or alumina (Fig. 2-3, *928*) or by extraction of the unsaponifiables while preparing methyl esters by the saponification–acidulation–esterification procedure (*567*). A blank area of the chromatogram should also be extracted and analyzed to demonstrate the absence of interfering contaminants.

The recovered triglycerides can be quantitated by most any of the micromethods listed in Table 2-3. The method of choice is usually GLC of the derived methyl esters with an internal standard (Table 2-4, *85, 176*), since this permits both the fatty acid composition and the amount of triglyceride to be determined simultaneously. If analogous studies on fractions recovered from Ag$^+$ TLC plates are taken as a guide (Table 4-2), the accuracy of the GLC internal standard method is ±0.9 absolute percent for major components *provided quadruplicate analyses are averaged*.

Vereshchagin and his co-workers (*926,928*) have measured the molar ratio of two adjacent triglyceride bands $X$ and $Y$ in preparative PC by determining the fatty acid composition of each individual band and also of the two bands combined. The relative amounts of $X$ and $Y$ can then be calculated from simultaneous equations summing the content of any major component acid:

Let $L_X$ = % 18:2 in band $X$, etc.
$X + Y = 1$
$L_X X + L_Y Y = L_{X+Y}$
$\therefore X = \dfrac{L_{X+Y} - L_Y}{L_X - L_Y}$ and $Y = 1 - X$

Triglyceride fractions recovered from thin-layer and paper partition chromatography have also been quantitated by triglyceride GLC with an internal standard (*567*) and the chromotropic acid color reaction (*949*). Gravimetric quantitation (*845*) is not recommended, since the recovered fractions are usually smaller than 5 mg and contain accompanying stationary phase.

Quantitation of triglyceride spots directly on the chromatogram is usually accomplished by combining densitometry with one of the color-producing reagents listed in Table 5-5. Chromatograms have been quantitated in this manner using iodine/cyclodextrin (*800,932*), Sudan black (*845,921*), KMnO$_4$ (*373,374*), and diphenylcarbazone (*678*). However, iodine/cyclodextrin, KMnO$_4$, and diphenylcarbazone all have the disadvantage that color density is dependent on the amount of unsaturation present. Triglyceride spots can also be quantitated by measuring spot width or spot area (*490,845*).

TABLE 5-5
REAGENTS FOR LOCATING TRIGLYCERIDE SPOTS AFTER THIN-LAYER OR PAPER PARTITION CHROMATOGRAPHY

| Reagent | Estimated sensitivity (μg) | | Used for | | | | | References |
|---|---|---|---|---|---|---|---|---|
| | Saturated triglycerides | Unsaturated triglycerides | Thin-layer chromatography | Paper chromatography | Non-volatile stationary phase | Volatile stationary phase | | |
| Fluorescein<br>Spray with 0.01% aqueous solution of fluorescein, view under long-wave ultraviolet light. Yields fluorescent yellow-green spots against purple background. Aqueous 2',7'-dichlorofluorescein solution slightly less sensitive. Spots fade in 5–10 minutes. | 1 | 1 | X | X | X[a] | X[b] | | 950<br>955 |
| Iodine<br>Place chromatogram in tank of iodine vapor until brown spots appear. Spots fade rapidly in air. | ~50 | 1 | X | X | X | X | | 499<br>567<br>800 |
| Iodine/starch<br>Iodine/cyclodextrin<br>Iodine/benzidine<br>Place chromatogram in tank of iodine vapor until brown spots appear, expose 20–30 seconds in air to lighten background, spray with solution of 1% starch in water/ethanol 7/3, 1% α-cyclodextrin in water/ethanol 7/3, or 0.5% benzidine in ethanol. Yields permanent blue spots on white background. | ~50 | 1 | X | X | X | X | | 27<br>140<br>857<br>951 |
| Cyclodextrin/iodine<br>Spray with 1% α-cyclodextrin in water/ethanol 7/3 solution, place in tank of iodine vapor. Produces violet spots against white background. | 5 | — | X | X | X | X | | 478 |
| Phosphomolybdic acid<br>Spray with solution of 3% phosphomolybdic acid in ethanol, heat at 120° until spots appear. Gives blue spots against a yellow background or vice versa. | 1[c] | 1 | X | X | X | X | | 485<br>498<br>951 |

# I. METHODS

| Method / Description | | | | | | | | Ref. |
|---|---|---|---|---|---|---|---|---|
| Iodine/phosphomolybdic acid<br>Place in tank of iodine vapor until brown spots appear, spray with 3% phosphomolybdic acid in ethanol solution, heat at 100° until spots appear. | ? | | X | | | | X | 485 |
| Rhodamine/KOH<br>Spray with 0.05% aqueous solution of rhodamine B, then spray liberally with 10 N aqueous KOH. After a few minutes, bright spots appear on a rose-red background; spots are often clearer when seen from reverse side. | 1 | — | X | X | | | X | 27 |
| Rhodamine<br>Spray with 0.005% rhodamine B in ethanol/water 1/1, view under long-wave ultraviolet light. Gives yellow spots on pink background. | 1 | 1 | X | X | X[a] | | X[b] | 140<br>483<br>485 |
| Radioactivity<br>Take autoradiogram to determine location of $^{14}$C. | <0.1 | <0.1 | X | X | | | X | |
| Dichlorodimethylsilane<br>Place in tank of $(CH_3)_2SiCl_2$ vapor for 60 minutes, dip in water/ethanol 4/1. Produces transparent spots on opaque background. | 2 | 2 | X | | | | X[b] | 486 |
| Fat-soluble dyes: sudan black, oil red, oil blue, etc.<br>Soak 3 hours in 0.1% sudan black in ethanol/water 1/1 solution, wash off excess dye in ethanol/water 1/1, dry. Yields dark blue spots on light background. Other dyes used similarly. | 1 | 1 | X[d] | X | X | | X | 480<br>482<br>921 |
| KOH | 5 | 5 | X | X | X | | X | 482 |
| Sulfuric acid | 10 | 10 | | X | X | | X | 838 |
| Potassium permanganate | — | 5 | X | X | | | | 373 |
| Diphenylcarbazone | — | 10 | X[d] | X | X | | X | 374<br>678 |
| KOH/Cu salt | 8 | 8 | | | X | | X[b] | 893<br>490 |
| Lipase/cyclodextrin/iodine | 200 | 200 | X | X | | | X | 604 |

[a] Only when a hydrocarbon stationary phase is coated on kieselguhr (140,950,955).
[b] Stationary phase must be evaporated by heating the chromatogram for 1 hour at 100–200° before applying reagent.
[c] When residual acetone/acetonitrile is left on the chromatogram, saturated triglycerides yield yellow spots on a blue background (951).
[d] Only when the solid support is not damaged by aqueous solutions, i.e., with silanized (486) or gypsum (480) supports.

## C. Column Chromatography

### 1. Solid Support

The solid support for a nonpolar stationary phase in column partition chromatography of triglycerides (Table 5-6) can be silanized Celite* (*80,673*), powdered cellulose (*858*), alkylated Sephadex (*247*), or solid particles of the polymer which also serves as the stationary phase (*338,378,894*). Kieselguhr would also serve for use in columns as it has in TLC. If the stationary phase is a polar liquid, it can be adsorbed directly on silicic acid (*971*). For maximum resolution, the smallest particle size compatible with acceptable column flow rates is usually selected (100 to 200 mesh is typical).

TABLE 5-6
Solvent Systems for the Separation of Triglycerides by Column Partition Chromatography

| Triglycerides separated | Solid support | Stationary phase | Mobile phase[a] (v/v) | References |
|---|---|---|---|---|
| Partition numbers 27–52 | Silanized Celite | Heptane + bonded silyl ethers | Acetonitrile/methanol 85/15 | 673, 686, 687 |
| | Silanized Celite | Heptane + bonded silyl ethers | Acetone/water 89/11 | 80 |
| | Cellulose | Paraffin oil | Acetone/methanol 60/40 | 858 |
| | Polymerized soybean oil factice | Polymerized soybean oil factice | Acetone/water 95/5 | 80, 378, 379, 518 |
| | Natural rubber | Natural rubber | Acetone/methanol 50/50 | 894 |
| | Alkoxypropyl/ hydroxypropyl Sephadex | $C_{11}/C_{18}$-alkoxypropyl ethers | Acetone/heptane/water 75/20/5 | 247 |
| Oxidized triglycerides | Natural rubber | Hexane + natural rubber | Acetone/water 50/50 to 100/0 in stepwise increments | 338 |
| | Silicic acid | Ethanol/water 90/10 | Petroleum ether (bp 58°–69°) then diethyl ether | 862, 971, 974 |

[a] Saturated with stationary phase where necessary.

* Celite 545, a calcined diatomaceous earth produced by Johns-Manville, is usually prewashed with concentrated HCl to remove any metal salts which might cause sample adsorption (*673*).

## 2. Solvent System

The same criteria used in selecting and preparing solvent systems for thin-layer partition chromatography (Section I,B,2) also apply to solvent systems for column operation. In addition to the usual hydrocarbon and silicone nonpolar phases, however, solid polymers such as natural rubber (*338,894*) and polymerized soybean oil factice (*378,379*) have also proven to be effective stationary phases in columns. Relatively few solvent systems have been tested for column partition chromatography of triglycerides (Table 5-6), but any solvents that have proven satisfactory for TLC and PC (Tables 5-1, 5-3, and 5-4), and CCD (Table 5-7) can generally be adapted to column operation.

It is desirable to choose a stationary phase that does not interfere with subsequent analyses of separated fractions. Paraffin oil, rubber, and polymerized soybean oil factice all elute to some extent with the mobile phase, necessitating a cleanup procedure such as adsorption chromatography (Chapter 2, Section II) or washing (*379*). A highly volatile stationary phase such as heptane presents no residue problem, but peak resolution is not as high us with longer chain hydrocarbons (Section I,B,2). Both these problems are avoided, however, if a long hydrocarbon chain is chemically bonded to an inert support; alkylsilyl silicates (*34,507*) and alkoxypropyl Sephadex (*247*) are two such nonpolar stationary phases which are nonextractable in column chromatography with lipid solvents.

## 3. Separation Procedure

It is difficult to describe a typical separation of triglycerides by column partition chromatography since so many diverse methods have been employed (Table 5-6). The best resolution of partition number 42–52 triglycerides has been achieved by Nickell and Privett (*673*) and Steiner and Bonar (*858*); the latter system will be described here since it can handle the larger sample sizes.

Steiner and Bonar (*94,858*) report a method for separating 300 mg samples of cocoa butter triglycerides using "dry column" chromatography with a paraffin oil stationary phase and an acetone/methanol 60/40 mobile phase:

> A cellulose powder impregnated with paraffin oil is prepared by mixing 250 g of Whatman cellulose powder (standard grade, dried 1 hour of 110°) with 45 g of paraffin oil (medicinal liquid paraffin) dissolved in 1 liter of $CCl_4$. The resulting slurry is allowed to stand at room temperature in a fume hood with occasional stirring until all odor of solvent is gone. The mixture is well stirred and then dried for 1 hour at 100°. The rather fluffy, slightly cream-colored powder is cooled and stored in a desiccator.

A 5–10 mm layer of nonabsorbent cotton wool is placed at the bottom of a 25 mm × 1.5 m chromatography tube. A tamping device consisting of a light rod attached to a cork which just passes through the column is used to pack down the column material. The cellulose-paraffin powder is added about 2 g at a time, and after each addition the material is pressed and tamped so as to form a uniform solid column 1.20 m long.

Three hundred milligrams of fat are melted with about 1 g of cellulose-paraffin powder. Enough acetone is added to make the material into a thin slurry, and the mixture is heated gently on a water bath until the solvent has evaporated. Frequent stirring is required. The prepared fat sample is then transferred to the top of the column and tamped level to form a uniform layer 2–3 mm thick. Enough cellulose-paraffin powder is added to form a 30 mm protective layer on top of fat sample. An acetone/methanol 60/40 mixture saturated with liquid paraffin oil is supplied continuously to the top of the column and 25 ml fractions are collected. StOSt is fully eluted from the column within 72 hours (3.5 liters of eluate).

The remarkable resolution of 300 mg samples (see Fig. 5-11) achieved by Steiner and Bonar was undoubtedly aided by their use of a long-chain hydrocarbon stationary phase, the 48:1 ratio of column height to diameter, and their method of uniform sample application. In addition, their use of a powdered cellulose solid support and of "dry column" operation may also have been contributing factors. Curiously, Bonar (*94*) reported that no appreciable separation could be obtained when the column was packed with a slurry of the impregnated cellulose in the mobile phase.

For consistent results, partition chromatography columns are maintained at a constant temperature by a jacket of circulating water (*379,673*) or by operating in a constant temperature room.

Column separations are conveniently monitored by examining an identical amount of every $n$th fraction on thin-layer partition plates. Spot positions show how effectively the original mixture was separated, and spot intensity is a rough guide to the amount of triglyceride in each fraction.

*4. Quantitation*

Most of the micromethods listed in Table 2-3 can be used to quantitate triglyceride separations by column partition chromatography. Automatic liquid chromatography monitors are the method of choice, since they produce a finished chromatogram which can then be quantitated on the basis of peak area. Since the refractive index of the mobile phase changes with triglyceride content, Trowbridge *et al.* (*894*) and Hirsch (*378,379*) have employed recording refractometers. Ellingboe *et al.* (*247*) used a moving chain flame ionization detector for detecting triglycerides in the column eluate. Other evaporative monitoring systems employing gas chromatography detectors (Table 2-3) should also prove useful, provided that any eluted stationary phase is sufficiently volatile for removal before sample detection.

If all the eluted solvents can be evaporated or removed by washing (*379,815*), then separations can be quantitated gravimetrically. A plot of weight per tube vs. fraction number yields a chromatogram describing the separation (see Fig. 5-12). Usually only selected fractions need to be processed, and the size of the recovered sample can be increased by combining adjacent tubes. The evaporation technique is quite tedious, but completely nondestructive.

Column separations have also been monitored by direct GLC of an aliquot of each fraction (*673,686*), GLC of the derived methyl esters (*673*), titrating the iodine value of each fraction (*858*), and titrating AAA, SAA, and SSA with NaOH (*338*). The titration techniques are not satisfactory for general application since they do not detect SSS and results depend on fatty acid composition.

## D. Countercurrent Distribution

### 1. *Apparatus*

The principle of liquid–liquid partition chromatography can also be applied on a much larger scale by repeated partitioning between two immiscible solvents in a series of separatory funnels. This process is called countercurrent distribution (CCD) and is schematically illustrated in Fig. 5-2.

Modern instruments for CCD consist of a train of glass cells connected in series in such a way that they are suitable for automatic operation. The cycle for an individual cell is illustrated in Fig. 5-3. Each cell contains sufficient lower-phase solvent (usually 20–40 ml) to bring the interface up to **a** in position **C**. The amount of upper-layer solvent is varied to give optimum resolution. Equilibration of the two phases is accomplished by rocking 10-15 times between positions **A** and **B**. The cells are then held in position **B** long enough for the two phases to separate. Tilting to position **C** decants the upper layer through **c** to **d**. Returning to position **A** causes the contents of **d** to flow through **e** to the adjoining cell. The process is repeated until the upper phase moves along the entire train of cells. A 200-cell automatic CCD apparatus is illustrated in Fig. 5-4.

### 2. *Solvent System*

Since the mobile phase in CCD must be the upper layer of the partition system, the nonpolar solvent is always the mobile phase and the polar solvent the stationary phase in CCD separations. The same criteria used to select solvent systems for thin-layer partition chromatography (Section I,B,2) also apply to CCD separations. However, the choice of nonpolar

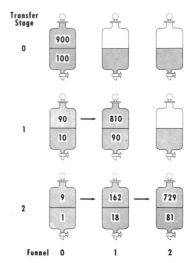

FIG. 5-2. Schematic model illustrating the movement of a solute through three transfer stages of a countercurrent distribution train. Each separatory funnel contains a given volume of lower-phase solvent. An equal volume of upper-phase solvent and 1000 units of solute are added to funnel 0, and the partition coefficient (concentration of solute in upper layer divided by concentration of solute in lower layer) is assumed to be 9. After equilibration of the solute between the two phases, 900 units are in the upper layer and 100 units are in the lower, as shown for transfer stage 0. The upper phase of funnel 0 is transferred to funnel 1, and fresh upper phase is added to funnel 0. After shaking and settling, the 900 units in funnel 1 and the 100 units remaining in funnel 0 are once again divided in a 9:1 ratio between the upper and lower layers as shown for transfer stage 1. Next, the upper phases of funnels 0 and 1 are transferred to funnels 1 and 2, respectively; and fresh upper phase is added to funnel 0. After once again shaking and settling, the new distribution is shown at transfer stage 2. When this procedure is repeated through 100–1000 transfer stages, triglycerides with different partition coefficients will move through the train of funnels at different rates, permitting them to be separated. From Scholfield (*812*).

phases (Table 5-7) is necessarily restricted to highly volatile hydrocarbons since small amounts of triglyceride must be recovered from large volumes of solvent.

Very few polar phases have been evaluated for separating triglycerides by CCD (Table 5-7); but any solvent mixture that has proven useful for TLC, paper chromatography (Tables 5-1, 5-3, and 5-4), and column (Table 5-6) partition chromatography should also be adaptable for CCD work. To date, the best CCD separations of natural fat triglycerides of partition numbers 36 to 48 have been obtained using nitroethane or a furfural/nitroethane 50/50 polar phase. One disadvantage of furfural, however, is its high boiling point (162°), which makes it necessary to re-

I. METHODS 83

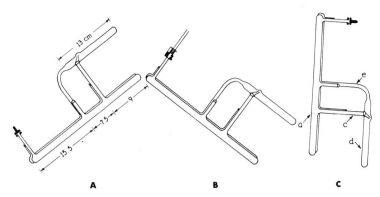

FIG. 5-3. Operating positions for a typical countercurrent distribution cell. Operating cycle consists of (i) rocking between **A** and **B** to equilibrate phases, (ii) holding at **B** to allow phases to separate, (iii) tilting to **C** to drain off upper phase, and (iv) returning to **A** to transfer upper phase to next cell. See text. From Scholfield (*811*).

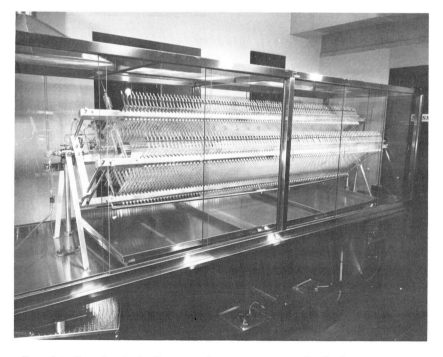

FIG. 5-4. Two hundred-cell automatic countercurrent distribution apparatus enclosed in a fume hood in a constant temperature room. From Scholfield (*811*).

TABLE 5-7
Solvent Systems for the Separation of Triglycerides by
Countercurrent Distribution

| Triglycerides separated | Stationary phase | Mobile phase[a] | References |
|---|---|---|---|
| **Natural fats** | | | |
| Linseed oil | 40 ml furfural | 2.5–5 ml petroleum ether (bp 35°–60°) | 238, 258 |
| Partition numbers 38–48 | 40 ml furfural/nitroethane 50/50 | 3–6 ml petroleum ether (bp 35°–60°) | 238, 241, 598, 641, 671, 813–815, 817 |
| Cottonseed oil | Nitroethane | Hexane | 641 |
| Cottonseed oil | Acetonitrile | Petroleum ether | 837 |
| Butterfat | 40 ml furfural/2-nitropropane 50/50 | 40 ml pentane | 334 |
| Butterfat | 40 ml 2-nitropropane/nitromethane/nitroethane 38/25/71 or 58/42/0 | 40 ml pentane | 334, 847 |
| Cottonseed oil, linseed oil | 40 ml methanol/aqueous 0.2 $N$ AgNO$_3$ 90/10 | 10 ml petroleum ether | 816, 837 |
| **Oxygenated triglycerides** | | | |
| Castor oil, oxidized triglycerides | 20 ml ethanol/water 90/10 | 20 ml petroleum ether (bp 58°–60°) | 3, 715, 971 |
| Oxidized triglycerides | methanol | Isooctane | 338 |
| Epoxy triglycerides | 39 ml acetonitrile | 20 ml hexane | 51a |

[a] Saturated with stationary phase.

move furfural from the recovered fractions by washing with ethanol/water 75/25 (*815*). On the basis of theoretical considerations, Hollingsworth *et al.* (*389*) have suggested that acetonitrile/chloroform and chloroacetonitrile/chloroform mixtures might also be effective polar phases for CCD triglyceride separations. Attempts to use a AgNO$_3$-containing polar phase in CCD to fractionate triglycerides by unsaturation (*816,837*) have so far produced rather poor separations, but resolution could undoubtedly be improved by further experimentation.

Both the polar and the nonpolar phases must be equilibrated with each other prior to use in CCD. This assures that the solvent interface will remain at the same level in each tube throughout the run.

Theoretical equations are available (*237,811,812*) for predicting the CCD separation of triglycerides by a specific solvent system if the partition coefficients of the triglycerides in that solvent system are known. However,

## 3. Separation Procedure

Before an analysis is started, each tube is filled to slightly above level a (Fig. 5-3) with equilibrated lower phase. The excess will flow from tube to tube with each transfer and will discharge from the instrument before the sample. Upper-phase solvent should be placed in several tubes directly ahead of the sample to ensure that the lower layer is saturated with upper layer before it comes in contact with the sample.

The sample is placed directly in the first tube unless it is so large that it would cause a significant shift in the position of the interface as the distribution proceeds. Such shifts can be minimized by equilibrating the sample between the two solvent phases and adjusting them to proper volume before putting the sample into the instrument (*334,814,815*). Some workers (*813,815*) have avoided interface shifts by dividing the sample among several of the initial tubes. The first three or four transfers are usually performed by hand, and any necessary adjustments of interface level are made by adding small amounts of lower phase to tube 0 or by removing small amounts of lower phase just ahead of the lead tube (*334,814,815*).

A countercurrent distribution apparatus can be operated by the fundamental, single-withdrawal, or recycle technique. Fundamental operation corresponds to that described for the separatory funnels in Fig. 5-2. The train of tubes is operated until the first upper phase has migrated to the last tube and the various sample components have moved partway along the train. The upper and lower layers are then removed from each tube, and the amount of solute in each tube is measured. A chromatogram plotting weight per tube vs. tube number is made to illustrate the distribution obtained.

Single-withdrawal operation is carried out in the same way as fundamental operation except that the distribution is continued after the upper phase reaches the last tube of the apparatus. The upper phase from the last tube is discharged into a fraction collector, and CCD is carried on until the last component has been eluted from the train of cells. The amount of triglyceride in each collection tube is then measured and plotted against the transfer number at which it left the instrument to construct a chromatogram (see Fig. 5-12).

If a mixture is not completely separated after one passage through the instrument, it may be recycled, i.e., each upper phase after removal from the last tube is reintroduced into tube 0. Recycling may be continued until the desired separation is obtained or until the leading and trailing edges of the band mix. Then the distribution is stopped as in fundamental operation or the upper layers are collected in a fraction collector as in single-withdrawal operation.

The temperature should be kept constant throughout a CCD run, since the position of the interface between the two phases can change significantly with temperature (*237,334,641*).

It is useful to note that the relative volumes of the two layers in CCD can affect peak resolution. Reducing the volume of the upper layer while the size of the lower layer remains constant will slow down the movement of solutes through the CCD train. With single-withdrawal operation, this will increase the number of transfer stages required to completely elute the sample and thereby improve peak resolution (*813*).

*4. Quantitation*

CCD separations of triglycerides can be monitored by most any of the micromethods listed in Table 2-3. The method of choice is an automatic recording refractometer of the type described by Butterfield and Dutton (*129,258*). However, most workers have quantitated the recovered fractions gravimetrically after removing the solvent by evaporation.

## II. APPLICATIONS

### A. Separation by Partition Number

*1. Elution Order*

The separation of triglycerides by liquid–liquid partition chromatography has been studied by a number of workers (*379,484,567,813,923*) and found to depend on both the molecular weight and the number of double bonds in the molecules. *Saturated* triglycerides are separated on the basis of their carbon number, with the lower molecular weights eluting first when the mobile phase is polar or last if the mobile phase is nonpolar. The introduction of a double bond into a saturated triglyceride molecule changes its partition coefficient so that it elutes with saturated triglycerides containing two less carbon atoms. This relationship is consistent for the triglycerides of common $C_{16}$–$C_{18}$ acids in all solvent systems evaluated so far and can be used to define a term *integral partition number*,* which can be

---

* Many names have been used to designate the groups of triglycerides which migrate together in liquid–liquid partition chromatography. Kaufmann and Makus (*484*) adopted *pc-wertzahl* or *paper chromatography value*, which was later modified to *chromatographische wertzahl* or *chromatography value* by Wessels

calculated for any triglyceride using the formula:

$$\text{INTEGRAL PARTITION NUMBER} = \text{CARBON NUMBER} - 2\begin{bmatrix} \text{NUMBER OF} \\ \text{DOUBLE BONDS} \end{bmatrix}$$

Triglycerides of the same integral partition number will travel together in the same band (i.e., as "critical partners") in liquid–liquid partition chromatography.

With the advent of higher-resolution techniques, it has been found (*478,494,673,950*) that the "critical partners" actually have slightly different mobilities and that one double bond is really equivalent to 2.2–2.8 carbon atoms, depending on the compounds compared and the operating conditions used. To accurately describe this relationship, a new definition of partition number is proposed here. Since there is a linear relationship between the carbon number of $n$-saturated triglycerides and the log of their elution volume in liquid–liquid partition chromatography (Fig. 5-5) just as in gas–liquid chromatography (Fig. 2-5), it would seem advantageous to define a term *equivalent partition number* analogous to the *equivalent chain length* concept used in GLC of fatty acid methyl esters (*655*):

$$\text{EQUIVALENT PARTITION NUMBER} = \begin{array}{l} \text{The carbon number of the equivalent} \\ n\text{-saturated triglyceride that will} \\ \text{elute at the same point as the} \\ \text{specified triglyceride} \end{array} = \text{EPN}$$

The equivalent partition numbers of unsaturated triglycerides are not whole numbers and must be established graphically by comparison with the elution volumes or $R_f$ values of $n$-saturated triglyccrides under identical experimental conditions. Figure 5-5 illustrates the linear relationship between the log of the elution volume and equivalent partition number for $n$-saturated triglycerides in column partition chromatography, while Fig. 5-6 gives the elution volumes of three unsaturated triglycerides under identical conditions. When the elution volume of PoPoPo (8.0 liters) is plotted on Fig. 5-5, tripalmitolein is found to have an EPN of 41.1. Hence one would predict that PoPoPo and MMM could be resolved by high-resolution partition chromatography since their EPN's differ by 0.9. One can also derive equivalent partition numbers by thin-layer or paper partition

---

and Rajagopal (*951*). McCarthy and Kuksis (*633*) used *polarity number*; Hirsch (*379*) employed *double-bond equivalence*. Vereshchagin (*922,923*) defined a *polarity constant* K; while Litchfield (*567*) suggested *partition number*. Of all these names, *partition number* seems to best describe the process by which the triglycerides are actually separated; and this term has been adopted here. *Integral* has been added to this nomenclature so that it can be distinguished from *equivalent partition number* defined below.

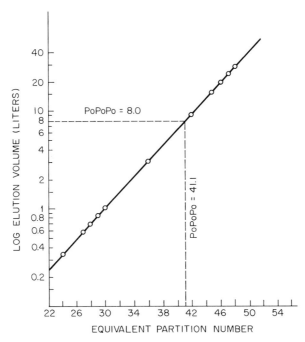

FIG. 5-5. Linear relationship between the log of the elution volume and the equivalent partition number for *n*-saturated triglycerides in liquid–liquid column partition chromatography. The equivalent partition number of an unsaturated triglyceride such as tripalmitolein is calculated graphically from its elution volume as shown. *Operating conditions:* same as Fig. 5-6. Adapted from Nickell and Privett (*673*) and reprinted by courtesy of Marcel Dekker Inc.

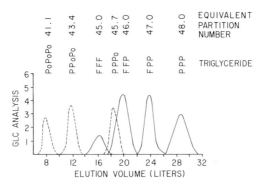

FIG. 5-6. Separation of a mixture of synthetic triglycerides by column partition chromatography. FFF, tripentadecanoin. *Operating conditions:* bonded dimethylsilyl ether + heptane stationary phase on Celite, acetonitrile/methanol 85/15 mobile phase, flow rate about 20 ml/hour (*686*), 25 × 1250 mm column, total sample size approximately 15 mg, separation monitored by GLC of derived methyl esters. Adapted from Nickell and Privett (*673*) and reprinted by courtesy of Marcel Dekker Inc.

chromatography using co-chromatography with suitable *n*-saturated triglyceride standards and interpolating the distances between spots. It should be realized that the EPN values of unsaturated triglycerides are known to vary slightly with chromatographic operating conditions, double-bond configuration, and placement of double bonds within the molecule (*950*); thus an EPN value must be defined for specific operating conditions and for specific compounds. As further EPN data become available, no doubt it will be possible to define relationships for predicting the EPN values of various homologous series of triglycerides, just as is done with the equivalent chain length values of fatty acid methyl esters in GLC (Fig. 2-5).

The use of partition numbers to predict the separation of triglycerides from a hypothetical natural fat containing palmitic, stearic, oleic, and linoleic acids is illustrated in Table 5-8. Integral partition numbers reveal that the 20 possible triglycerides (positional isomers not considered) can

TABLE 5-8
Use of Partition Numbers to Predict the Separation of Triglycerides Containing Palmitic, Stearic, Oleic, and Linoleic Acid

| Band | Triglycerides | Integral partition number |
|---|---|---|
| I | LLL | 42 |
| II | OLL | 44 |
|  | PLL | 44 |
| III | OOL | 46 |
|  | StLL | 46 |
|  | POL | 46 |
|  | PPL | 46 |
| IV | OOO | 48 |
|  | StOL | 48 |
|  | PStL | 48 |
|  | POO | 48 |
|  | PPO | 48 |
|  | PPP | 48 |
| V | StOO | 50 |
|  | StStL | 50 |
|  | PStO | 50 |
|  | PPSt | 50 |
| VI | StStO | 52 |
|  | PStSt | 52 |
| VII | StStSt | 54 |

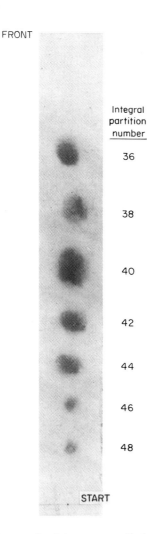

Fig. 5-7. Analytical separation of linseed oil triglycerides by paper partition chromatography. *Operating conditions:* 200 μg linseed oil, Schleicher and Schüll 2043b paper impregnated with tetradecane, one development (12 hours) with acetone/acetonitrile 80/20, spots located with iodine vapor. Adapted from Kaufmann *et al.* (*499*) with spots identified by comparison with other analyses (*238, 926*).

be resolved into seven distinct bands by partition chromatography. If EPN values were available, however, they would indicate moderate subfractionation of molecular species within each band.

## 2. Thin-Layer and Paper Chromatography

Since the polar solvent is the mobile phase in TLC and PC, lower partition number triglycerides move ahead of higher partition number molecules, i.e., LLL migrates further than OOO and MMM further than PPP. Resolution becomes poorer as the partition number increases. Hence

## II. APPLICATIONS

highly unsaturated natural fats are resolved better than the more saturated fats of similar molecular weight, and good resolution above partition number 52 is difficult to achieve with large samples.

Natural fat triglyceride mixtures containing only even carbon number acids can be fully resolved into fractions of different integral partition number (as in Table 5-8) by thin-layer and paper partition chromatography. From 5 $\mu$g to 40 mg can be separated into five to nine spots or bands on a single chromatogram. Figures 5-7 and 5-8 illustrate typical analytical separations of linseed, cottonseed, olive, and soybean triglycerides by paper chromatography and TLC. Using TLC with multiple development, it is sometimes possible to separate synthetic triglycerides having the same integral partition numbers. Figure 5-9 shows such a separation of PPP, PPO, POO, and OOO by Kaufmann and Das (*478*). Similarly, the critical partners in natural fat triglycerides can sometimes be partially separated if the mixture is not too complex (note subfractionation of spots 44 and 46 in cottonseed and soybean oils in Fig. 5-8). However, triglyceride isomers such as PPP and LaStSt (*485*) or *sn*-PPO and *sn*-POP (*494*) cannot be resolved by partition chromatography.

Typical preparative separations of olive oil and soybean oil triglycerides

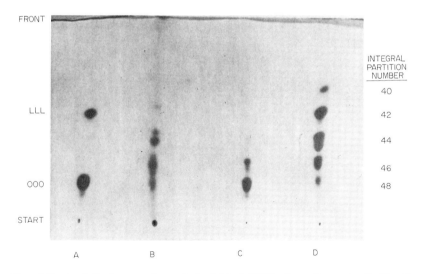

Fig. 5-8. Analytical separation of **A**, LLL + OOO; **B**, cottonseed oil; **C**, olive oil; and **D**, soybean oil triglycerides by thin-layer partition chromatography. Note partial resolution of critical partners in **B** and **D**. *Operating conditions:* sample size probably 10–30 $\mu$g, 0.25 × 200 × 200 mm layer of silanized silicic acid impregnated with 8% (w/w) paraffin oil, triple development with acetone/acetonitrile 70/30, spots located with iodine vapor. From Applied Science Laboratories Inc. (*30*).

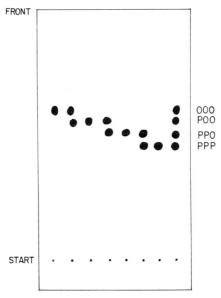

FIG. 5-9. Analytical separation of the critical partners PPP, PPO, POO, and OOO by thin-layer partition chromatography with multiple development. *Operating conditions:* 3 μg of each component triglyceride, kieselguhr layer impregnated with paraffin oil, triple development (40 minutes each) with acetone/acetonitrile 80/20, spots located with cyclodextrin/iodine reagent. From Kaufmann and Das (*478*).

are shown in Fig. 5-10. With a 1.0 × 200 × 200 mm TLC layer and using multiple development, up to 2 mg per band can be resolved. Kaufmann *et al.* (*499*) have fractionated as much as 40 mg of linseed oil triglycerides on a single paper partition chromatogram using circular development (Fig. 5-1). After the individual bands have been isolated and analyzed for fatty acid composition, the component triglycerides in each are calculated using the technique outlined in Table 5-9.

### 3. Column Chromatography

The polar solvent is the mobile phase in column partition chromatography; therefore, triglycerides are eluted in ascending order of their partition numbers, i.e., LLL before OOO and MMM before PPP. As with TLC and PC, resolution decreases as the partition number becomes greater.

Column partition chromatography can serve for both high-resolution analyses and for preparative work: Nickell and Privett (*673*) employed an ~15 mg sample and a very slow solvent flow rate to fully resolve PPP and pentadecanodipalmitin (which have a 1.0 difference in equivalent partition numbers) and to partially resolve the critical partners palmitoleodi-

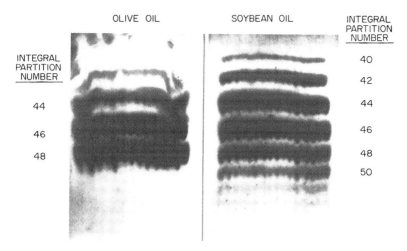

FIG. 5-10. Preparative separation of olive oil and soybean oil triglycerides by thin layer partition chromatography. *Operating conditions:* 2 mg triglyceride, $1.0 \times 90 \times 150$ mm layer of kieselguhr impregnated with paraffin oil, triple development (30 minutes each) with acetone/acetic acid 70/30, bands located with iodine vapor. From Kwapniewski and Sliwiok (547) with bands identified by comparison with other analyses (30,815,951).

palmitin and palmitodipentadecanoin (Fig. 5-6). However, these analyses required 4–8 weeks to complete under the conditions employed. On the other hand, Steiner and Bonar (858) used a 300 mg sample size and a fairly rapid flow rate to achieve full fractionation of cocoa butter triglycerides, including partial resolution of the critical partners StOO and POSt, in only 72 hours (Fig. 5-11).

After the triglyceride fractions have been isolated by column partition chromatography and their fatty acid composition determined by GLC, their triglyceride composition is calculated in the same manner as for CCD (Table 5-9).

## 4. Countercurrent Distribution

Since the nonpolar solvent is the mobile phase in CCD, triglyceride fractions are eluted in descending order of their partition numbers, i.e., StStSt before OOO and PPP before MMM. Resolution becomes poorer as the partition number increases; thus highly unsaturated natural fats are fractionated better than the more saturated types of similar molecular weight.

Large samples (1–10 g) of natural fat triglycerides up to partition number 52 can be partially to fully resolved into fractions of different integral partition number by CCD in 40–60 hours. The definitive work of Dutton, Scholfield, and co-workers (238–241,811–817) has indicated that

TABLE 5-9
METHOD FOR CALCULATING THE TRIGLYCERIDE COMPOSITION OF PEAK 46
FROM COUNTERCURRENT DISTRIBUTION OF WHEAT SEED OIL (FIG. 5-12)

**Fatty acid composition** (671)

| | Integral partition number of fatty acid | Mole % | |
|---|---|---|---|
| 18:0 | 18 | 1.3 | |
| 16:0 | 16 | 36.5 | } 64.3 |
| 18:1 | 16 | 27.8 | |
| 14:0 | 14 | 0.7 | |
| 16:1 | 14 | 2.3 | } 34.4 |
| 18:2 | 14 | 31.4 | |

**Checking purity of fraction**
(a) This fraction is mainly POL which has an integral partition number of 46.
(b) The integral partition number of this fraction calculated from its fatty acid composition is

$$3\left[\frac{1.3(18) + 64.3(16) + 34.4(14)}{100}\right] = 46.01$$

(c) Since the theoretical and experimental values check closely, this fraction is of high purity and an accurate triglyceride composition can be calculated.

**Calculating component triglycerides**
(a) Assume 14:0, 16:1, and 18:0 are present in such small amounts that they are unlikely to appear more than once in any single triglyceride molecule.
(b) Then the only possible triglycerides in this fraction are [St + (M,Po,L) + (M,Po,L)] and [(P,O) + (P,O) + (M,Po,L)].
(c) All the 18:0 is in [St + (M,Po,L) + (M,Po,L)] = 3(1.3) = 3.9% = StLL + StML + StPoL.
(d) Assume StLL/StPoL/StML = 31.4/2.3/0.7. Then StLL ≈ 3.6%, StPoL ≈ 0.3%, and StML ≈ trace. A slight error is introduced by making this assumption, but this error is small (probably <1.0 absolute percent) in terms of the overall composition.
(e) All remaining 14:0 is in [(P,O) + (P,O) + M], so PPM + POM + OOM = 3(0.7) = 2.1%.
(f) Remaining 16:1 is all in [(P,O) + (P,O) + Po], so PPPo + POPo + OOPo = 3(2.3 − 0.1) = 6.6%.
(g) Then PPL + POL + OOL = 100.0 − 3.9 − 2.1 − 6.6 = 87.4%.
(h) Thus the probable composition of Peak 46:

$$\begin{aligned}
PPL + POL + OOL &= 87.4\% \\
PPPo + POPo + OOPo &= 6.6\% \\
StLL &= 3.6\% \\
PPM + POM + OOM &= 2.1\% \\
StPoL &= 0.3\% \\
StML &= trace
\end{aligned}$$

**Checking results vs. original sample**

$$18:0 = \frac{3.6 + 0.3}{3} = 1.3\% \text{ vs. } 1.3\% \text{ found}$$

$$16:0 + 18:1 = \frac{2(87.4) + 2(6.6) + 2(2.1)}{3} = 64.1\% \text{ vs. } 64.3\% \text{ found}$$

$$14:0 = \frac{2.1}{3} = 0.7\% \text{ vs. } 0.7\% \text{ found}$$

$$16:1 = \frac{6.6 + 0.3}{3} = 2.3\% \text{ vs. } 2.3\% \text{ found}$$

$$18:2 = \frac{87.4 + 2(3.6) + 0.3}{3} = 31.6\% \text{ vs. } 31.4\% \text{ found}$$

## II. APPLICATIONS

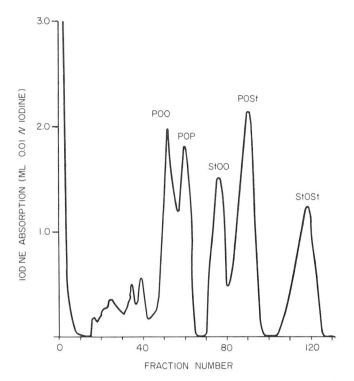

FIG. 5-11. Separation of Brazilian cocoa butter unsaturated triglycerides by column partition chromatography. *Operating conditions:* paraffin oil stationary phase coated 9/50 (w/w) on cellulose powder, acetone/methanol 60/40 mobile phase, 25 × 1200 mm column, 300 mg cocoa butter, separation monitored by measuring iodine absorption of alternate fractions. From Steiner and Bonar (*858*).

800–1000 transfers are usually necessary on a 200-tube apparatus to separate the triglycerides of safflower, soybean, and linseed oils using the single-withdrawal procedure. Figure 5-12 illustrates an efficient separation of 4.3 g of wheat seed triglycerides by CCD.

The component triglycerides in each CCD fraction can often be calculated from the fatty acid composition as outlined in Table 5-9. Difficulties are encountered, however, if adjacent fractions overlap. This problem is partially overcome if only the central portion of each peak is analyzed, but the resultant answer is an approximation since CCD peaks from natural fats are not homogeneous (Section II,A,1).

### B. Separation by Unsaturation

Partition chromatography can also separate triglycerides on the basis of unsaturation rather than partition number. This is accomplished either

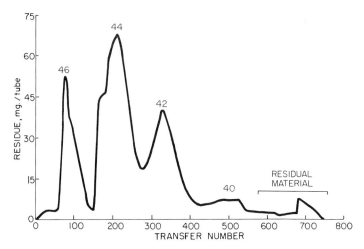

FIG. 5-12. Countercurrent distribution of wheat seed triglycerides showing resolution of four distinct peaks of different integral partition numbers. *Operating conditions:* 40 ml furfural/nitroethane 50/50 stationary phase, 5 ml petroleum ether (30°–60° bp) mobile phase, 4.3 g wheat seed triglycerides, 200 tubes, single-withdrawal operation, 927 transfers. From Nelson et al. (*671*) with additional identification of peak 40 based on its fatty acid composition.

by using a polar phase containing silver nitrate or by adding bromine or mercuric acetate at the double bonds of the molecules. The $AgNO_3$-containing solvent is the preferable procedure, since it involves no derivatization reaction.

### 1. $AgNO_3$-Containing Solvent Systems

The addition of silver nitrate to the polar phase in partition chromatography causes $Ag^+$/olefin $\pi$-complexing to occur and increases the solubility of the unsaturated molecules in the polar solvent. This dramatically alters the partition coefficients of the triglycerides, producing a separation based on the number of double bonds per molecule with some subfractionation by partition number (*683,684,696,924,926*). Figure 5-13 illustrates the separation of two synthetic triglyceride mixtures using this technique with TLC. The more unsaturated molecules have the higher $R_f$ values; and OOO/POO/PPO, which were formerly critical partners, are now widely separated. Operating conditions are the same as discussed above for non-$AgNO_3$ solvent systems, except that any recovered samples must be washed to remove traces of $AgNO_3$.

Vereshchagin and his co-workers (*683,684,924,926*) have demonstrated how successive partition separations, first in a normal solvent system and then in a $AgNO_3$-containing system, can be used for a detailed analysis of natural fat triglyceride mixtures. Molecules are initially separated into integral partition number fractions and then subfractionated on the basis of unsaturation. Table 5-10 shows the analysis of *Monarda fistulosa* seed triglycerides by these combined procedures.

Silver ion partition chromatography has not yet been applied to unfractionated natural fat triglyceride mixtures, and its usefulness for this purpose remains to be seen. The results in Fig. 5-13 and Table 5-10 imply that different solvent systems would be required for separating 0–4 and 5–9 double bond triglycerides, just as in $Ag^+$ adsorption chromatography. There is some evidence (Table 5-10) that highly unsaturated triglycerides may be resolved more effectively by $Ag^+$ partition than by $Ag^+$ adsorption chromatography. If future work confirms this, then partition chromatog-

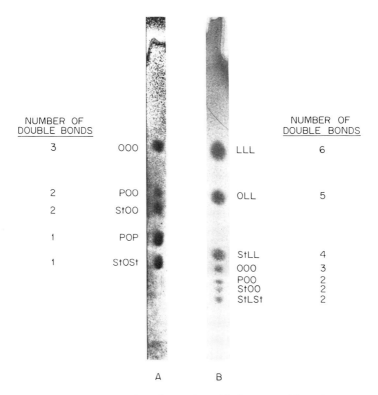

FIG. 5-13. Separation of triglycerides using thin-layer partition chromatography with a $AgNO_3$-containing polar phase. *Operating conditions:* 1–2 μg per component, 200 × 200 mm plates coated with 0.25 mm thick layer of methanol-washed silanized kieselguhr, developed (90 minutes) in acetone/ethanol/water/acetonitrile/$AgNO_3$ (**A**) 72/18/8/2/saturated or (**B**) 83/8/7/2/saturated, spots located with phosphomolybdic acid reagent. Adapted from Ord and Bamford (*696*).

TABLE 5-10

SEPARATION OF *Monarda fistulosa* SEED TRIGLYCERIDES BY CONSECUTIVE PAPER PARTITION CHROMATOGRAPHY IN NORMAL AND AgNO$_3$-CONTAINING SOLVENT SYSTEMS[a]

|  | First separation by partition number[b] | | Second separation by unsaturation[d] | | |
|---|---|---|---|---|---|
| Triglyceride | Integral partition number | $R_h$[c] | Number of double bonds | $R_h$[c] | Mobile phase (methanol/water/AgNO$_3$) |
| LnLnLn | 36 | 0.78 | 9 | — | — |
| LLnLn | 38 | 0.63 | 8 | — | — |
| LLLn + OLnLn | 40 | 0.42 | 7 | 2.51 | 80/20/sat.[e] |
| PLnLn |  |  | 6 | 0.74 |  |
| LLL + OLLn |  |  | 6 | 1.02 |  |
| StLnLn | 42 | 0.32 | 6 | 0.79 | 80/20/sat. |
| PLLn |  |  | 5 | 0.71 |  |
| OLL + OOLn |  |  | 5 | 0.93 |  |
| StLLn | 44 | 0.25 | 5 | 0.71 | 95/5/sat. |
| POLn |  |  | 4 | 0.58 |  |
| OOL |  |  | 4 | 0.51 |  |
| StOLn | 46 | 0.21 | 4 | 0.38 | 95/5/sat. |
| POL |  |  | 3 | 0.27 |  |
| OOO |  |  | 3 | 0.29 |  |
| POO | 48 | 0.18 | 2 | 0.17 | 98/2/sat. |
| StOO | 50 | 0.16 | 2 | 0.07 | 98/2/sat. |

[a] From Novitskaya and Mal'tseva (684).

[b] *Operating conditions:* paper impregnated with aliphatic hydrocarbons (bp 260°–310°), developed in acetone/acetic acid 50/50, spots located with sudan black reagent.

[c] $R_h = \dfrac{\text{distance travelled by given triglyceride}}{\text{distance travelled by butyl hexabromostearate}}$.

[d] *Operating conditions:* paper impregnated with dodecane, developed in methanol/water/AgNO$_3$, spots located with sudan black reagent.

[e] Saturated.

raphy with AgNO$_3$-containing solvents might prove very useful for analysis of the highly unsaturated triglycerides from marine animals.

## 2. Brominated Triglycerides

Bromination is an effective means of changing the partition number of unsaturated triglycerides so that critical partners can be easily separated on the basis of unsaturation. Vereshchagin *et al.* (*921,928*) and Vorob'ev

(*938*) have utilized this technique to examine the triglyceride composition of poppyseed, cottonseed, and soybean oils. After a first separation of triglycerides by integral partition number, the recovered fractions are quantitatively brominated (Chapter 3, Section I,E) and then refractionated in the same solvent system to give a separation based on bromine content (i.e., unsaturation). Table 5-11 illustrates the type of subfractionation of complex triglyceride mixtures that can be achieved in this manner. The same separation can also be accomplished by two-dimensional TLC using a normal moble phase for the first development and the same solvent contaning 0.5% bromine for the second development (*487*), but the quantitative nature of such *in situ* bromination has not been tested.

The addition of bromine to the double bonds in an unsaturated triglyceride decreases its partition number in a manner dependent on its fatty acid composition. Note, for example, that the critical partners StOL and OOO

TABLE 5-11

SEPARATION OF POPPYSEED TRIGLYCERIDES BEFORE AND AFTER BROMINATION USING PAPER PARTITION CHROMATOGRAPHY[a]

| Triglyceride | Separation before bromination | | Separation after bromination | |
|---|---|---|---|---|
| | Integral partition number | $R_h$ [b] | Number of double bonds | $R_h$ [b] |
| LLL | 42 | 0.32 | 6 | 0.69 |
| OLL | 44 | 0.25 | 5 | 0.57 |
| PLL | | | 4 | 0.47 |
| OOL | 46 | 0.21 | 4 | 0.47 |
| StLL | | | 4 | 0.42 |
| POL | | | 3 | 0.36 |
| PPL | | | 2 | 0.29 |
| OOO | 48 | 0.18 | 3 | 0.34 |
| StOL | | | 3 | 0.30 |
| POO | | | 2 | 0.27 |
| PStL | | | 2 | 0.25 |
| PPO | | | 1 | 0.21 |

[a] *Operating conditions:* paper impregnated with aliphatic hydrocarbons (bp 260°–310°), developed in acetone/acetic acid 85/15, spots located with sudan black reagent. Data from Vereshchagin (*921*).

[b] $R_h = \dfrac{\text{distance travelled by given triglyceride}}{\text{distance travelled by butyl hexabromostearate}}$.

are both converted to hexabromides but they can still be separated after bromination.

### 3. Mercurated Triglycerides

Mercuric acetate adducts of unsaturated triglycerides (Chapter 3, Section I,F) have been fractionated by Noda and Hirayama (678,679) using a methanol/acetic acid 83/17 mobile phase on alkane- or tetralin-impregnated paper. Results show a clearcut separation of triglycerides containing one, two, three, four, five, and six double bonds with some subfractionation of molecules such as POO and StOO by partition number. Long-chain SSS molecules presumably remain at the origin with such a polar mobile phase.

## C. Oxidized Triglycerides

Oxidized triglycerides are readily resolved into SSS, SSA, SAA, and AAA fractions using partition chromatography techniques. Haighton et al. (338) accomplished this by countercurrent distribution between isooctane and methanol, while Youngs (971) utilized 90 CCD transfers between petroleum ether and ethanol/water 90/10. Haighton et al. (338) have also resolved oxidized triglycerides into four fractions on columns of powdered rubber, eluting with various acetone/water mixtures in stepwise increments (Fig. 5-14). Youngs (971) has reported the resolution of SSS/SSA and SAA/AAA fractions by column partition chromatography using ethanol/water 90/10 as the polar stationary phase with petroleum ether and diethyl ether as the eluting solvents. Undoubtedly SSS, SSA, SAA, and AAA could also be separated by thin-layer and paper partition chromatography.

Mangold (603) has suggested adding 10% peracetic acid to the usual polar mobile phase so that unsaturated triglycerides will be quantitatively oxidized directly on the chromatogram during partition chromatography. The resulting oxidized molecules are highly polar and move with the solvent front, leaving only the saturated triglycerides in their normal locations on the chromatogram.

## D. Other Functional Groups and Derivatives

### 1. Trans Unsaturation

Triglycerides containing *trans* double bonds have slightly higher equivalent partition numbers than their corresponding *cis* isomers. Kaufmann and Das (478) reported that ElElEl has a lower $R_f$ value than OOO in partition TLC, and Nickell and Privett (673) found trilinolelaidin eluting after

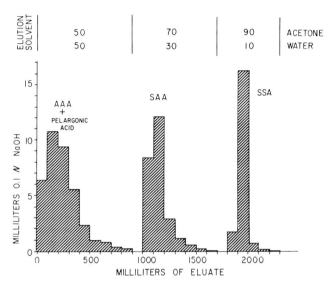

FIG. 5-14. Separation of oxidized palm oil triglycerides by column partition chromatography. SSS (not shown) is eluted last using 100% acetone. *Operating conditions:* 2–3 g permanganate-oxidized palm oil, powdered rubber stationary phase, acetone/water mobile phase, separation monitored by titration with 0.1 N NaOH. Redrawn from Haighton *et al.* (*338*).

LLL in column partition chromatography. In general, however, geometrical isomers are poorly separated by liquid–liquid partition chromatography, and isomer pairs such as OOO and OOEl cannot be resolved.

*2. Hydroxy Triglycerides*

Triglycerides containing free hydroxyl groups are very easily separated according to the number of –OH groups per molecule using partition chromatography. For example, Achaya *et al.* (*3,715*) successfully fractionated castor oil triglycerides into tetra-, tri-, di-, mono-, and nonhydroxy molecules using only 103 CCD transfers (Fig. 5-15).

*3. Hydrogenated Triglycerides*

Hydrogenation can be used to change the integral partition numbers of unsaturated triglycerides so that critical partners are separated. For example, the critical partners PPO and POO isolated from cocoa butter by partition chromatography could be converted into PPSt and PStSt by hydrogenation and then resolved by a second partition separation. Such analyses are conveniently carried out using two-dimensional TLC by hydrogenating the sample directly on the chromatogram between the first and second developments (Chapter 3, Section I,A; *487*). Some compositional informa-

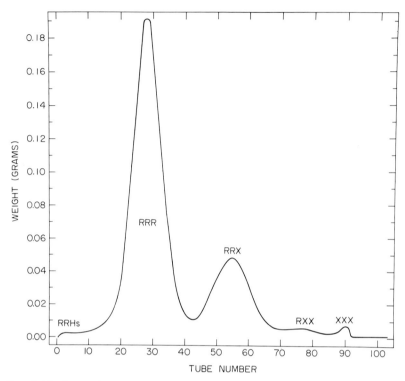

FIG. 5-15. Separation of castor oil triglycerides by countercurrent distribution. R, ricinoleic acid; Hs, dihydroxystearic acid; X, nonhydroxy acids. *Operating conditions:* 20 ml ethanol/water 90/10 stationary phase, 20 ml petroleum ether (bp 58°–69°) mobile phase, 3 g castor oil triglycerides, fundamental operation with 103 transfers. From Achaya et al. (*3*).

tion is lost by hydrogenation, however. Any PStL present in the PPO/POO fraction from cocoa butter would be hydrogenated to PStSt, which would be indistinguishable from the PStSt produced from POO.

### E. Derived Diglycerides

Diglycerides can be effectively fractionated by liquid–liquid partition chromatography in the same manner as triglycerides. The same partition number relationships apply, except that diglyceride integral partition numbers are computed on the basis of two rather than three acyl groups. Kaufmann and Makus (*483*) separated dilaurin, dimyristin, dipalmitin, and distearin by TLC using methanol/chloroform/water 71/24/5 as the mobile phase, while Hirayama and Inouye (*375*) achieved similar separations of unsaturated diglycerides by paper chromatography with a variety of solvent systems. Neither of these publications reports any effect of isomeric diglycerides on the separations. Hirsch (*379*) has fractionated saturated diglycerides by chain length on columns employing soybean oil factice as the stationary phase. With an acetone/water 87.5/12.5 mobile phase, the *sn*-1,2(2,3)-diglycerides were eluted before the corresponding *sn*-1,3-isomers; but diglyceride isomers were very poorly separated with an acetone/water 95/5 mobile phase. Apparently the amount of separation of diglyceride isomers in partition chromatography varies with the polarity of the mobile phase. For best resolution, therefore, it would seem desirable to separate the *sn*-1,2(2,3)- and *sn*-1,3-isomers by TLC on silicic acid (Chapter 8, Section I,H) prior to using partition chromatography. Possible isomerization of diglycerides during liquid–liquid partition separations has apparently not been investigated, although one would expect it to be minimal if separations were carried out rapidly enough and if the solid support were made inert by silanization.

Diglycerides are often acetylated before chromatographic fractionation to avoid any possibility of isomerization during handling. Diglyceride acetates (which are actually triglycerides) with integral partition numbers between 26 and 38 have been resolved by Mangold (*603*) by developing silicone-impregnated paper with methanol/chloroform/water 71/24/5 or acetic acid/water 90/10. Similar separations have been reported by Åkesson (*13*) using acetone/acetonitrile/water 65/35/5 with undecane-impregnated TLC plates.

*sn*-1,3-Diglycerides can also be separated by integral partition number using the dihydroxyacetone derivatives (*479*).

# 6

# GAS–LIQUID CHROMATOGRAPHY

Gas–liquid chromatography (GLC) has established itself over the past 15 years as an indispensable tool for lipid research. The high resolving power, speed of analysis, automated chart readout, and extreme sensitivity of GLC have led to its universal adoption for fatty acid analysis. The application of GLC to the analysis of intact triglycerides had produced a very rapid technique for characterizing natural fat triglycerides by molecular weight.

The possibility of using GLC for triglyceride analysis was first explored in 1959 by Huebner (*399–401*), who discovered that intermediate molecular weight triglycerides such as LaLaLa, MMM, and PPAc could be eluted from short GLC columns containing a silicone liquid phase. Subsequent work by Martin *et al.* (*610*), Fryer *et al.* (*290*), and Pelick *et al.* (*28,706*) demonstrated that pure PPP and StStSt were eluted as individual peaks and that characteristic "fingerprint chromatograms" could be obtained from butterfat, olive oil, cottonseed oil, etc. In 1961, Huebner (*402*) achieved the first clearcut resolution of GLC peaks with natural fat triglyceride mixtures, indicating the true analytical capabilities of the method. Kuksis and McCarthy placed the technique on a firm quantitative basis in 1962 when they defined the molecular weight distribution of triglycerides in butterfat (*540,634*). Since then, Litchfield *et al.* (*346,578*), Kuksis and co-workers (*109,530,535,539*), and others (*134,793,943*) have extensively investigated the variables affecting quantitative GLC of triglycerides; and optimum operating conditions have been defined. The technique is now widely used to determine the molecular weight distribution of triglycerides in natural fats.

GLC on short silicone columns separates triglyceride molecules according to the number of carbon atoms they contain, and each peak in the chromatogram is conveniently referred to by its "carbon number":

$$\text{CARBON NUMBER} = \text{the number of carbon atoms in the acyl chains of a triglyceride}$$

The carbon atoms in the glycerol moiety are not counted so that the carbon numbers of triglycerides and their component fatty acids will be analogous. For example, the carbon number of oleodipalmitin is $18 + 16 + 16 = 50$, which is often abbreviated to "$C_{50}$." Molecules such as PPP, MPO, and MMAd, which all have a carbon number of 48, will all elute together in the same $C_{48}$ peak.

This chapter describes the analytical methodology required for quantitative GLC of triglycerides and discusses feasible applications to various types of samples. The reader may also wish to consult two excellent reviews on this subject by Kuksis (*531,533*).

## I. METHODS

### A. Apparatus

The gas chromatograph (Fig. 6-1) is a familiar instrument to all lipid chemists, for it is indispensable in the analysis of fatty acids. For general background information on its use, the reader should refer to one of the many books on the theory and practice of GLC (*126,255,347,582,635, 891*).

The application of GLC to the separation of high-boiling compounds such as triglycerides requires several instrumental design features beyond those necessary for fatty acid methyl ester analysis. Automatic temperature programming up to 400° is required, and all parts that come in contact with the sample should be made of glass or stainless steel.

Litchfield *et al.* (*578*) and Kuksis and Breckenridge (*531,535*) have found that on-column injection is essential for accurate, quantitative GLC of triglycerides. Figure 6-2 shows a typical flash heater designed for on-column injection. The syringe needle is inserted through the septum into the heated open part of the column. The preheated carrier gas sweeps across the inner face of the septum with extremely high velocity, carrying the sample immediately to the top of the column packing and helping to

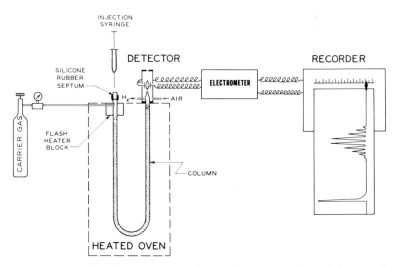

FIG. 6-1. Schematic diagram of typical gas chromatograph used for gas–liquid chromatography of triglycerides.

control "flash-back" into the carrier gas line. All "dead spaces" (regions of low gas velocity where sample vapor could accumulate) which might broaden peaks or lead to peak tailing are eliminated.

Another critical point in gas chromatograph design is the connection between the column and the detector. To prevent any sample condensation in this area, the column/detector connection should be as short as possible, contain no dead spaces, and should be separately heated outside the column oven.

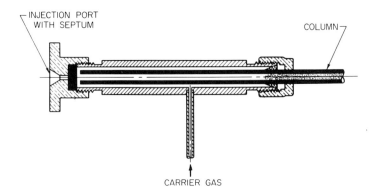

FIG. 6-2. Cross section of typical flash heater used for on-column injection of triglycerides.

The hydrogen flame detector has been universally adopted for GLC of triglycerides. It is 10–100 times more sensitive to carbon compounds yet less affected by temperature changes than a thermal conductivity detector. Moreover, the hydrogen flame detector is not as sensitive to silicone column bleed as the argon ionization and thermal conductivity detectors since the Si in a silicone polymer is already partially oxidized.

The very high flash heater temperatures used for triglyceride GLC cause rapid degradation of normal silicone rubber injection septa. This degradation sometimes leads to the elution of a series of extraneous small peaks between 200° and 300° *(131,520,578,846,880)*. Such peaks appear even during a blank programmed run when no sample is injected and are due to material being vaporized out of the septum as column temperature increases. These extraneous peaks can be eliminated by purchasing special septa of high temperature stability (Applied Science Laboratories, State College, Pa.; Supelco Inc., Bellefonte, Pa.; Canton Biomedical Products Inc., Boulder, Colo.; *131*), by preconditioning septa at high temperatures before use *(131,242,520,578,579,880)*, or by external cooling of the septum during use *(579,846,899)*.

## B. Column

GLC of triglycerides requires packed columns. Attempts to separate high molecular weight triglycerides on capillary columns have so far proven unsuccessful *(577)*.

### 1. Solid Support

Standard diatomaceous earth supports are used for triglyceride GLC, but special chemical treatment is necessary to reduce adsorption effects which cause peak tailing. This treatment includes washing with concentrated HCl to remove metal impurities followed by reaction with dimethyldichlorosilane to convert surface –SiOH groups to silyl ethers *(393,868)*. Commercial solid supports such as Gas Chrom Q (Applied Science Laboratories), Chromosorb W/AW-DMCS (Johns-Manville), and Supelcoport (Supelco) are prepared in this manner and have proven suitable for triglyceride separations.

The particle size of the solid support affects both column efficiency and carrier gas flow rate. A very finely divided solid support of uniform size produces maximum column resolution, but the fine particles also impede passage of gas through the column. The high carrier gas flow rates needed for triglyceride GLC (100 ml/minute for a 2.4 mm i.d. column) necessitate a compromise between high column resolution and gas flow rate in selecting the particle size of the solid support. Supports of 60/80, 80/100,

TABLE 6-1
LIQUID PHASES FOR GAS–LIQUID CHROMATOGRAPHY OF TRIGLYCERIDES

| Type | Proprietary name | Temperature limit | Polarity | References |
|---|---|---|---|---|
| Long-chain hydrocarbons | Apiezon L | 325° | Least | 577 |
| Silicone polymers | | | | |
|   Dimethyl polysiloxane | OV-1 | 375° | | 269, 962 |
| | OV-101 | 375° | | 651 |
| | JXR | 375° | | 535, 578 |
| | SE-30 | 350° | | 530, 974 |
|   Carborane dimethyl siloxane copolymer | DEXSIL-300 | 500° | | 265a |
|   Phenyl methyl polysiloxane | OV-17 | 375° | | 545, 961 |
| | SE-52 | 350° | | 35, 565 |
|   Trifluoropropyl methyl polysiloxane | QF-1 | 325° | Most | 530, 918, 943 |

and 100/120 mesh size have been most widely used for triglyceride GLC. The difference in column efficiency between 60/80 and 100/120 mesh supports is small, but the finer 100/120 mesh size does yield slightly better peak resolution (577).

## 2. Liquid Phase

The choice of liquid phases for GLC separation of triglycerides is severely limited by temperature requirements. Only those liquids which are stable to 325° can be used. Table 6-1 lists the various liquid phases which have been evaluated so far for GLC separation of triglycerides. All of these, except Apiezon L, are silicone polymers which are known for their high temperature stability. With the usual short columns (450–600 mm), all the liquid phases listed in Table 6-1 will separate triglycerides only on the basis of carbon number (i.e., molecular weight). Separation by unsaturation or of positional isomers is not generally practical (Sections II,B and II,C,2) even with the most polar liquid phase on much longer columns. Therefore, the major criterion in selecting a liquid phase for triglyceride analysis is the thermal stability of the liquid. Hence OV-1, OV-101, JXR, Dexsil-300, and OV-17 are the liquid phases of choice since they will produce the least baseline rise during temperature programming. Another interesting possibility would be the use of silicones chemically bonded to the solid support (2,34,349,349a,508) to minimize column bleed at high temperatures.

The amount of liquid phase coated on the support is generally 1–3%

(w/w). Although early workers used much higher levels, elution temperatures can be significantly reduced by using less than 3% liquid phase (*531,540*). Considerable liquid is removed from the support during column conditioning, and the final loading of liquid phase on the support is a matter of speculation. The amount remaining after conditioning will vary with the amount originally applied, the severity of the conditioning process, and the thermal stability of the original silicone. Kuksis (*529,530*) has estimated that over half the original silicone on an SE-30 column is removed during extensive conditioning at 330°–350°.

The liquid phase is coated on the solid support using the standard filtration technique (*393*), preferably with fluidized drying (*527*). The solvent evaporation coating technique (*540*) has been reported to give poorer quantitative results with high molecular weight triglycerides (*794*). Most laboratories prefer to purchase the solid support already coated with the desired liquid phase, however, since excellent column packings suitable for triglyceride GLC are now available commercially at relatively low cost.

## 3. Column Tubing Material

Both glass and stainless steel tubing have been used to prepare columns for GLC of high molecular weight triglycerides. Although both tubing materials are satisfactory, they are not equivalent in their performance. Litchfield *et al.* (*578*) have shown that glass tubing gives significantly better peak resolution than steel tubing when compared under identical operating conditions (Fig. 6-3). Apparently stainless steel exhibits an adsorption effect with triglycerides, but glass does not. This adsorption effect decreases peak resolution, but appears to be reversible since the calibration factors for PPP and StStSt are practically the same on both glass and steel columns (Table 6-2). Another advantage of glass columns is the ability to see gaps or discoloration in the column packing should such occur during operation. In addition, resolution is less dependent on carrier gas flow rate (Fig. 6-3) and temperature programming rate (see Fig. 6-6) with glass tubing than with steel columns. Kuksis and Breckenridge (*535*) silanized the interior surfaces of column tubing before use, but the advantages of such treatment have not been defined experimentally.

Installation of steel columns in a gas chromatograph is easily accomplished with standard Swagelok tubing fittings. With glass columns, however, it is difficult to maintain a leak-free, glass-to-metal seal at the ends of the column. Silicone O-rings crack after a few days of use above 300° (*578,793*) and often stick tightly to the glass and metal surfaces, making it difficult to replace the leaking O-rings without breaking the fragile glass column. One solution to this problem is the Kovar glass-to-metal seal used by Litchfield *et al.* (*579*) for triglyceride GLC on glass columns at tem-

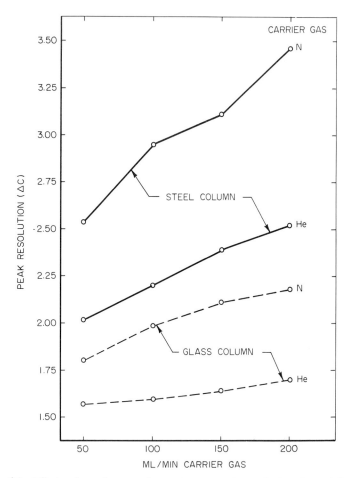

Fig. 6-3. Effect of carrier gas flow rate on peak resolution in stainless steel and glass columns. $\Delta C$ is the minimum carbon number difference between two saturated triglycerides which could be separated with baseline resolution in the $C_{42}$–$C_{48}$ region of the chromatogram. *Operating conditions:* $610 \times 2.5$ mm i.d. columns packed with 3.0% JXR on 100/120 mesh Gas Chrom Q; column programmed $170° \rightarrow 325°$ at $3°$ minute; flash heater, $325°$; detector base, $300°$–$340°$; sample, equal weights of LaLaLa, MMM, PPP, and StStSt. From Litchfield *et al.* (578).

peratures up to 400°. However, Kovar seals are quite fragile and can be easily broken during column installation. A more practical answer has recently been reported by Beroza and Bowman (74), who used asbestos gaskets impregnated with Dexsil 300 to obtain gas-tight, glass-to-metal seals during GLC up to 400°.

## I. METHODS

TABLE 6-2

EFFECT OF CARRIER GAS FLOW RATE ON CALIBRATION FACTORS FOR TRIPALMITIN AND TRISTEARIN IN GLASS AND STAINLESS STEEL COLUMNS[a,b]

| Carrier gas flow rate (ml/min) | Weight calibration factor ($f_w$) | | | |
|---|---|---|---|---|
| | Tripalmitin | | Tristearin | |
| | Glass column | Steel column | Glass column | Steel column |
| 50 | 1.00 | 1.01 | 1.14 | 1.12 |
| 100 | 0.98 | 0.96 | 1.08 | 1.08 |
| 150 | 0.97 | 0.95 | 1.01 | 1.02 |
| 200 | 0.98 | 0.99 | 1.07 | 1.06 |

[a] From Litchfield et al. (578).
[b] *Operating conditions:* 0.61 m × 2.5 mm i.d. columns packed with 3.0% JXR on 100/120 mesh Gas Chrom Q with 100 ml/minute He carrier gas; column programmed 170° → 325° at 3°/minute; flash heater, 325°; detector base, 300°–340°; sample, equal weights of LaLaLa, MMM, PPP, and StStSt.

### 4. Column Size

The internal diameter of columns used for GLC of triglycerides is usually 2.5–3.5 mm for analytical separations (531,533,578,579) and 5–10 mm for preparative work (225,539).

Column length in triglyceride GLC depends on the separation desired. For triglycerides above $C_{40}$, 0.5–0.7 m columns will resolve molecules differing by two carbon atoms while 1.5–1.9 m columns are required to resolve triglycerides differing by only one carbon atom (346,535,540,578, 579). Columns longer than 2.0 m are impractical for separating triglycerides above $C_{48}$, since the very high temperatures necessary for sample elution will cause excessive bleed of the liquid phase and thermal degradation of the sample (579).

Sample losses increase significantly with column length (530,578,793). Sato et al. (793) compared the ratio of tristearin and trilaurin peak areas when the same sample was chromatographed on 0.35, 1.00, and 1.80 m columns under identical operating conditions (Fig. 6-4). Losses were much greater on the 1.8 m column than on the short 0.35 m column. Litchfield et al. (578) reported similar findings, but later work (579) indicated that extensive conditioning can improve sample recovery to some extent (see Section I,B,6).

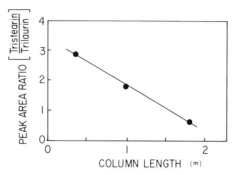

FIG. 6-4. Effect of column length on recovery of tristearin. *Operating conditions:* 0.35, 1.00, and 1.80 m × 4 mm i.d. glass columns packed with 2% JXR on 60/80 mesh Shimalite W; column programmed 100° → 325° at 4°/minute with 100 ml/minute N carrier gas; flash heater, 450°; detector base, 370°; sample, 20 μg LaLaLa and 100 μg StStSt. Redrawn from Sato et al. (*793*).

## 5. Single vs. Dual Columns

All GLC analyses of triglycerides described in this chapter can be accomplished with a single-column gas chromatograph. Dual-column instruments are sometimes used to extend the upper temperature limit of a liquid phase, but this is unnecessary when separating triglycerides by carbon number. Well-conditioned OV-1 and JXR columns can be used up to 375°, allowing the elution of $C_{66}$ molecules without column bleed causing excessive baseline rise (*579*).

Analysis of triglycerides by dual-column GLC has been described by Kuksis *et al.* (*545*) and Watts and Dils (*943*). This technique would prove useful if less thermostable liquid phases or extra long columns should find application in triglyceride GLC.

## 6. Column Preparation

Two different methods have been used to pack columns for triglyceride GLC. Kuksis and his co-workers (*531,535*) employed the following procedure to prepare tightly packed columns:

> The outlet end of a suitably shaped tube is closed with a compact plug of siliconized glass wool extending about 6 mm from the end of the tube to the column interior. This plugged end is attached to a water pump, and the column is filled (a few centimeters at a time) under suction with the help of a small funnel connected to the column inlet by Tygon tubing. During packing, the tube is mechanically vibrated. Violent vibration of the column should be absolutely avoided, as this fractures the support particles and exposes adsorptive sites. The column is uniformly packed to about 35 mm from the inlet end, and a small siliconized glass wool plug is pushed down the tube to rest against the packing.

On the other hand, Litchfield and his co-workers (*577–579*) have successfully used a loose-packing technique in which the column is filled without suction using only light tapping with a small rubber mallet to achieve gravity settling of the packing in the U-tube. There are advantages to both methods. Tight packing gives slightly higher resolution, but loose packing requires less conditioning to achieve high recoveries of $C_{46}$–$C_{66}$ triglycerides (see below). Satisfactory columns can be prepared by either method. Kuksis (*531*) expresses a preference for tight packing with coarse (60/80 mesh) supports and for loose packing with fine (100/120 mesh) supports.

Opinions vary as to whether the part of the column that is in the flash heater should be empty (*578,579*) or partially filled with packing (*535*). Theoretically, the front of the column should be empty so that the vaporized sample is blown directly onto the front of the packing to achieve a sample band of minimum width. In practice, however, excellent peak resolution has been obtained either by injection into the heated empty front of the column (*579*), by injection into the actual packing contained in the flash heater (*535*), or even by cold injection where the septum was removed and the sample solution placed directly in the glass wool below the flash heater (*579*). It appears, therefore, that the exact location of the front of the column packing is not particularly significant.

GLC columns for triglyceride analysis must be thoroughly conditioned for good recovery of the high molecular weight compounds. During the conditioning of silicone columns, the calibration factors for tripalmitin and tristearin slowly decrease until they become fairly constant (*530*). The amount and temperature of conditioning is related to the column length and tightness of packing. With a loosely packed 3% JXR, 0.56 m × 2.4 mm i.d. column, Litchfield *et al.* (*578*) reported that after 2 hours of conditioning at 350° (150 ml/minute carrier gas) and two to six trial runs, typical low calibration factors for PPP and StStSt were obtained. With a tightly packed 3% JXR, 0.5 m × 3.0 mm o.d. column, Kuksis and Breckinridge (*535*) recommended 6–8 hours or more of "thermal stripping" at 350° (100 ml/minute carrier gas) for good recovery of StStSt. Longer columns require more extensive conditioning procedures. Litchfield *et al.* (*579*) conditioned loosely packed 3% JXR and 1% OV-17 1.83 m × 2.4 mm i.d. columns for 4 hours at 350°, 1 hour at 375°, and 1 hour at 400° (100 ml/minute carrier gas) and then used the columns for 5–10 days before stable calibration factors were obtained. In any case, it is essential that conditioning at the highest usage temperature be continued until the calibration factors for the $>C_{42}$ triglycerides are stabilized.

It has been suggested (*402,539,839*) that conditioning silicone GLC columns in the presence of air will improve their thermal stability, possibly by increasing the cross linking in the polymer. Other laboratories

(*344,577*) have not found any advantage in this conditioning procedure so perhaps results may depend on the particular batch of silicone polymer evaluated.

## C. Operating Conditions

### 1. Sample Injection

A 1–5% solution of the sample is injected through the septum into the column using a standard microsyringe. On-column sample injection is essential for accurate results (*531,535,578*). Approximately 10–20 μg of triglyceride is required to produce a full scale peak on a gas chromatogram with a 1.0 mV recorder (*578*). Triglyceride samples for GLC analysis are often dissolved in carbon disulfide, since the hydrogen flame detector is fairly insensitive to this solvent, and any degradation products appearing near the solvent front are more easily detected. However, fully saturated $C_{48}$–$C_{66}$ triglycerides are more easily dissolved in chloroform or benzene than carbon disulfide.

The flash heater temperature must be maintained high enough to assure instantaneous sample vaporization but also low enough to avoid any thermal degradation. Flash heater temperatures between 300° and 350° are most commonly employed, and studies by Litchfield *et al.* (*578*) and Carracedo and Prieto (*134*) have confirmed that maximum sample recovery occurs in this range.

It is important to avoid any pyrolysis of the sample while it is being vaporized in the flash heater. The presence of unidentified peaks following the solvent front is often an indication of fragmentation in the flash heater (*530*). These pyrolysis products are most easily seen when the sample is dissolved in $CS_2$. Extraneous peaks at a point in the chromatogram where diglycerides would elute can also indicate degradation in the flash heater (*579*). Three causes of sample pyrolysis in the flash heater have been encountered in the author's laboratory: flash heater temperature is too high, carrier gas contains reactable impurities, and flash heater contains catalytically active sites (i.e., nonvolatile residues from previous samples, nonsilanized glass wool, copper or other active metals).

### 2. Carrier Gas

The type of carrier gas used for triglyceride GLC influences peak resolution. Litchfield *et al.* (*578*) have demonstrated that helium gives significantly better peak resolution than nitrogen in both glass and stainless steel columns (Fig. 6-3). Calibration factors are equivalent with either carrier gas under optimum operating conditions (*578*). The reason for superior peak resolution with helium is not fully understood, although a similar effect has been noted by Barr and Sawyer (53) in the GLC of 3-pentanone.

Carrier gas flow rates for triglyceride GLC are considerably higher than those used for fatty acid methyl esters because triglycerides are much less volatile than methyl esters. Typical gas flow rates (measured at room temperature) for triglyceride GLC are 100 ml/minute with 2.4–3.0 mm i.d. columns (578) and 180–300 ml/minute with 4.0–6.0 mm i.d. columns (539). Gas chromatographs designed for programmed temperature operation are normally equipped with a constant mass flow controller to keep carrier gas flow constant over a wide temperature range.

Peak resolution improves as the carrier gas flow rate decreases (Fig. 6-3; 134). This improvement is most pronounced on a steel column with a nitrogen carrier and is rather small on a glass column with a helium carrier. The higher the carrier gas flow rate, the lower the elution temperature and the sharper the peak shape for any particular triglyceride. Under optimum conditions calibration factors for tripalmitin and tristearin are not much affected by the type of carrier gas (578) or by the flow rate in the 50–200 ml/minute range (Table 6-2), although this has not proven true for some gas chromatographs (134,578).

*3. Column Temperature*

For best results, temperature programming is recommended for all GLC of triglycerides. If a sample contains only a limited range of triglyceride carbon numbers, quantitative separations are possible with isothermal operation. When a wider range of carbon numbers is present, however, the column temperature must be programmed to elute all components as sharp, well-defined peaks. Isothermal analysis of coconut oil is compared with programmed temperature conditions in Fig. 6-5. With the column temperature fixed at 250°, only the $C_{30}$ to $C_{40}$ peaks are well-defined and can be accurately quantitated. By programming column temperature from 200° to 335°, all the peaks from $C_{28}$ to $C_{54}$ are sharp and well-defined.

For best resolution, the initial column temperature should be 25°–50° below the elution temperature of the most volatile component. Linear programming rates of 2° to 5°/minute are most commonly used for GLC of triglycerides. Litchfield and co-workers (578) compared peak resolution at different program rates on glass and stainless steel columns (Fig. 6-6). Slower program rates gave significantly better resolution on steel columns, but peak resolution was almost independent of rate on glass columns. Varying the program rate had no effect on the quantitative recovery of tristearin in one study (578) but produced significant changes in the $C_{54}$ calibration factor in another study (134). Carracedo and Prieto (134) found that the recovery for $C_{52}$ and $C_{54}$ triglycerides increased when the

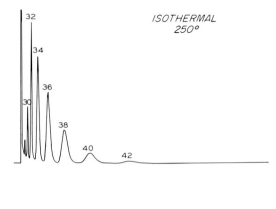
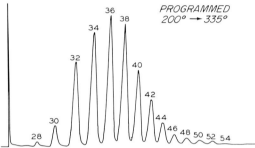

FIG. 6-5. Comparison of isothermal and programmed column temperatures for gas–liquid chromatography of coconut oil triglycerides. *Operating conditions for isothermal run:* 0.40 m × 2.8 mm i.d. glass column packed with 2.25% SE-30 on Chromosorb W; column temperature, 250°; 75 ml/minute N carrier gas; flash heater, 320°. From Leegwater and van Gend (*564a*). *Operating conditions for programmed run:* 0.46 m × 3.2 mm o.d. stainless steel column packed with 2.25% SE-30 on 60/80 mesh Chromosorb W; column temperature programmed 200° → 335° at 3°/minute; 150 ml/minute N carrier gas; flash heater, 325°. From Kuksis et al. (*542*).

initial column temperature was raised from 150° to 250°; the calibration factors for $C_{30}$–$C_{48}$ molecules were the same under both conditions, however.

Kuksis and co-workers (*531,534,539*) have recommended nonlinear program rates for GLC of triglycerides on stainless steel columns. As the molecular weight of triglycerides increases, vapor pressure differences between consecutive members of the homologous series decrease. Triglyceride peaks are eluted at progressively closer intervals as column temperature is raised. Under nonlinear programming conditions, rapid program rates (15°–20°/minute) at 200°–230° are decreased to slow rates (1°–2°/minute) at 300°–330° to maintain an even spacing of peaks at all temperatures. The data in Fig. 6-6 suggest that nonlinear temperature program rates are advantageous on stainless steel but not on glass columns.

## 4. Detector

The air and hydrogen flows to the flame ionization detector are adjusted to give maximum detector response at the carrier flow rate selected; a 10:1:1 ratio of air, hydrogen, and carrier flows is typical. It is not uncommon for the high carrier gas flow rates used in triglyceride GLC to blow out the flame in a standard hydrogen flame detector, but this can be corrected by using a flame tip with a larger internal diameter. Silicone column bleed deposits on the detector electrodes and will decrease sensitivity if not removed. To avoid this problem, the detector electrodes should be soaked in $N,N$-dimethylformamide after every 15–30 analyses.

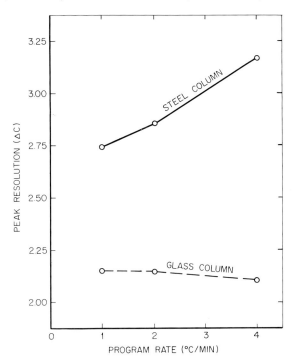

FIG. 6-6. Effect of temperature program rate on peak resolution in glass and steel columns. $\Delta C$ is the minimum carbon number difference between two saturated triglycerides which could be separated with baseline resolution in the $C_{42}$ to $C_{48}$ region of the chromatogram. *Operating conditions:* 0.61 m × 2.5 mm i.d. columns packed with 3.0% JXR on 100/120 mesh Gas Chrom Q; column programmed 170° → 325° with 100 ml/minute N carrier gas; flash heater, 325°; detector base, 300°–340°; sample, equal weights of LaLaLa, MMM, PPP, and StStSt. From Litchfield *et al.* (*578*).

TABLE 6-3
Typical Operating Conditions for Analytical GLC of $C_{30}$–$C_{60}$ Natural Fat Triglycerides

| | |
|---|---|
| Gas chromatograph | Instrument designed for steroid analysis with on-column injection, equipped with flame ionization detector and temperature programming |
| Column | 0.55 m × 2.5 mm i.d. glass or stainless steel |
| Packing | 3.0% OV-1 on 100/120 mesh Gas Chrom Q |
| Column conditioning | 2–8 hours at 350° |
| Carrier gas | 100 ml/minute helium |
| Column temperature | 170° → 350° at 2°–4°/minute |
| Flash heater temperature | 300°–350° |
| Detector base temperature | 300°–340° |
| Sample size | 10–20 μg triglyceride for each full scale peak on a 1 mV recorder |

## 5. Optimum Operating Conditions

Experience has shown that the typical operating conditions listed in Table 6-3 give excellent results for GLC of triglycerides.

## D. Quantitation

Triglycerides are among the highest molecular weight compounds that can be effectively analyzed by GLC. Their exceedingly low vapor pressures, even at elevated temperatures, make quantitative analysis difficult. Even under optimum operating conditions, not all of the high molecular weight triglyceride injected into a gas chromatograph can be made to elute from the column. Hence special precautions are necessary to produce meaningful quantitative results.

## 1. Peak Identification

Natural fat triglycerides usually contain an homologous series of regularly spaced peaks representing successive increments of two carbon atoms (see Fig. 6-11). Co-chromatography of known and unknown samples is the most convenient and accurate method for peak identification. The unknown is first chromatographed by itself. For a second run, the syringe is successively filled with the same size sample of unknown, a small air bubble, and then a few microliters of a known mixture (LaLaLa, MMM, PPP, and StStSt, for example). Comparison of the two chromatograms identifies which peaks have increased their height when the known mixture was added. The in-between peaks are then assumed to be members of the same homologous series of even-carbon-number triglycerides, provided no obvious shoulders or irregular peak spacings are noted.

Schmit and Wynne (*810*) have described an alternative method for the identification of peaks in programmed temperature GLC based on elution temperatures. Although absolute elution temperatures are dependent on operating conditions, the relative elution temperature, $T_{RE}$, defined as

$$T_{RE} = \frac{\text{elution temperature of peak X in °C}}{\text{elution temperature of standard in °C}}$$

is practically independent of initial column temperature and program rate. Watts and Dils (*943*) have found $T_{RE}$, values to be quite reproducible for $C_6$–$C_{54}$ triglycerides on silicone columns.

Under isothermal conditions, a linear relationship exists between the log of the retention volume and the carbon number for an homologous series of triglycerides (Fig. 6-7). If the retention volumes of two members of the series are known, then the retention volumes of other members can be estimated graphically. This procedure is seldom used for peak identification, however, since triglyceride GLC usually requires temperature programming for best results (Section I,C,3).

## 2. *Linearity of Detector Response*

For quantitative GLC of any compound, it is essential to establish the linearity of detector response to variations in sample size. For triglyceride

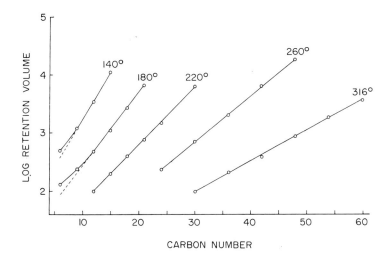

FIG. 6-7. Linear relationship between the log of the retention volume and carbon number for saturated, monoacid triglycerides. *Operating conditions:* 0.53 m × 6.5 mm o.d. stainless steel column packed with .10% SE-30 on 100/120 mesh Gas Chrom Q; isothermal column temperature; 35 ml/minute N carrier gas; sample load, 5–25 µg per peak. From Watts and Dils (*943*).

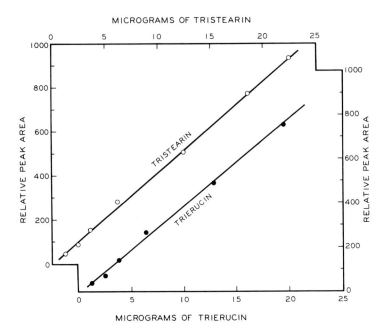

FIG. 6-8. Linear relationship between GLC peak area and the amount of tristearin or trierucin injected. *Operating conditions:* 0.53 m × 2.4 mm i.d. stainless steel column packed with 3.0% JXR on 100/120 mesh Gas Chrom Q; column temperature programmed 200° → 360° at 4°/minute; 100 ml/minute He carrier gas; flash heater, 350°; detector base, 310°–350°. From Harlow et al. (*346*).

GLC, the entire gas chromatographic system must be evaluated, since there is some loss of $C_{48}$ and larger molecules during passage through the column (Section I,D,3). Litchfield and co-workers (*346,578*) and Watts and Dils (*943*) have studied hydrogen flame ionization detector response to $C_6$–$C_{66}$ triglycerides on different gas chromatographs. Their resulting plots of peak area vs. amount of triglyceride injected (Fig. 6-8) were linear over the normal operating range of 0–20 μg for each molecular species tested except for triacetin and tripropionin (*943*). Thus the losses of $C_{48}$ and higher triglycerides during passage through the column are proportional to the amount of sample injected, indicating that these losses can be accurately compensated for by using proper calibration procedures.

### 3. Calibration

A mixture of monoacid triglycerides is usually employed for calibrating a gas chromatograph since these compounds are commercially available in 99% purity. A typical calibration chromatogram employing $C_{24}$–$C_{54}$

standards is shown in Fig. 6-9. Peak areas are measured by triangulation, planimetry, an integrator, cutout weight, or other accurate method. Quantitative weight response factors ($f_w$) and molar response factors ($f_m$) for individual triglycerides are then calculated by the internal normalization technique (*126,491*):

$$f_w = \frac{\text{weight }\%}{\text{area }\%} \qquad f_m = \frac{\text{mole }\%}{\text{area }\%}$$

A value of 1.00 is assigned to $f_w$ and $f_m$ for a low molecular weight primary standard (usually trilaurin) which is assumed to be completely recovered from the column. This primary standard can then be included in all calibration mixtures so that the calibration factors from all GLC runs will be comparable. Comparison of calibration factors under different operating conditions gives an accurate indication of sample recovery and permits optimum operating conditions to be selected. The factors are then directly usable for the quantitative analysis of unknown triglyceride mixtures. If the area of each peak is multiplied by its respective $f_w$ or $f_m$ calibration factor, then the relative weight or molar amounts represented by each peak are obtained.

Theoretical $f_w$ calibration factors for specific triglycerides can be calculated as shown in Table 6-4, assuming that all the injected sample reaches the flame ionization detector. A plot of theoretical $f_w$ vs. carbon number would approximate a horizontal line with a slight negative slope. This negative slope is due to the decreasing percent of oxygen in the molecule as the carbon number increases (*8*). On the other hand, as the molecular

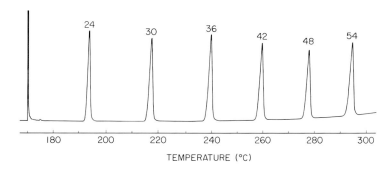

FIG. 6-9. Typical calibration chromatogram for a mixture of trioctanoin, tridecanoin, trilaurin, trimyristin, tripalmitin, and tristearin. *Operating conditions:* 0.56 m × 2.4 mm i.d. glass column packed with 3.0% JXR on 100/120 mesh Gas Chrom Q; column temperature programmed 170° → 305° at 3°/minute; 100 ml/minute He carrier gas; flash heater, 335°; detector base, 300°–340°. From Litchfield *et al* (*578*).

TABLE 6-4
Typical Weight Calibration Factors for GLC of Monoacid Triglycerides

| | | Weight calibration factor[a] ($f_w$) | | | |
| --- | --- | --- | --- | --- | --- |
| Triglyceride | $\left(\begin{array}{c}\text{Carbon}\\\text{number}\end{array}\right):\left(\begin{array}{c}\text{Double}\\\text{bonds}\end{array}\right)$ | Theory[b] | Litchfield et al. (578) | Watts and Dils (943) | Kuksis (531) |
| Triocatanoin | 24:0 | 1.10 | 1.12 | 1.13 | 1.00 |
| Tridecanoin | 30:0 | 1.04 | 1.04 | 1.08 | 1.00 |
| Trilaurin | 36:0 | 1.00 | 1.00 | 1.00 | 1.00 |
| Trimyristin | 42:0 | 0.97 | 0.96 | 1.01 | 1.00 |
| Tripalmitin | 48:0 | 0.95 | 0.98 | 1.09 | 1.00 |
| Tripalmitolein | 48:3 | 0.95 | 1.01 | 1.12 | — |
| Tristearin | 54:0 | 0.93 | 1.09 | 1.52 | 1.00 |
| Triolein | 54:3 | 0.93 | 1.03 | 1.35 | 1.00 |
| Trilinolein | 54:6 | 0.93 | 1.10 | — | 1.05 |
| Trilinolenin | 54:9 | 0.93 | 1.12 | — | — |
| Triarachidin | 60:0 | 0.92 | 1.21 | — | 1.10 |
| Tri-11-eicosenoin | 60:3 | 0.92 | 1.10 | — | — |
| Tribehenin | 66:0 | 0.91 | 1.43 | — | — |
| Trierucin | 66:3 | 0.91 | 1.34 | — | — |

[a] $f_w$ = weight percent/area percent. The $f_w$ for trilaurin is arbitrarily chosen as 1.00.
[b] Calculated assuming that the flame ionization detector response is proportional to the hydrocarbon content of each triglyceride [i.e., that the 12 C—O linkages in each molecule are incapable of combustion (see Ackman and Sipos, 8)].

weight of the triglycerides increases, the vapor pressure decreases. At some point the molecules must become so large and so nonvolatile that they can not fully participate in the gas–liquid partition effect necessary for GLC. One would expect this problem to cause differences between theoretical and experimental $f_w$ values for very high molecular weight triglycerides.

Typical experimental $f_w$ values for various monoacid triglycerides on 0.45–0.56 m columns are listed in Table 6-4 for comparison with the theoretical values. Litchfield et al. (578) found good agreement between theoretical and experimental $f_w$ values for $C_{24}$–$C_{42}$ triglycerides. Above $C_{42}$, the experimental $f_w$ value increased when in theory it should have decreased. These results indicate negligible losses up through $C_{42}$; while at $C_{48}$ and above, losses occur and become greater as carbon number increases. Watts and Dils (943) confirmed these findings, although losses became apparent as low as $C_{42}$ under their operating conditions. The calibration factors published by Kuksis and Breckenridge (531) reflect their assumption that weight percent equals area percent for $C_{24}$–$C_{54}$ triglycerides on thoroughly conditioned columns. While this assumption may be approximately true over a narrow range of carbon numbers, the use of actual $f_w$ calibration factors will clearly produce more accurate analytical results, particularly on chromatograms containing more than four or five peaks.

Harlow et al. (346) have estimated that approximately 6% of the tristearin and 24% of the trierucin injected is lost during analysis under the conditions listed in Table 6-4. This lost triglyceride is apparently retained unaltered on the column, since experiments with $^{14}$C-labeled tristearin and trierucin have shown that radioactive triglyceride can be recovered from the column packing after the run (109). Preparative GLC of $^{14}$C-trilaurin (125) has shown that this retained material will later bleed very slowly off the column if a high enough temperature is maintained. For quantitative analysis, however, such losses can be accurately compensated for by using calibration factors, since detector response curves show that the loss is proportional to the amount of sample injected (Fig. 6-8).

In theory, the calibration factors of unsaturated triglycerides should approximately equal those of saturated molecules of the same carbon number. This has not always proven true in practice, however (Table 6-4).

Tripalmitin and tripalmitolein have approximately equal calibration factors. So do StStSt, LLL, and LnLnLn, although OOO may have a slightly lower $f_w$ value. Tri-11-eicosenoin has a lower calibration value than triarachidin; and trierucin shows a better recovery than tribehenin. This may indicate a tendency for unsaturated triglycerides of high carbon number to show lower losses than their corresponding saturated compounds. Other workers have also reported problems with the GLC of unsaturated triglycerides, indicating that sample losses are strongly influenced by the chromatograph, column, and operating conditions employed. Jurriens and Kroesen (443) hydrogenated natural fat triglycerides because unsaturated molecules suffered severe degradation during GLC analysis. Kuksis and Breckenridge (535) reported that GLC recovery of triolein was 5% higher than that of tristearin, while the recovery of trilinolein was 25–50% lower. Thus sample hydrogenation before GLC analysis is often desirable for best quantitative results.

In actual practice, triglyceride compositions are most frequently determined in mole percent values. A plot of $f_m$ vs. carbon number is made for the saturated monoacid triglycerides, and the approximate curve is drawn (Fig. 6-10). The $f_m$ values for mixed-acid triglycerides (i.e., $C_{50}$, $C_{52}$, etc.) are read from this graph. All saturated triglycerides of the same carbon number are assumed to have the same $f_m$ value. Where the $f_m$ values for saturated and unsaturated triglycerides of the same carbon number are different, an average $f_m$ value is assigned to each peak based on its estimated fatty acid composition.

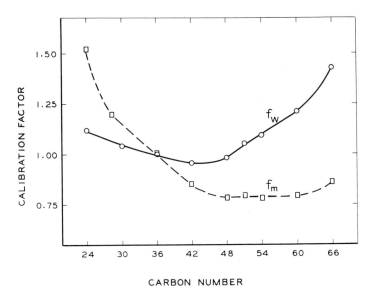

Fig. 6-10. Variation of weight ($f_w$) and molar ($f_m$) calibration factors with carbon number for saturated, monoacid triglycerides. *Operating conditions:* 0.56 m × 2.4 mm i.d. stainless steel column packed with 3.0% JXR on 100/120 mesh Gas Chrom Q; column temperature programmed 170° → 350° at 3°/minute; 100 ml/minute He carrier gas; flash heater, 320°–350°; detector base, 300°–340°; sample, equal weights of OcOcOc, DDD, LaLaLa, MMM, PPP, trimargarin, StStSt, AdAdAd, and BeBeBe. From Litchfield et al. (578).

The above evidence emphasizes that *regular calibration is essential for quantitative GLC of triglycerides.* Possible degradation of unsaturated triglycerides should always be checked before quantitative analysis of unsaturated samples is attempted. Since calibration factors vary substantially with the column, the operating conditions, and the chromatograph used, the $f_w$ and $f_m$ values listed in Table 6-4 and Fig. 6-10 cannot be used directly in other laboratories. Even with the same instrument, column, and operating conditions, calibration factors vary over a period of weeks and must be checked almost daily (578).

### 4. Accuracy

A thorough study of the accuracy of triglyceride GLC data using known-composition mixtures has never been attempted. One would expect triglyceride GLC data to be less accurate than methyl ester GLC data because of the difficulties involved in the GLC of higher molecular weight compounds. Some judgment on the method's reproducibility can be gained from Table 6-5, which compares four consecutive GLC analyses of

TABLE 6-5
COMPARISON OF FOUR CONSECUTIVE GLC ANALYSES (MOLE PERCENT) OF RAPESEED OIL TRIGLYCERIDES[a,b]

| Analysis | Carbon number | | | | | | | |
|---|---|---|---|---|---|---|---|---|
| | 50 | 52 | 54 | 56 | 58 | 60 | 62 | 64 |
| A | 0.4 | 1.7 | 3.9 | 10.4 | 18.7 | 25.3 | 38.9 | 0.7 |
| B | 0.4 | 1.6 | 3.8 | 11.0 | 18.7 | 24.7 | 39.1 | 0.7 |
| C | 0.6 | 1.7 | 4.0 | 10.6 | 19.0 | 24.6 | 39.0 | 0.5 |
| D | 0.4 | 1.8 | 4.2 | 9.3 | 19.9 | 25.1 | 38.9 | 0.4 |
| Range | 0.2 | 0.2 | 0.4 | 1.7 | 1.2 | 0.7 | 0.2 | 0.3 |

[a] From Litchfield and Harlow (577).
[b] *Operating conditions:* 0.56 m × 2.4 mm i.d. stainless steel column packed with 3.0% JXR on 100/120 mesh Gas Chrom Q; column temperature programmed 200° → 350° at 1°/minute; 100 ml/minute He carrier gas; flash heater, 335°; detector base, 320°–360°.

rapeseed oil triglycerides. The four values for any specific carbon number lie within a range of 0.2–1.7% absolute.

Kuksis et al. (*531,541*) have devised a useful technique for checking the accuracy of GLC analyses of natural fat triglycerides when the fatty acid composition is known. The average fatty acid chain length (or average triglyceride carbon number) is readily calculated from both the methyl ester GLC data and the triglyceride GLC results *when they are expressed in mole percent*. If these two independent calculations agree within a few percent, then the triglyceride carbon number distribution found is probably

TABLE 6-6
COMPARISON OF AVERAGE FATTY ACID CHAIN LENGTHS CALCULATED FROM METHYL ESTER AND TRIGLYCERIDE GLC DATA

| Source of triglycerides | Average fatty acid chain length | | Fatty acid carbon recovery | Reference |
|---|---|---|---|---|
| | Calculated from methyl ester data | Calculated from triglyceride data | | |
| Watercress seed fat | 18.74 | 18.79 | 100.3% | *346* |
| Rapeseed oil | 19.80 | 19.85 | 100.3% | *346* |
| Cocoa butter | 17.43 | 17.49 | 100.3% | *578* |
| Lindera praecox seed fat | 11.35 | 11.33 | 99.8% | *580* |
| Butterfat | 14.63 | 14.06 | 96.2% | *541* |
| Rat adipose tissue fat | 16.86 | 17.00 | 100.8% | *578* |
| Tuna muscle fat | 18.34 | 18.31 | 99.8% | *579* |

close to the correct values. Kuksis *et al.* express this agreement in terms of "% fatty acid carbon recovery":

$$\frac{\% \text{ fatty acid}}{\text{carbon recovery}} = \frac{\begin{bmatrix} \text{average fatty acid chain length} \\ \text{calculated from triglyceride data} \end{bmatrix}}{\begin{bmatrix} \text{average fatty acid chain length} \\ \text{calculated from methyl ester data} \end{bmatrix}} \times 100$$

Typical results with seven natural fats are listed in Table 6-6. This method of checking triglyceride GLC analyses is most effective when a wide range of fatty acid chain lengths is present in the sample.

## II. APPLICATIONS

### A. Separation by Carbon Number

Triglycerides differing by two carbon atoms are readily separated on 0.45–0.60 m GLC columns using the operating conditions listed in Table 6-3. Molecular weights up through $C_{66}$ can be resolved (*346*). Natural fat triglycerides containing only even-chain-length acids will produce an homologous series of even-carbon-number peaks which are well separated and easily quantitated. Fats containing short (*Lindera praecox*), average (rat adipose tissue), and long (*Crambe abyssinica*) chain acids have been analyzed by this technique, as illustrated in Fig. 6-11. More peaks are resolved when the sample contains a wide range of fatty acid chain lengths (*Lindera praecox* with $C_8$, $C_{10}$, $C_{12}$, $C_{14}$, $C_{16}$, and $C_{18}$ acids) than when only a few chain lengths are present (rat adipose tissue with only $C_{14}$, $C_{16}$, and $C_{18}$ acids).

Triglycerides differing by one carbon atom have been separated on a 1.83 m, 3.0% JXR column by Litchfield *et al.* (*579*). Both even and odd carbon numbers up through $C_{64}$ have been resolved by programming to 375°. Hydrogenated fish oil triglycerides containing 2.4–20.4 mole % odd-chain-length fatty acids were separated into an homologous series of peaks by this method (Fig. 6-12).

Saturated and unsaturated triglycerides of the same carbon number have slightly different retention times during GLC on a silicone liquid phase (Section II,B). Although natural fat triglycerides cannot be separated on the basis of unsaturation by GLC, this effect does broaden peak width appreciably. Better peak resolution is obtained if a sample is completely hydrogenated before analysis (*579*), especially when triene, tetraene, pentaene, and hexaene acids are present. No information is lost by hydrogenation,

## II. APPLICATIONS

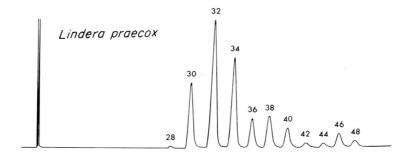

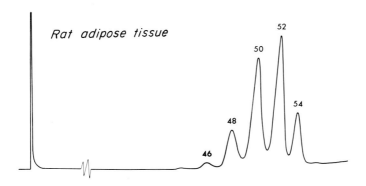

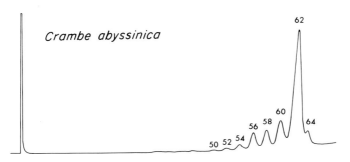

FIG. 6-11. Gas chromatograms of unhydrogenated *Lindera praecox* seed, rat adipose tissue, and *Crambe abyssinica* seed triglycerides showing resolution of molecules differing by two carbon atoms. *Operating conditions:* 0.56 m × 2.5 mm i.d. glass (rat) or stainless steel (*Lindera* and *Crambe*) columns packed with 3.0% JXR on 100/120 mesh Gas Chrom Q; column temperature programmed 170° → 340° (*Lindera* and rat) or 220° → 350° (*Crambe*) at 4°/minute; 100 ml/minute He carrier gas; flash heater, 350°; detector base, 300°–340° (*Lindera* and rat) or 320°–360° (*Crambe*). From Litchfield et al. (577,578).

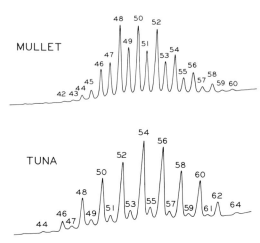

FIG. 6-12. Gas chromatograms of hydrogenated tuna and mullet oil triglycerides showing resolution of molecules differing by only one carbon atom. *Operating conditions:* 1.83 m × 2.4 mm i.d. glass column packed with 3.0% JXR on 100/120 mesh Gas Chrom Q; column temperature programmed 210° → 375° at 4°/minute, 100 ml/minute He carrier gas; flash heater, 350°; detector base, 320°–360°. From Litchfield et al. (579).

since GLC separations are on the basis of carbon number only. Hydrogenation also prevents any thermal decomposition of polyunsaturated fatty acids at the high temperatures required for triglyceride GLC.

### B. Separation by Unsaturation

Since GLC can separate fatty acid methyl esters on the basis of unsaturation, efforts have been made to do the same with triglycerides. Unfortunately, the polar polyester liquid phases commonly used for methyl ester analysis will bleed heavily above 250° and cannot be used for triglycerides. Alternative organic liquid phases which are stable above 300° (Table 6-1) are of fairly low polarity.

Litchfield et al. (577,579) have attempted to separate tristearin, triolein, and trilinolein on 1.83 m JXR, OV-17, and Apiezon L columns. Results (Fig. 6-13) show some resolution based on unsaturation in each case. The unsaturated triglycerides elute before tristearin on the JXR and Apiezon L columns, while the reverse order is observed on OV-17. Analogous results on an SE-30 column have been reported by Watts and Dils (943). To be useful in the analysis of unsaturated triglycerides, however, GLC must resolve molecules of the same carbon number which differ by only one double bond; i.e., StStSt, StStO, StOO, and OOO. The results in

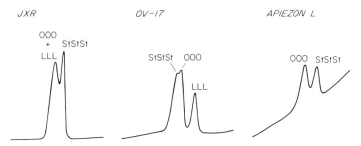

FIG. 6-13. Gas chromatograms showing partial separation of tristearin, triolein, and trilinolein on JXR, OV-17, and Apiezon L liquid phases. *Operating conditions:* 1.83 m × 2.4 mm i.d. glass column packed with 3.0% JXR, 1.0% OV-17, or 3.0% Apiezon L on 100/120 mesh support; column programmed 200° → 350° at 4°/minute; 100 ml/minute He carrier gas; flash heater, 330°; detector base, 320°–360°. From Litchfield and Harlow (*577*).

Fig. 6-13 indicate that molecules must differ by at least three double bonds before they can be separated. Apparently one double bond makes too little difference in the gas–liquid partition characteristics of a molecule as large as triolein. It is evident, therefore, that present GLC methodology cannot be used to resolve natural fat triglycerides on the basis of unsaturation.

## C. Separation of Isomers

### 1. Isomeric Fatty Acids

Triglycerides containing isomeric branched-chain acids elute slightly before straight-chain compounds containing an equal number of carbon atoms (*531*). Hence the presence of substantial amounts of branched-chain acids will decrease resolution in triglyceride GLC so that molecules differing by only one carbon atom cannot be fully separated (*579*). The complex elution patterns of cetacean triglycerides containing isovaleric acid have been studied by Litchfield *et al.* (*575*).

GLC of triglycerides containing fatty acids with midchain cyclopropane rings has been attempted (*510,579*), but the precise influence of the cyclopropane ring on elution order has not been determined.

The presence of a terminal cyclopentane ring in a fatty acid, however, makes its triglyceride elute later than the straight-chain analog (*579*). Resolution is sufficient to completely separate hydrogenated palmitodihydnocarpin from hydrogenated trihydnocarpin, even though both molecules have the same theoretical carbon number. This phenomenon makes GLC a useful tool for the triglyceride analysis of cyclopentene acid seed fats such as *Hydnocarpus wightiana* (Fig. 6-14).

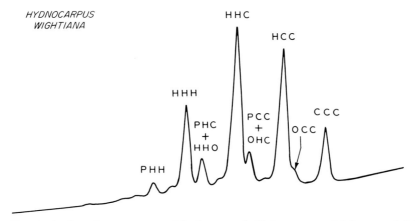

Fig. 6-14. Gas chromatogram of hydrogenated *Hydnocarpus wightiana* seed fat showing separation of cyclopentene acid triglycerides. P, 16:0; O, 18:1; H, hydnocarpic acid; C, chaulmoogric and gorlic acids. *Operating conditions:* 1.83 m × 2.4 mm i.d. glass column packed with 3.0% JXR on 100/120 mesh Gas Chrom Q; column temperature programmed 210° → 375° at 4°/minute; 100 ml/minute He carrier gas; flash heater, 350°; detector base, 320°–360°. From Litchfield *et al.* (*579*).

## 2. Triglyceride Positional Isomers

Triglyceride positional isomers can be separated using GLC only when they contain two fatty acids of greatly different chain length. Early work by Huebner (*399*) showed that triglycerides containing one or two acetate esters can be resolved into two positional isomer peaks on a 0.61 m, 23% silicone grease column. In all cases, the isomer having the acetate in the 2-position elutes after the isomer having a longer chain acid in the 2-position. Separations of diacetotriglyceride isomers ranged from β-AcOcAc/β-OcAcAc through β-AcOAc/β-OAcAc. Separations of positional isomers containing one acetate ester included β-AcDD/β-DAcD and β-AcMM/β-MAcM. Kuksis and Breckenridge (*534*) have resolved 2-palmito-1,3-dibutyrin and 1-palmito-2,3-dibutyrin on a 2.44 m, 5% SE-30 column (Fig. 6-15), and similar separations have also been reported by Watts and Dils (*943*). Like the acetotriglycerides, the isomer having the shorter butyrate chain in the 2-position eluted last. These workers also attempted to separate 2-palmito-1,3-dihexanoin from 1-palmito-2,3-dihexanoin on the same column but without success.

So far, only the separation of acetate and butyrate positional isomers has proven possible with present GLC techniques. This may be extended to hexanoates and octanoates as higher resolution columns and more polar, thermostable liquid phases are tested; but it seems very doubtful that triglycerides such as β-POP and β-PPO will ever be resolved by GLC.

## 3. Triglyceride Chain-Length Isomers

The distribution of carbon atoms between the three acyl chains in a triglyceride has a small effect on the elution temperature of the molecule. This sometimes permits the separation of synthetic triglycerides of the same carbon number if fatty acids of greatly different chain length are present. Thus Watts and Dils (*943*) have reported good resolution of trilaurin and butyrodipalmitin, both $C_{36}$ triglycerides, on a 0.53 m, 10% SE-30 column. Huebner (*401*) resolved tridecanoin and acetodimyristin under similar conditions. In both cases, the monoacid triglyceride eluted before the mixed-acid species. Such separations are not found during GLC of natural fat triglyceride mixtures, however.

## D. Hydroxy and Epoxy Triglycerides

Hydroxy triglycerides have been analyzed by GLC of their acetate esters by Powell *et al.* (*728*). The exact increase in carbon number due to the acetoxy group is not reported but is apparently less than 4.

Fioriti *et al.* (*269*) have studied the GLC of epoxy triglycerides and their 1,3-dioxolane derivatives. Underivatized trivernolin elutes in the same peak as nonoxygenated $C_{60}$ triglycerides, indicating that each epoxy group adds the equivalent of two carbon numbers to the elution times of its triglyceride. Conversion of the epoxy groups to 1,3-dioxolane derivatives of cyclopentanone increases the molecular weights of the epoxy triglycerides so that they all elute after the $C_{54}$ peak, allowing epoxy and nonepoxy triglycerides to be clearly distinguished. However, the very low conversions

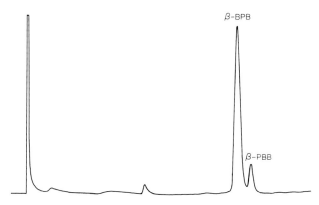

FIG. 6-15. Gas chromatogram showing separation of the positional isomers 2-palmito-1,3-dibutyrin and 1-palmito-2,3-dibutyrin. *Operating conditions:* 2.44 m × 3.2 mm o.d. stainless steel column packed with 5% SE-30 on 60/80 mesh Chromosorb W; column temperature programmed 200° → 300°; 100 ml/minute N carrier gas. From Kuksis and Breckenridge (*534*).

in the dioxolane reaction make these derivatives unsuitable for quantitative analysis.

### E. Oxidized Triglycerides

Youngs and Subbaram (*974*) separated saturated and unsaturated triglycerides of the same carbon number after oxidizing the unsaturated molecules to lower molecular weight derivatives. The triglyceride sample was first oxidized by a modified $KMnO_4/KIO_4$ procedure (Chapter 3, Section I,B) and then reacted with diazomethane to esterify the free carboxyl group of the azelaic acid. They found that triazelain trimethyl ester eluted at the same point as trilaurin on a 1.22 m, 2% SE-30 column, indicating that methyl azelate has an equivalent carbon number of 12. By this technique, a sample containing StStSt, StStO, StOO, and OOO which would normally elute together in one peak is converted into a mixture of StStSt, StStA, StAA, and AAA which have GLC carbon numbers of 54, 48, 42, and 36, respectively. Figure 6-16 shows typical gas chromatograms of oxidized triglycerides from a synthetic mixture, cocoa butter, and olive oil. Proper calibration with model mixtures of oxidized triglycerides is essential for quantitative interpretation.

GLC of oxidized triglycerides has the advantage of being able to distinguish between saturated and unsaturated triglycerides of the same carbon number, which cannot be done by GLC of the original sample. However, the method has several limitations:

(a) Palmitoleic, oleic, linoleic, and linolenic acids cannot be distinguished since they are all converted into azelaic acid by the oxidation procedure.

(b) If unsaturated acids are present which do not have the first double bond in the 9-position (i.e., vaccenic, petroselinic, 20:1-*11c*, 22:6-*4c,7c,10c,13c,16c,19c*, etc.), the number of possible oxidized triglycerides increases substantially, resulting in overlapping of many peaks.

(c) The $KMnO_4/KIO_4$ oxidation procedure does not quantitatively convert oleate, linoleate, and linolenate to azelate. Some overoxidation and/or double-bond migration occurs during oxidation (Chapter 3, Section I,B) causing 1–5% $C_7$ and $C_8$ dibasic acids to occur in the expected $C_9$ azelate. These unwanted $C_7$ and $C_8$ dibasic acids create oxidized triglycerides of unexpected molecular weights, leading to the appearance of artifact peaks in the gas chromatogram (peaks labeled "?" in Fig. 6-16). These artifact peaks are most prominent just preceding the AAA, PAA, and

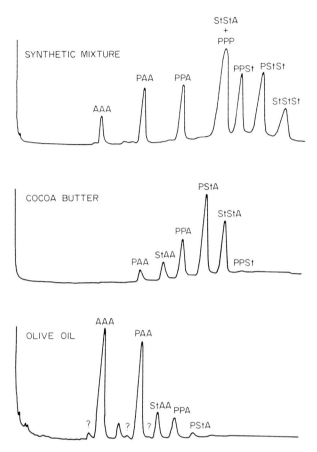

FIG. 6-16. Gas chromatograms showing separation of oxidized triglycerides from a synthetic mixture (originally OOO, POO, PPO, StStO, PPP, PPSt, PStSt, and StStSt), cocoa butter, and olive oil, ?, probable artifacts from $KMnO_4/KIO_4$ oxidation (see text). *Operating conditions:* 1.22 m × 4.8 mm o.d. stainless steel column packed with 2.0% SE-30 on 60/70 mesh Anakrom ABS; column temperature programmed 260° → 325° at 3°/minute and then held at 325°; 100 ml/minute He carrier gas; flash heater, 385°; detector base, 355°. Redrawn from Youngs and Subbaram (*974*).

StAA peaks and are usually included with the presumed parent peak for quantitative calculations.

## F. Preparative Separations

Natural triglyceride mixtures are so complex that each GLC peak represents a mixture of molecules having the same carbon number. If each peak were collected and analyzed further by other techniques, additional data

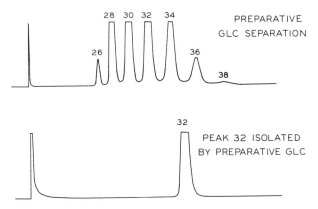

FIG. 6-17. Gas chromatograms showing preparative GLC separation of butterfat fraction and rechromatography of the carbon number 32 peak isolated. *Operating conditions:* 0.61 m × 6.3 mm o.d. stainless steel column packed with 5% SE-30 on 60/80 mesh silanized Chromosorb W; column temperature programmed 195° → 285° at nonlinear rate; 200 ml/minute N carrier gas; flash heater, 325°; detector base, 325°. Redrawn from Kuksis and Breckenridge (534).

on the triglycerides present could be obtained. For example, the monounsaturated triglycerides MStO, PPO, and PPoSt would all elute together in a carbon number 50 peak; but the relative amounts of these three triglycerides could be determined from the fatty acid composition of the collected peak.

Kuksis and co-workers (531,534,539) have described a semiautomatic system for the GLC separation and recovery of triglycerides in milligram amounts. They employed a 0.61 m × 6.3 mm o.d., 5% SE-30 column in an Aerograph Autoprep gas chromatograph (Varian Aerograph, Walnut Creek, California) equipped with a stream splitter so that column effluent could be monitored with a hydrogen flame detector. With 10–20 injections of 5–10 mg samples, 20–50 mg of each peak could be collected in the $C_{18}$ through $C_{38}$ range. Figure 6-17 shows a typical preparative GLC separation of the $C_{32}$ peak from a butterfat fraction. Eluted triglycerides were condensed in special collection vials maintained at room temperature and partially filled with glass wool. It was essential to maintain the tubing connection between the column oven and the condensing vials at 325°–350° to prevent any sample condensation before reaching the collection vial. Sample recoveries of monoacid triglycerides were >90% up through tripalmitin, but substantial losses of tristearin and triolein were reported. Collection of pure $C_{48}$, $C_{50}$, $C_{52}$, and $C_{54}$ peaks from natural fat mixtures was not accomplished. Small amounts of silicone column bleed were present in the collected material, but this was easily removed by thin-layer chromatography. Examination of the recovered triglycerides by thin-layer chromatography, infrared spectroscopy, GLC of derived butyl esters, and pancreatic lipase hydrolysis showed expected triglyceride behavior in all cases.

Similar systems for preparative GLC of triglycerides have been described by Lefort et al. (565), Dixon and Schmit (225), and Bugaut and co-workers (76,125). These authors also report good recoveries of high-purity fractions up to $C_{42}$. At higher carbon numbers, however, severe contamination with triglycerides from adjacent peaks and (surprisingly) with triglycerides of widely different carbon numbers was encountered. Apparently triglyceride molecules previously retained on the column (Section I,D,3) are eluted fairly rapidly above 300° and contaminate the collected material.

At the present state of the art, therefore, it is not possible to collect pure $C_{46}$, $C_{48}$, $C_{50}$, and higher peaks from natural triglyceride mixtures and obtain completely representative sampling. Repurification of the collected material by additional preparative GLC runs is one means of removing most of the contaminants (125). However, it must be remembered that the GLC peaks of triglyceride mixtures are nonhomogeneous (Sections II,A, and II,B); and any discarding of the front and tail portions of a peak probably produces a nonrepresentative sample.

## G. Radioisotope Detection

A procedure for the simultaneous measurement of both mass and $^{14}$C-content of triglycerides eluted from GLC columns has been developed by Breckenridge and Kuksis (109) using a modified Barber-Colman radioactivity monitoring system. The effluent from the column was split with 10% going to a flame ionization detector for mass measurement and the other 90% flowing through a combustion train and then to a proportional radioactive gas counter. The combustion train (CuO at 700°) converted $^{14}$C into $^{14}$CO$_2$ before entry into the counter. The counting efficiency of the entire system was better than 90%. Under optimum conditions, as little as 500 cpm/peak could be detected with a relative error of 10%; and at higher counting rates the error was less than 5%. Typical mass and $^{14}$C-radioactivity chromatograms of the same sample are shown in Fig. 6-18.

Simultaneous mass and radioisotope monitoring systems of this type are particularly suited for metabolic studies where the specific activities of several triglycerides of different carbon number must be determined (389a). The radioactivity monitoring system used by Breckenridge and Kuksis can also be adapted to measure $^3$H content of the eluate (869a). However,

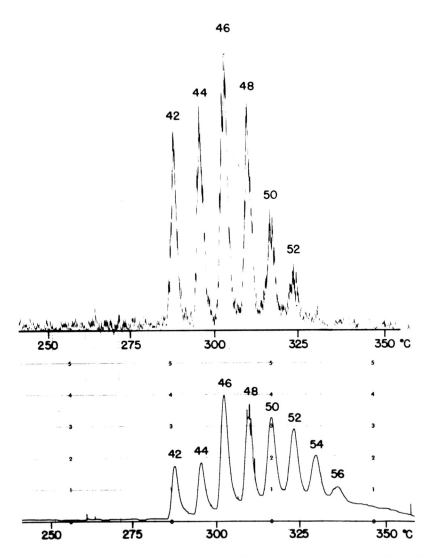

FIG. 6-18. Simultaneous measurement of $^{14}C$-content and mass of triglyceride peaks separated by gas–liquid chromatography. *Operating conditions:* dual 0.61 m × 6.3 mm o.d. glass columns packed with 3% JXR on 100/120 mesh Gas Chrom Q; column temperature programmed 200° → 355° at 5°/minute; 115 ml/minute Ar carrier gas; flash heater, 310°; flame ionization detector, 340°; column effluent split with 90% going to Barber-Colman radioactivity monitoring system and 10% to flame ionization detector; sample, rat intestine triglycerides containing $^{14}C$-labeled 14:0 and 16:0. From Breckenridge and Kuksis (109).

practical applications of the technique are limited by the need for samples of relatively high specific activity (500 dpm/50 µg per peak).

## H. Derived Diglycerides

Diglycerides can be effectively separated by GLC in essentially the same manner as triglycerides. However, free diglycerides undergo transesterification at the high flash heater and column temperatures used (537,729) so they must be converted into acetate (537,764,959) or trimethylsilyl ether (524,690,778,944) derivatives prior to analysis. Trifluoroacetate derivatives are not sufficiently thermostable for diglyceride GLC (546).

Gas chromatography of diglyceride derivatives on the short (0.45–0.65 m), nonpolar silicone columns used for triglyceride analysis yields a simple separation by carbon number (Fig. 6-19). Different species of the same carbon content [sn-1,2(2,3)- and sn-1,3-isomers, molecules of different unsaturation] have very similar elution volumes on such columns (537,690,778) so that well-separated peaks (possibly with shoulders) are obtained for all even-carbon-number diglycerides. If odd chain length fatty acids were present, however, peak overlap could be a problem. The separation of diglycerides containing hydroxy, epoxy, and nonoxygenated fatty acids has been studied by Tallent et al. (877,878) using trimethylsilyl ether derivatives.

It is also possible to separate different diglyceride species of the same carbon number by GLC on longer columns and more polar liquid phases. Several laboratories (399,531,690) have resolved corresponding

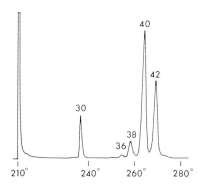

Fig. 6-19. Separation of diglyceride acetates by carbon number on a short silicone column. *Operating conditions:* 0.46 m × 3.2 mm o.d. stainless steel column packed with 3% JXR on 100/120 Gas Chrom Q; 150 ml/minute N carrier gas; column temperature programmed 210° → 280°; flash heater, 300°; detector, 350°; sample tridecanoin internal standard (peak 30) plus hexaene diglyceride acetates derived from rat liver lecithins. From Kuksis et al. (537).

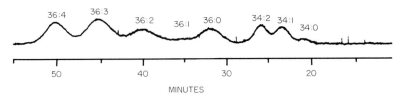

FIG. 6-20. Resolution of the trimethylsilyl ethers of $sn$-1,2(2,3)-diglycerides by unsaturation on a polyester column. Peaks are labeled in the same manner as fatty acids, i.e., (carbon number):(number of double bonds). *Operating conditions:* 1.20 m × 3 mm i.d. glass column packed with 10% EGSS-X on 100/120 mesh Gas Chrom Q; 30 ml/minute N carrier gas; isothermal column temperature, 265°; flash heater, 280°; detector, 290°; sample, trimethylsilyl ethers of $sn$-1,2(2,3)-distearin plus $sn$-1,2(2,3)-diglycerides derived from corn oil. Redrawn from Kuksis (*533*).

$sn$-1,2(2,3)- and $sn$-1,3-isomers on 1.5–2.0 m silicone columns and found that the $sn$-1,3-species has the greater retention volume. Kuksis (*531a,533,533a*) has recently reported the effective resolution of diglyceride mixtures by unsaturation using a polyester liquid phase such as EGSS-X (Applied Science Laboratories) or SP-1000 (Supelco) operated briefly at higher than normal temperatures. The chromatogram shown in Fig. 6-20 indicates that diglyceride derivatives differing by only one double bond can be partially resolved, making it possible to analyze an homologous series such as StSt–, StO–, OO–, OL–, and LL–, provided only one positional isomer is present.

# 7

# FRACTIONAL CRYSTALLIZATION

The earliest technique used to separate natural fats into their component triglycerides was fractional crystallization from solvents. An unsaturated molecule such as PPO is more soluble than PPP in acetone; hence PPP can be precipitated when an acetone solution of the two is cooled to 20°. Application of this principle to natural fats allows triglyceride molecules to be partially fractionated according to the number of saturated acyl groups they possess. Although complete separation of SSS, SSU, SUU, and UUU is not achieved, the major groups of triglycerides can be concentrated and identified in the more saturated natural fats.

During the nineteenth and early twentieth centuries, workers such as Chevreul (159), Duffy (234), Heise (351,352), Amberger (20,21), Klimont (516,517), and Bömer (92,93) utilized repeated fractional crystalization to isolate major component triglycerides from natural fats and obtain a qualitative idea of their composition. It was not until the 1930s, however, that Hilditch and his associates at the University of Liverpool (364) perfected fractional crystallization as a semiquantitative method for determining component triglycerides. They showed that a series of recrystallizations could separate solid and semisolid fats into fairly simple fractions whose triglyceride composition could be approximated from their fatty acid composition. This technique was widely used for triglyceride analysis until the early 1960s, when it was replaced by the more accurate chromatographic separation methods. A new approach was introduced in 1964 when Gunstone et al. (325,330) found that crystallizing triglyceride mixtures from Ag⁺-containing solvents produced very efficient separations on the basis of the number of double bonds per molecule.

Fractional crystallization is no longer used for quantitative analysis, but it is still quite useful for the large scale (>2 g) fractionation of triglyceride mixtures to concentrate various components for subsequent isolation and analysis. This chapter reviews fractional crystallization techniques from the standpoint of preparing such concentrates. Brief comments on former analytical procedures are also included to aid the researcher in interpreting the older literature.

## I. METHODS

### A. Solvent

Two types of solvents have been used for fractional crystallization of triglyceride mixtures: (i) solvents such as acetone or diethyl ether that separate triglycerides according to the number of saturated acyl groups that they possess; and (ii) Ag⁺-containing solvents which separate molecules according to the number of double bonds they contain.

Anhydrous acetone has been the most widely used solvent for separating triglyceride mixtures by fractional crystallization (*367*). Triglycerides containing different numbers of saturated acyl groups have markedly different solubilities in acetone, and the crystals formed are reasonably well defined and easily filtered. Unfortunately, acetone is miscible with the water condensate formed at subambient temperatures, and the presence of a small amount of water in acetone has a marked effect on the solubility of triglycerides. This problem can be avoided, however, by conducting the crystallization in a closed, nitrogen-blanketed vessel. Petroleum ether (*973*) and diethyl ether (*52,359*) have also been successfully used for fractional crystallization of triglycerides, but they usually require a higher ratio of solvent to sample. Ethanol (*886*) and methanol (*105*) are useful when more polar triglycerides are to be separated, i.e., molecules containing short-chain or oxygenated fatty acids.

Triglycerides can also be separated according to the number of *cis* double bonds per molecule by crystallization from solvents containing $AgNO_3$. The Ag⁺ complexes with the $\pi$-electrons of the double bonds to alter normal triglyceride solubilities and allow fractionation by unsaturation. Gunstone and co-workers (*325,330*) have demonstrated this technique with natural fat triglycerides using $AgNO_3$-saturated methanol/acetone 70/30. The amount of $AgNO_3$ used was twice the amount required to form $\pi$-complexes with all the double bonds present.

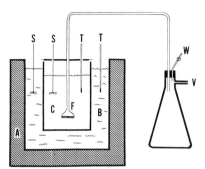

FIG. 7-1. Typical bench apparatus for large-scale crystallization of triglyceride mixtures. **A**, insulated container; **B**, constant-temperature bath; **C**, solution of triglycerides; **F**, filter stick with fritted glass disk on bottom; **S**, stirrer; **T**, thermometer; **V**, to vacuum; **W**, to atmosphere. Redrawn from Doerschuk and Daubert (226).

## B. Procedure

The triglyceride mixture to be separated is dissolved in the appropriate solvent using heat if necessary. The ratio of solvent to sample is usually in the range of 5/1 to 15/1 (v/w) depending on the nature of the sample. If constant-temperature cabinets are not available, equipment of the type shown in Figs. 7-1 and 7-2 (see also *156,287,798*) can be used for large- and small-scale crystallizations. Both are simply a crystallization vessel placed in a constant-temperature bath with provision for filtration at the same temperature at which crystallization has taken place. A water bath can be used for temperatures above $-10°$, while dry ice in acetone will provide temperatures between $-10°$ and $-78°$. Temperature control by the direct addition of dry ice pellets to the sample solution should be avoided since this causes localized supercooling. When crystallizing large

FIG. 7-2. Simple apparatus for small-scale crystallization of triglyceride mixtures. Solution to be crystallized (**A**) is placed in the bottom of the V-tube, which is then immersed in a constant-temperature bath. Crystals are recovered by tipping the V-tube to the right, applying nitrogen pressure at (**B**), and filtering the solution through fritted glass disk (**C**). The V-tube is easily constructed from a standard glass chromatography tube. Redrawn from Nickell and Privett (674).

samples, both the sample solution and the constant-temperature bath should be stirred to assure a uniform temperature throughout. The sample solution should be stirred very gently, however, so as not to fracture the triglyceride crystals. Equilibrium between the dissolved and precipitated triglycerides is achieved very slowly, and slow cooling is desirable for good crystal formation and maximum resolution. The solution is usually held at the final temperature for 3–24 hours before filtering off the precipitate.

After crystallization is finished, the precipitate is allowed to settle, and the remaining solution is filtered off. It is essential that filtration take place at the same temperature as the crystallization to avoid redissolving any precipitate. This is easily accomplished by using a precooled filter stick with large batches (Fig. 7-1) or by filtering through the fritted glass disk in the small-scale crystallization apparatus (Fig. 7-2). To avoid contamination of the precipitate by the mother liquor, the crystals should be washed twice with fresh solvent precooled to the temperature of the crystallization bath. Solvent is then removed from the filtrate and precipitate by evaporation, and the yield of each is determined by weighing.

## II. APPLICATIONS

### A. Separation by Number of Saturated Acyl Groups

*1. Solubility Considerations*

Fractional crystallization separates triglycerides mainly on the basis of their saturated acid content when solvents such as acetone, diethyl ether, or petroleum ether are used. This was clearly illustrated by Youngs and Sallans (973) who fractionated a mixture of StStO, StStL, and StOO by crystallizing from acetone or petroleum ether (Fig. 7-3). StStO and StStL have the same saturated acid content, while StStL and StOO have the same number of double bonds. If the triglycerides were separated on the basis of their unsaturation, the StStO would be separated from the other two, and the compositions of the fractions would lie on a line through StStO and the original mixture. If the saturated acid content of the triglycerides were the determining factor, however, then StOO would be separated from the other two, and the compositions of the fractions would lie on a line through StOO and the original mixture. The points representing the experimental fractions lie very close to the latter line, indicating that the saturated acid content of the triglycerides is the predominant factor in determining their relative solubilities.

In theory, therefore, it should be possible to resolve triglyceride mixtures into SSS, SSU, SUU, and UUU by fractional crystallization. Unfortunately,

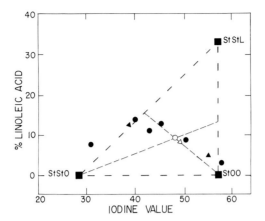

FIG. 7-3. Fractional crystallizaton of StStO, StStL, and StOO. ■, orginal triglycerides in mixture; ○, △, original composition of mixture; ●, fractions crystallized from acetone; ▲, fractions crystallized from petroleum ether. *Crystallization conditions:* solvent/sample 10/1 (v/w), samples chilled until an estimated 50% of material had crystallized. From Youngs and Sallans (*973*).

this cannot be achieved in practice. Crystallization can be used to concentrate each of these triglyceride groups, but complete resolution into discrete groups of predictable composition is not possible. Natural fat triglyceride mixtures are quite complex, and undesirable "mutual solubility effects" (*367*) cause more of a compound to be retained in solution than would be predicted from its individual solubility at that temperature. Triglycerides form true crystals very slowly, and equilibrium conditions are probably never reached in normal laboratory operations. Hence mixed crystals are often encountered. Furthermore, mutual solubility and mixed-crystal effects increase with increasing unsaturation, so that solid fats can be fractionated fairly well, while liquid oils are very poorly resolved. Additional difficulties arise if $C_4$–$C_{12}$ acids are present, since their solubility properties resemble those of long-chain unsaturated acids. As a result of these problems, useful fractionations by crystallization are only obtained with semisolid fats of simple fatty acid composition, i.e., triglyceride mixtures containing mostly 16:0, 16:1, 18:0, 18:1, and less than 30% 18:2.

## 2. Crystallization Sequence

The proper crystallization sequence for fractionating triglyceride mixtures varies widely with the composition of the sample. In most cases, a crystallization procedure for a similar fat can be found in the literature and adapted to the test sample. Otherwise, preliminary tests are advisable to ensure efficient separation. Figure 7-4 illustrates a typical crystallization

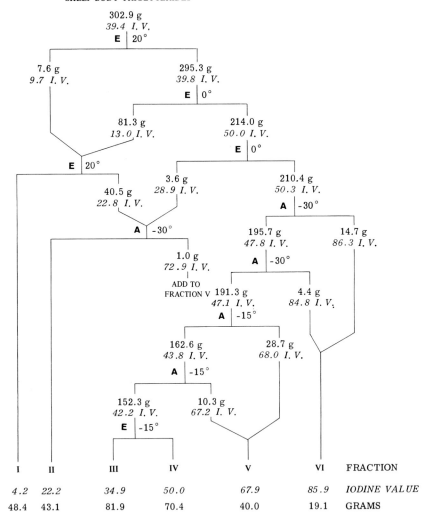

FIG. 7-4. Crystallization sequence for separating sheep body triglycerides into six fractions of widely different iodine value (I.V.). Samples were crystallized from 10% (w/v) solutions in acetone (**A**) or diethyl ether (**E**) at the temperatures indicated. Fatty acid compositions of the six final fractions are given in Table 7-1. Redrawn from Hilditch and Shrivastava (*360*).

sequence for separating sheep body triglycerides into fractions having as widely different triglyceride compositions as possible.

Hilditch (*360,367*) recommends that the SSS and SSU be precipitated in the first few crystallizations (i.e., at 20° and 0°) so that the SUU and

## II. APPLICATIONS

UUU can be processed in a more concentrated form. The SSS- and SSU-rich precipitates are then recrystallized at the same or at higher temperatures, while the SUU- and UUU-rich mother liquor is recrystallized at lower temperatures. To minimize mutual solubility effects, it is best to start the SUU/UUU crystallizations at the lowest temperature and then use progressively higher temperatures. Best resolution is obtained by crystallizing each sample twice under identical conditions before proceeding to different conditions. When the full crystallization sequence has been completed, fractions of similar composition are combined.

### 3. Calculation of Triglyceride Composition

Since the complete resolution of SSS, SSU, SUU, and UUU cannot be achieved by crystallization, early workers had to make certain simplifying assumptions before they could calculate the composition of the fractions. The most popular approach was that of Hilditch (Table 7-1) who assumed that the SSS, SSU, SUU, and UUU present in the original sample had been sufficiently separated so that only two or three of these triglyceride groups would be present in each fraction. Thus the amounts of SSS and SSU in the highly saturated Fraction I could be calculated from the amounts of S and U acids present in that fraction. Furthermore, if the

TABLE 7-1
APPROXIMATE TRIGLYCERIDE COMPOSITION OF SHEEP BODY FAT DETERMINED BY
FRACTIONAL CRYSTALLIZATION SEQUENCE GIVEN IN FIG. 7-4[a]

|  | Fractions (mole %) | | | | | | |
| --- | --- | --- | --- | --- | --- | --- | --- |
|  | I | II | III | IV | V | VI | Total |
| Component acids | | | | | | | |
| Saturated | 94.8 | 75.6 | 63.1 | 49.8 | 36.0 | 18.2 | 60.8 |
| Unsaturated | 5.2 | 24.4 | 36.9 | 50.2 | 64.0 | 81.8 | 39.2 |

Component triglycerides—calculated from
(1) fatty acid composition of individual fractions
(2) SSS content of several fractions determined by $KMnO_4$ oxidation
(3) assumption that components marked * are absent in the fractions indicated

| | | | | | | | |
| --- | --- | --- | --- | --- | --- | --- | --- |
| SSS | 84.4 | 35.2 | 23.2 | 13.0 | * | * | 28.0 |
| SSU | 15.6 | 56.4 | 42.3 | 23.4 | 8.1 | * | 28.5 |
| SUU | * | 8.4 | 34.5 | 63.9 | 91.9 | 54.6 | 40.7 |
| UUU | * | * | * | * | * | 45.4 | 2.8 |
| Total | 16.0 | 14.5 | 27.1 | 23.0 | 13.2 | 6.2 | 100.0 |

[a] After Hilditch and Shrivastava (360).

SSS content were determined independently by permanganate oxidation, then the SSU and SUU content could be directly calculated, provided UUU was assumed to be absent (Fractions II, III, and IV).

Modern chromatographic studies (*328,329,331,641*) have shown that the above assumptions were not wholly justified, but the error varies with the type of sample. Triglyceride mixtures containing mostly SSS and SSU (i.e., solid and semisolid fats) are moderately well resolved by fractional crystallization; and SSS/SSU/SUU compositional data obtained with this technique can be regarded as semiquantitative. On the other hand, liquid fatty oils containing no SSS are very poorly separated by low-temperature crystallization, and SSU/SUU/UUU data from crystallization analyses on such triglyceride mixtures are definitely unreliable. Gupta and Hilditch (*333*) also attempted to extend these calculation methods for SSS, SSU, SUU, and UUU to other acids. To do this, they presumed that PPP, PPX, PXX, and XXX or OOO, OOX, OXX, and XXX are effectively separated by crystallization. This assumption is highly questionable, since the data in Fig. 7-3 indicate that oleo- and linoleotriglycerides are very poorly fractionated by crystallization. It is also doubtful that PPO and PStO could be separated anywhere near as easily as PPO and POO.

### B. Separation by Number of Double Bonds

Early attempts by Hilditch and his co-workers (*52,199,326,359*) to separate the triglycerides of highly unsaturated natural fats by crystallization from acetone at $-10°$ to $-70°$ were unsuccessful. Modern silver ion TLC analyses (*328,329,331,641*) reveal that these low-temperature crystallizations do not give an efficient separation of triglycerides on the basis of unsaturation as earlier presumed.

More recent worp by Gunstone *et al.* (*325,330*), however, has demonstrated that triglyceride molecules can be effectively separated according to the number of *cis* double bonds they contain when crystallized in solvents containing $AgNO_3$. The $Ag^+$ complexes with the $\pi$-electrons of the double bonds to alter normal triglyceride solubilities and allow fractionation by unsaturation. Figure 7-5 illustrates the effectiveness of this technique. One crystallization of *Jatropha curcas* seed triglycerides from $AgNO_3$-saturated methanol/acetone 70/30 at $-10°$ produced a precipitate (**A**) that was mainly SSU containing one or two double bonds per triglyceride. Cooling the filtrate to $-70°$ yield crystals (**B**) that were predominantly SUU having three or four double bonds. The remaining mother liquor (**C**) was almost entirely UUU containing five or six double bonds per triglyceride.

It is obvious that two crystallizations in the presence of silver ion have produced a much better fractionation of *Jatropha curcas* triglycerides (Fig.

## II. APPLICATIONS

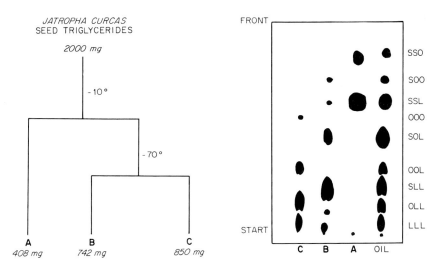

FIG. 7-5. *Left.* Separation of *Jatropha curcas* seed triglycerides by number of *cis* double bonds using fractional crystallization in the presence of Ag⁺. *Crystallization sequence:* 2000 mg sample crystallized at $-10°$ for 24 hours from 10 ml AgNO₃-saturated methanol/acetone 70/30. Filtrate then recrystallized at $-70°$ for 24 hours. All fractions were subsequently dissolved in petroleum ether and washed with distilled water to remove AgNO₃. *Right.* Analysis of three crystallization fractions by Ag⁺ TLC. *Operating conditions:* silicic acid impregnated with 17% AgNO₃, developing solvent benzene/diethyl ether 90/10, spots visualized by charring with glass-blowing torch. After Gunstone *et al.* (*325*).

7-5) than ten crystallizations from plain solvent have done for sheep body triglycerides (Fig. 7-4). The efficient resolution of SUU and UUU molecules is particularly noteworthy. Since only one example of silver ion crystallization has been published so far, the exact potential of this technique for triglyceride analysis remains to be established. It seems certain, however, to be the method of choice when large amounts ($>2$ g) of a triglyceride mixture are to be fractionated to obtain particular components for animal feeding studies or for further analysis.

### C. Oxidized Triglycerides

Fractional crystallization of oxidized triglycerides has been employed in a number of laboratories. It is difficult to evaluate the published results, however, since the procedure has always been coupled with the controversial permanganate oxidation reaction (Chapter 3, Section I,B); and the probable presence of unwanted reaction by-products (diglycerides, unoxidized SSU, "non-triglyceride neutral material," acetate esters, and short-chain monocarboxylic acids) undoubtedly makes subsequent fractional crystallization more difficult.

Assuming pure SSS/SSA/SAA/AAA mixtures as starting materials, useful separations can be achieved using either the neutral oxidized triglycerides or their metal salts. Desnuelle and Naudet (*215*) have published a procedure for isolating SSS and concentrating SSA from oxidized triglyceride mixtures using fractional crystallization from diethyl ether.

Kartha (*451,455,460,461,470*) has utilized the relative solubilities of the magnesium salts of oxidized triglycerides to resolve SSS/SSA/SAA/AAA mixtures into two simpler, more easily analyzed fractions. In his procedure the ammonium salts of the oxidized triglycerides are dissolved in water, and $MgSO_4$ is added until no further precipitate forms. The mother liquor and precipitate are separated by filtration, acidified, evaporated to constant weight, and both are analyzed for fatty acid composition. According to Kartha, the precipitate will contain only SSS, SSA, and SAA, while the filtrate contains only SAA and AAA. If the SSS content of the original sample is determined independently, then the glyceride composition of each crystallization fraction can be determined from its S and U content; and the SSS/SSA/SAA/AAA content of the original sample is easily calculated. Two problems have been reported in the application of this procedure: some SSA may appear in the filtrate if samples contain less than 30% S (*470*), and SAA molecules are easily hydrolyzed to diglycerides during Mg salt precipitation (*455,463*). Suggestions for minimizing these difficulties have been presented (*455,460,461,463,470*).

Mixtures of oxidized triglyceride have also been fractionated by their relative abilities to form sodium or potassium salts and dissolve in an aqueous solution. When a diethyl ether solution of SSS/SSA/SAA/AAA is washed with $NaHCO_3$, $Na_2CO_3$, $KHCO_3$, or $K_2CO_3$, AAA is extracted more rapidly than SAA, SAA more rapidly than SSA, and SSS is not extracted at all (*215,355,461,462,552*). However, severe emulsion problems and some hydrolysis of azelaic acid (Chapter 9, Section I,B) are usually encountered in such a procedure.

**D. Other Derivatives**

A number of other chemical reactions at the double bonds of triglycerides were explored by early workers in attempts to improve crystallization separations, but none yielded satisfactory results. Hydrogenation (*354,362*) was not particularly useful because so much compositional information was lost and because fractional crystallization of fully saturated molecules proved so difficult. Other workers (*246,348,787,869,919,931*) brominated natural triglyceride mixtures expecting to enhance the solubil-

## II. APPLICATIONS

ity differences in highly unsaturated fats. However, more recent quantitative studies by Mhaskar *et al.* (*645*) indicate that the bromo derivatives are poorly resolved by crystallization and extraction procedures. Piguelevsky and his co-workers (*720,721*) showed that the epoxy derivatives of oleic acid triglycerides were easily separated by crystallization, but the nonquantitative nature of the epoxidation reaction (Chapter 3, Section I,D) limits its analytical applications. Fractional crystallization of *cis–trans* isomerized triglycerides (*245,322*) proved very unsatisfactory due to the very complex nature of the reaction products (Chapter 3, Section I,G).

# 8

# OTHER SEPARATION TECHNIQUES

A number of other separation techniques that are not generally applicable to natural triglyceride mixtures can be quite useful when certain unusual fatty acids are present. These auxiliary separation methods will be discussed in this chapter.

## I. SILICIC ACID ADSORPTION CHROMATOGRAPHY

Silicic acid, the most versatile adsorbent for separating lipid classes, can also be used to separate groups of triglycerides having major differences in polarity. Molecules containing short-chain or oxygenated fatty acids are often resolved by this method.

### A. Methods

General methods for chromatography on columns and thin layers of silicic acid are described in Chapter 2, Section II and need not be repeated here. Appropriate solvent systems will be cited as each type of separation is considered. Quantitation procedures for the triglyceride fractions separated by silicic acid are discussed in Chapter 2, Section III.

### B. Separation by Molecular Weight

Larger molecules travel ahead of smaller molecules in the chromatography of triglycerides on silicic acid. Figure 8-1 shows the $R_f$ values for

# I. SILICIC ACID ADSORPTION CHROMATOGRAPHY

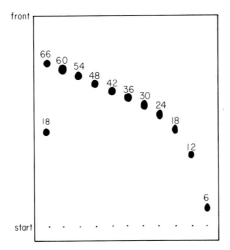

FIG. 8-1. Analytical separation of saturated, monoacid triglycerides by carbon number using thin-layer chromatography on silicic acid. 66, BeBeBe; 60, AdAdAd; 54, StStSt; 48, PPP; 42, MMM; 36, LaLaLa; 30, DDD; 24, OcOcOc; 18, HHH; 12, BBB; 6, AcAcAc. *Operating conditions:* 200 × 200 mm TLC plate coated with 0.25 mm layer of Adsorbosil-1; sample size, ∼5 μg per spot; developed with petroleum ether/diethyl ether 80/20; spots visualized by using rhodamine 6G-impregnated adsorbent.

a series of saturated, monoacid triglycerides during TLC. A difference of 12 carbon atoms is required for the complete resolution of $C_{24}$ to $C_{66}$ triglycerides (StStSt vs. MMM, for example), while molecules differing by only 2 to 6 carbons can be separated below $C_{24}$ (i.e., HHH vs. BBB). Blank and Privett (*82*) used this approach to separate butterfat triglycerides into long-chain and short-chain fractions by column chromatography on silicic acid (Table 8-1), and analogous results can also be achieved by TLC (*82,236,260,304,536*). Triglycerides containing both long-chain and acetic acids (i.e., "acetin fats") are easily separated according to acetate content by TLC (Fig. 8-2). Similarly, beluga whale head triglycerides containing isovaleric acid have been fractionated by TLC on the basis of the number of isovalerate moieties per molecule by Litchfield *et al.* (*575*).

The presence of phytanic acid (3,7,11,15-tetramethylhexadecanoic acid) in a triglyceride decreases its adsorption by silicic acid so that triglycerides containing no, one, two, or three phytanate groups can be resolved. Karlsson *et al.* (*447,448*), Laurell (*559*), and Sezille *et al.* (*833,834*) have achieved this separation on thin layers of silicic acid (Fig. 8-3).

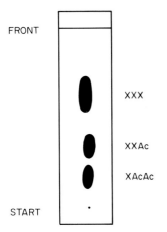

Fig. 8-2. Analytical separation of triglycerides containing acetic acid using thin-layer chromatography on silicic acid. Ac, acetic acid; X, fatty acids of herring oil. *Operating conditions:* 200 × 200 mm TLC plate coated with silicic acid; sample size, ∼50 μg per spot; developed with petroleum ether/diethyl ether 92/8; spots visualized with iodine vapor. From Gruger et al. (*318*).

TABLE 8-1

Fatty Acid Composition of Long-Chain and Short-Chain Fractions of Butterfat Separated by Column Chromatography on Silicic Acid[a,b]

|  | Composition (mole %) | |
| --- | --- | --- |
| Fatty acid | Long-chain fraction | Short-chain fraction |
| 4:0 | tr | 25.9 |
| 6:0 | 0.3 | 6.3 |
| 8:0 | 0.5 | 4.5 |
| 10:0 | 1.7 | 4.2 |
| 12:0 | 2.4 | 4.6 |
| 14:0 | 10.5 | 11.9 |
| 15:0 (?) | 3.0 | 2.9 |
| 16:0 | 32.7 | 19.7 |
| 16:1 | 3.4 | 2.0 |
| 18:0 | 15.4 | 3.4 |
| 18:1 | 30.1 | 13.1 |
| 18:2 | tr | 1.5 |
| Weight % | 61.7 | 38.3 |

[a] From Blank and Privett (*82*).

[b] *Operating conditions:* 25 × 300 mm column of silicic acid; sample, 1.0 g butterfat; triglycerides eluted with petroleum ether/diethyl ether 97/3; 10 ml fractions collected; separation monitored by TLC and appropriate fractions combined.

## I. SILICIC ACID ADSORPTION CHROMATOGRAPHY

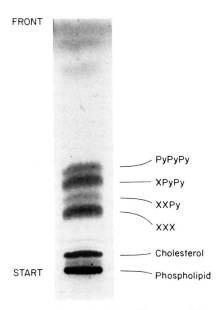

FIG. 8-3. Analytical separation of triglycerides containing phytanic acid using thin-layer chromatography on silicic acid. Py, phytanic acid; X, other fatty acids. *Operating conditions:* 200 × 200 mm TLC plate coated with silicic acid; sample, heart lipids from rabbit fed 2% phytol in diet; developed with petroleum ether/diethyl ether/acetic acid 90/10/1; spots visualized by spraying with a solution of ammonium sulfate in dilute $H_2SO_4$ and charring. From Sezille *et al.* (*833,834*).

### C. Separation by Unsaturation

Among triglycerides containing fatty acids of the same chain length, the more unsaturated molecules are adsorbed more strongly on silicic acid than the more saturated species (*777*). This difference is not great enough to separate triglycerides according to the number of double bonds they contain, but the effect is sufficient to create nonhomogeneous spots or peaks during thin-layer or column chromatography. Marinetti (*606*) has noted this effect in eluting egg yolk triglycerides from a column of silicic acid (Fig. 8-4). The early triglyceride fractions are richer in 16:0, 18:0, and 18:1, while 16:1, 18:2, and 18:3 are found in greatest concentration in the later fractions.

### D. Triglyceride Positional Isomers

Triglyceride positional isomers can be separated on silicic acid only when they contain two fatty acids of greatly different chain length. Renkonen (*759*) has resolved β-PAcP and β-PPAc by TLC (Fig. 8-5)

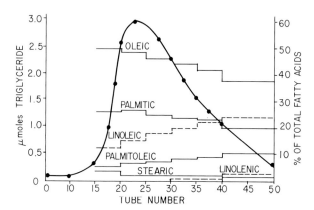

FIG. 8-4. Subfractionation of egg yolk triglycerides during column chromatography on silicic acid. The skewed curve represents the total triglyceride eluted. Fatty acid composition of each fraction is given by the horizontal lines. *Operating conditions:* column containing 20 g of silicic acid; sample, 180 mg of egg yolk triglycerides; eluted with 250 ml *n*-heptane/diethyl ether 96/4; eluate collected in 5 ml fractions and monitored with hydroxamate/ferric chloride reactions. From Marinetti (*606*).

with the symmetrical β-PAcP isomer having the higher $R_f$ value. Similar results have been reported by Åkesson *et al.* (*14*) and Kleiman and co-workers (*511,512*). The analogous butyrate compounds might possibly be separated in the same manner, but long-chain positional isomers such as β-PMP and β-PPM cannot be resolved on silicic acid.

### E. Oxygenated Triglycerides

Triglycerides containing hydroxy, epoxy, ketone, aldehyde, ozonide, or ester groups on the fatty acid chains can be separated according to polarity by chromatography on silicic acid. Natural fats can be fractionated according to the number of oxygenated fatty acids per molecule using this technique.

Epoxy triglycerides from *Euphorbia lagascae* have been separated into bands containing zero, one, two, and three epoxy fatty acids per triglyceride molecule using TLC (Fig. 8-6). Similar results have been reported by others (*267,268,717,934*). The same separation has also been achieved on a column of silicic acid using hexane/diethyl ether 80/20 as the eluting solvent (*875*).

Hydroxy triglycerides from castor oil, *Lesquerella*, or other seed fats can be fractionated into normal, monohydroxy, dihydroxy, and trihydroxy bands on silicic acid. Suitable solvent systems for TLC are hexane/diethyl ether 60/40 or 70/30 (*332,653*), benzene/diethyl ether 75/25 (*262*), or chloroform/methanol 99/1 (*832*). Analogous column chromatography separations of hydroxy triglycerides have also been described (*652,725*).

Evans and co-workers (*256*) have resolved the hydroxy triglycerides of isano oil on columns of partially deactivated silicic acid using benzene/methanol 98/2 as the eluting solvent. Triglycerides of di- and trihydroxy fatty acids can also be separated with silicic acid, but the solvent system used and the resolution obtained depend on the fatty acid composition of the original mixture (*649,725*).

Keto triglycerides from oiticica oil have been separated into normal, monoketo, and diketo fractions on a column of partially deactivated silicic acid using benzene/methanol 98/2 for elution (*256*). Franzke *et al.* (*283*) have also resolved PPP and 2-(2-ketomyristo)-1,3-dipalmitin by column chromatography on silicic acid. The same separations could no doubt also be achieved on TLC plates using a hexane/diethyl ether solvent mixture.

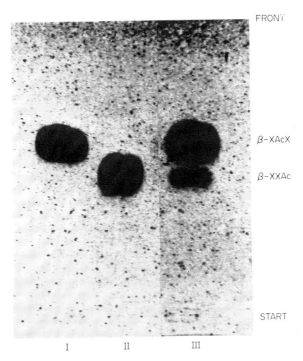

Fig. 8-5. Analytical separation of triglyceride positional isomers by thin-layer chromatography on silicic acid. Ac, acetic acid; X, normal long-chain fatty acid. *Operating conditions:* TLC plate coated with thin layer of Kieselgel G; sample: I, β-OAcO; II, β-XXAc derived from egg lecithin; III, β-PAcP + β-PPAc; developed with petroleum ether/diethyl ether 80/20; spots visualized by charring with sulfuric acid. From Renkonen (*759*).

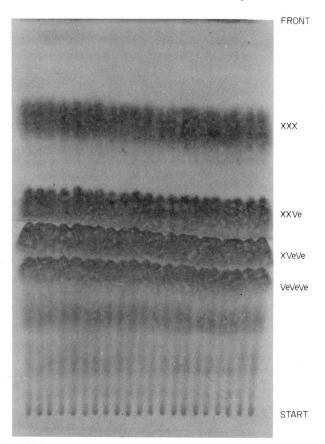

Fig. 8-6. Preparative separation of epoxy triglycerides by thin-layer chromatography on silicic acid. Ve, vernolic acid; X, normal long-chain fatty acids. Unidentified bands in chromatogram are nontriglyceride lipids. *Operating conditions:* 200 × 200 mm TLC plate coated with 0.5 mm layer of silicic acid; *Euphorbia lagascae* seed oil; developed once with 80/20/1 and then with 90/10/1 petroleum ether/diethyl ether/acetic acid; spots visualized with iodine vapor. From Kleiman *et al.* (*513*).

Ozonization of triglyceride molecules converts each double bond into an ozonide group $(-\mathrm{HC} \underset{O-O}{\overset{O}{\diagup\diagdown}} \mathrm{CH}-)$. The products can then be separated on silicic acid in two ways: as the ozonides, or after reduction to aldehydes. Privett and Blank (*733*) have demonstrated that triglyceride ozonides can be resolved according to the number of —CHOOOCH— groups they contain (i.e., according to the number of double bonds in the original molecules). Such a TLC separation of ozonides is shown in Fig.

8-7. On the other hand, the ozonides may be reduced with triphenylphosphine or by hydrogenation so that an aldehyde replaces each unsaturated fatty acid chain. The resultant "aldehyde cores" can then be separated into SSS, SSU, SUU, and UUU fractions by TLC (732). Privett et al. (735) have resolved the reduced ozonides from egg triglycerides by this technique (Fig. 8-8). These two separation methods can also be employed consecutively for a more detailed characterization (733): (i) preparative separation of ozonides according to —CHOOOCH— content, (ii) reduction of each fraction to aldehydes, and (iii) TLC separation of each fraction according to aldehyde content. However, because of the nonquantitative nature of the ozonization reaction (Chapter 3, Section I,C), ozonide separations are rarely used for triglyceride analysis.

Oxidized triglycerides can be directly resolved into SSS, SSA, SAA, and AAA bands by TLC on silicic acid using hexane/diethyl ether 60/40 as the developing solvent (439,874). Alternatively, Youngs and Baker (972) have reacted oxidized triglycerides with diazomethane and fractionated the resultant methyl esters by TLC into four bands having three, four, five, and six ester groups per molecule. A double development procedure was used for this separation: an initial development with diethyl ether until the solvent front had traveled 1 cm above the point of sample application, followed by full development with benzene/diethyl ether 90/10.

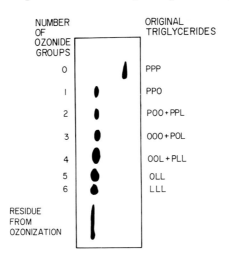

FIG. 8-7. Analytical separation of triglyceride ozonides according to —HCOOOCH— content using thin-layer chromatography on silicic acid. *Operating conditions:* 50 × 200 mm TLC plate coated with silicic acid; sample 50 μg ozonized corn oil triglycerides plus PPP; developed with petroleum ether/diethyl ether 80/20; spots visualized by spraying with 80% $H_2SO_4$ saturated with $K_2Cr_2O_7$ and charring. From Privett and Blank (733).

158                8. OTHER SEPARATION TECHNIQUES

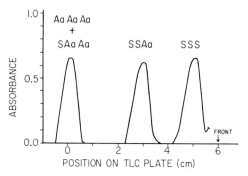

FIG. 8-8. Analytical separation of aldehyde cores obtained by ozonization and reduction of egg triglycerides using thin-layer chromatography on silicic acid. Aa, azeleoyl semialdehyde; S, saturated fatty acid. *Operating conditions:* TLC plate coated with silicic acid; sample, ∼50 μg reduced ozonides from egg triglycerides; developed with petroleum ether/diethyl ether 85/15; spots visualized by charring with $H_2SO_4/K_2Cr_2O_7$ and measured by densitometry. From Privett *et al.* (735), copyright 1962 by the Institute of Food Technologists.

Mixtures of estolide triglycerides can also be separated on silicic acid according to the number of ester groups per molecule. Ergot oil, for example, contains ricinoleic acid in which the hydroxyl group is acylated by normal long-chain fatty acids. The resulting tri- (normal), tetra-, penta-, and hexaester glycerides have been separated by Morris and Hall (664) by silicic acid TLC (Fig. 8-9). Similar separations of estolide triglycerides have been reported by others (328,598,650).

## F. Brominated Triglycerides

Although methyl stearate, oleate, linoleate, and linolenate cannot be resolved on silicic acid, their bromine additon products (methyl stearate, dibromostearate, tetrabromostearate, and hexabromostearate) can be completely separated (697,835). Triglycerides containing zero, one, two, and three double bonds could probably also be brominated and resolved by TLC on silicic acid, but no experiments attempting this have been reported.

## G. Mercurated Triglycerides

The addition of mercuric acetate to the double bonds of triglycerides (Chapter 3, Section I,F) enhances their polarity so that the molecules can be separated on silicic acid according to the number of double bonds they originally contained. Hirayama (371) has utilized this method to separate the mercuric acetate adducts of stillingia tallow triglycerides (*Sapium sebiferum* fruit coat fat) on a column of silicic acid (Fig. 8-10). Triglycerides containing zero, one, two, and three double bonds were completely resolved with Hirayama's pattern of elution solvents. However, molecules containing more than four double bonds were so strongly adsorbed by the silicic acid that they could not be quantitatively eluted from the column. The original triglycerides were regenerated from the eluted mercuric acetate adducts by reaction with dilute hydrochloric acid (Chapter 3, Section I,F). Similar separations of mercurated triglycerides have also been obtained using columns of Florisil (Section II) or alumina (Section III).

## H. Derived Diglycerides

The resolution of *sn*-1,2(2,3)- and *sn*-1,3-diglyceride isomers by preparative TLC (Chapter 2, Section II,C) is readily accomplished using a

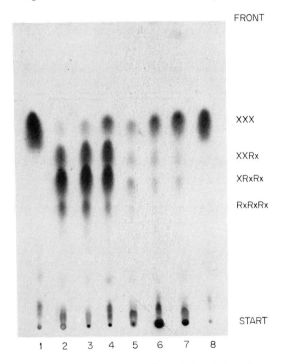

FIG. 8-9. Analytical separation of ergot oil estolide triglycerides containing three, four, five, and six ester groups per molecule using thin-layer chromatography on silicic acid. Rx, ricinoleate with 12-hydroxyl esterified to a normal long-chain fatty acid; X, normal long-chain fatty acid. *Operating conditions:* TLC plate coated with silicic acid; samples 1 and 8, corn oil; samples 2–7, different ergot oils; developed three times with hexane/diethyl ether 92.5/7.5; spots visualized by charring with $H_2SO_4$. From Morris and Hall (664).

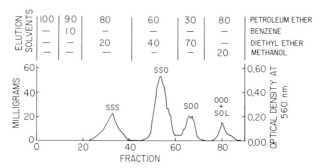

FIG. 8-10. Separation of mercuric acetate adducts of *Sapium sebiferum* fruit coat triglycerides by column chromatography on silicic acid. *Operating conditions:* 21 × 300 mm column containing 15 g silicic acid; sample, mercuric acetate adducts prepared from 508 mg *Sapium sebiferum* fruit coat triglycerides; stepwise elution at 0.5 ml/minute using solvent mixtures shown above; 100 drop fractions collected; SSS peak monitored gravimetrically, other peaks monitored spectrophotometrically after reaction with diphenylcarbazone. Redrawn from Hirayama (*371*).

petroleum ether/diethyl ether 50/50 or 60/40 solvent system (*170,217,975*). Prolonged contact between silicic acid and partial glycerides causes acyl migration, but this can be minimized by impregnating the adsorbent with 8% (w/w) boric acid (*975*).

Separations of diglyceride species by silicic acid adsorption chromatography closely parallel those described above for triglycerides. Diglycerides containing short-chain acids are adsorbed more strongly than species having only long-chain acids; this allowed Kleiman *et al.* (*512*) to separate *sn*-XX– from *sn*-–XAc (X = $C_{16}$ or $C_{18}$) on silicic acid TLC plates developed with hexane/diethyl ether 70/30. Fioriti *et al.* (*267*) resolved diglycerides containing zero, one, and two epoxy acids by TLC using petroleum ether/diethyl ether/acetic acid 60/40/1 as the solvent system.

## II. FLORISIL ADSORPTION CHROMATOGRAPHY

Florisil (nominally magnesium silicate) has the same general adsorption characteristics for triglycerides as silicic acid. The only major chromatographic difference between the two adsorbents is the stronger adsorption of free carboxyl groups by Florisil (Chapter 2, Section II,A). Any of the chromatographic separations described in the above section for silicic acid could probably also be accomplished with Florisil, except those with oxidized triglycerides containing —COOH groups.

Kerkhoven and deMan (*505*) have reported the separation of SSS from natural fats by chromatography of their mercuric acetate adducts on columns of partially inactivated Florisil. The unmercurated SSS molecules were eluted with hexane/diethyl ether 80/20, and then the unsaturated triglycerides were regenerated and eluted with ethanol (95%)/chloroform/conc. HCl 9/8/1. The SSS content of natural fats could be determined in this manner with an accuracy of ±0.6 absolute percent.

## III. ALUMINUM OXIDE ADSORPTION CHROMATOGRAPHY

Activated aluminum oxide (alumina) possesses triglyceride adsorption characteristics similar to silicic acid, except that free carboxyl groups are retained more strongly. Some hydrolysis of triglyceride ester bonds may occur during chromatography on alumina (*563,890*), but this loss can be minimized by keeping adsorbent/sample contact time less than 60 minutes and by performing separations at a subambient temperature (*563,758*). Most triglyceride separations obtained with aluminum oxide can be duplicated on silicic acid or Florisil, which do not hydrolyze esters. For this reason, alumina has not been widely used for triglyceride separations in recent years.

Several workers have attempted to use the adsorption of double bonds by $Al_2O_3$ as a means of separating triglycerides by unsaturation. Walker and Mills (*939–941*) have used a form of displacement chromatography to segregate linseed triglycerides of different iodine values. Linseed oil was placed at the head of a column of alumina, and the column was developed with hexane which is not polar enough to elute triglycerides. The column was then blown dry with $CO_2$, extruded, and cut into 1.3 cm sections from which the triglycerides were recovered and analyzed. Four zones of iodine values 203, 175, 146, and 116 were obtained with the more unsaturated triglycerides appearing near the top of the column. The exact resolution cannot be judged from the data presented, but it seems doubtful that pure fractions containing four, five, six, and seven double bonds per molecule were separated as claimed. Reinbold and Dutton (*753*) have tried to fractionate soybean triglycerides on a column of aluminum oxide by elution with petroleum ether/diethyl ether 65/35. Some separation by unsaturation was possible, but no discrete fractions were obtained.

Low molecular weight triglycerides are adsorbed more strongly by $Al_2O_3$ than high molecular weight molecules. Thus Pokorny and Prochazkova (*726*) have separated PPP, PPAc, PAcAc, and AcAcAc on thin layers of alumina using benzene as the developing solvent. Kaufmann and Wolf (*500*) isolated pure MMM from a mixture of BBB and MMM by column chromatography on aluminum oxide using benzene as the eluting solvent.

Inkpen and Quackenbush (*405*) have fractionated the mercuric acetate derivatives of synthetic triglycerides and of olive oil on columns of alumina using various diethyl ether/acetic acid mixtures for stepwise elution. Their separations resembled those of Hirayama on silicic acid (Fig. 8-10) with incomplete separation between the various peaks.

The strong adsorption of free carboxyl groups on alumina has been used to separate the SSS from oxidized triglycerides. Michel (*648*) applied 1.8 g of oxidized *Mycobacterium marianum* triglycerides to a column of 70 g $Al_2O_3$ and eluted the SSS with benzene. Similar techniques have been described by Lakshminarayana and Rebello (*551,552*) and Sylvester *et al.* (*871,872*) using chloroform or diethyl ether to elute the SSS.

## IV. CHARCOAL ADSORPTION CHROMATOGRAPHY

The separation of model mixture of triglycerides by column chromatography on activated charcoal was investigated many years ago by Claesson (*178*) and Hamilton and Holman (*339*). The adsorbent was evaluated using either frontal analysis or displacement analysis procedures with ethanol (for $C_{12}$ to $C_{36}$ triglycerides), benzene, or diethyl ether (for $C_{36}$ to $C_{54}$ triglycerides) as the eluting solvent. Saturated triglycerides eluted in the order of their solubilities in the mobile phase (i.e., lower carbon numbers first); and triolein eluted with trimyristin. The resolution achieved was insufficient for the fractionation of natural fat triglyceride mixtures.

## V. PAPER CHROMATOGRAPHY

Triglyceride separations on cellulose have not been successful except for molecules containing free hydroxyl groups. Early work by Kaufmann and co-workers (*477*) separated mixtures of castor oil and linseed triglycerides by paper chromatography. Development with methanol moved the castor oil triglycerides and fractionated them according to hydroxyl content, while linseed triglycerides remained at the origin.

## VI. ION-EXCHANGE CHROMATOGRAPHY

Neutral triglycerides are readily separated from molecules containing ionic bonds by ion-exchange treatment. Savary and Desnuelle (*797*) oxidized palm oil and then passed the oxidized triglycerides through a column of Amberlite IRA-400 ion-exchange resin in hydroxyl form using diethyl ether as the eluting solvent. Molecules having free carboxyl groups were retained by the resin, and the neutral SSS was recovered from the eluate. Eshelman *et al.* (*254*) added mercaptoacetic acid to the double bonds in unsaturated fats and then passed the ammonium salts of the reaction products through a column of DEAE-cellulose (*N,N*-diethylaminoethylcellulose). Petroleum ether eluted the SSS, while the addition products remained on the column.

## VII. PERMEATION CHROMATOGRAPHY

Permeation chromatography on a column of hydrophobic gel can separate triglycerides on the basis of molecular weight or carbon number. Suitable stationary phases are cross-linked polystyrene or methylated cross-linked dextran of known pore size, while the mobile phase is a good lipid solvent such as tetrahydrofuran or chloroform. A high-resolution separation of saturated, monoacid triglycerides is illustrated in Fig. 8-11. Similar results with synthetic mixtures have been reported in other papers (*89,90,436,675,689*). Since the log of the elution volume is proportional to the molecular weight in permeation chromatography (*91*), saturated and unsaturated triglycerides of the same carbon number would have approximately the same elution volumes. In the $C_{30}$–$C_{60}$ range, molecules must differ by at least six carbon atoms to obtain baseline resolution (Fig. 8-11). Since separation of two-carbon-number differences is essential for useful separations of natural triglyceride mixtures, permeation chromatography has not found wide application in the field of triglyceride analysis.

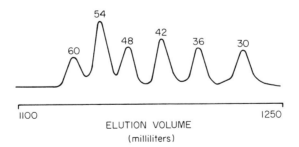

FIG. 8-11. Separation of saturated, monoacid triglycerides by carbon number using permeation chromatography. *Operating conditions:* 48.8 m × 7.8 mm i.d. column packed with 500 Å beads of cross-linked polystyrene; solvent flow, 0.4 ml/minute tetrahydrofuran; sample, 2.5 mg each of AdAdAd, StStSt, PPP, MMM, LaLaLa, DDD; column eluate monitored by differential refractometer detector. From Bombaugh *et al.* (*91*).

## VIII. THERMAL GRADIENT CHROMATOGRAPHY

Thermal gradient chromatography is a process for the repeated recrystallization of a solute as its solution moves through a column having a temperature gradient. The different sample components initially crystallize in different parts of the column according to their solubilities. As the composition of the mobile solvent and/or the temperature gradient of the column are slowly changed, components are eluted from the column in the order of their solubilities.

Van den Tempel et al. (906) have described a thermal gradient apparatus for the separation of triglycerides (Fig. 8-12). The column consists of a thick-walled brass tube having water jackets at both ends. A temperature gradient is established by passing warm water through the upper jacket and cold water through the lower jacket. The column is packed with an inert, porous material allowing movement of liquid but not of solid particles. Glass beads, steel wool, sand, and cotton have been used as packings. The column is initially filled with a poor triglyceride solvent (usually acetone), and a solution of the sample in this solvent is applied at the top of the packing. A gradient mixture of the poor solvent with a good triglyceride solvent (hexane, benzene, or carbon tetrachloride) is then passed through the column to elute the triglycerides in the order of their solubilities. If desired, the temperature of the bottom jacket can be raised during the run to speed the elution of high-melting triglycerides from the column (435). The eluate is collected with an automatic fraction collector.

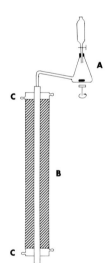

FIG. 8-12. Apparatus used for thermal gradient chromatography. (A) Mixing vessel containing a predetermined volume of acetone; hexane is added from the dropping funnel to create a solvent gradient. (B) Thick-walled, 21 mm i.d., 400 mm long brass column packed with a mixture of steel wool with coarse sand or cotton wool. (C) Water jackets for heating top of column and cooling bottom of column to create thermal gradient. From van den Tempel et al. (906).

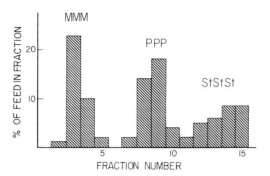

FIG. 8-13. Separation of trimyristin, tripalmitin, and tristearin by thermal gradient chromatography. *Operating conditions:* column as described in Fig. 8-12; sample, ~1 g of each component; thermal gradient of 40° (top) to 15° (bottom); gradient elution with 80 ml of acetone to which 50 ml of heptane and then 50 ml of carbon tetrachloride were slowly added; 28 mm/minute solvent velocity through column; separation monitored gravimetrically by collecting and evaporating 12 ml fractions. Redrawn from van den Tempel *et al.* (*906*).

Thermal gradient chromatography can resolve triglycerides which differ greatly in solubility, but the complex mixtures found in natural fats cannot be fully separated by this technique. Van den Tempel *et al.* (*906*) have successfully resolved MMM, PPP, and StStSt by thermal gradient chromatography (Fig. 8-13), and Magnusson and Hammond (*597*) have separated PPO and PPP. The SUU and SSU of cocoa butter could only be partially resolved, however (*435*). Thermal gradient chromatography has not found wide application in the analysis of triglyceride mixtures because other separation techniques are more effective.

## IX. DISTILLATION

Natural fat triglycerides can be distilled at relatively low temperatures without decomposition by performing a short-path distillation from a thin film in a "molecular still" at a pressure of a few microns. Molecules are separated on the basis of molecular weight, but resolution is inadequate for the analysis of natural fats. Augustin (*35*) has fractionated coconut oil triglycerides by distillation, and his results (Table 8-2) illustrate the type of separation obtained. GLC analyses of the various fractions show that a wide range of carbon numbers is present in each. McCarthy *et al.* (*634*) found similar carbon number separations after distillation of butterfat triglycerides.

TABLE 8-2
FRACTIONATION OF COCONUT OIL TRIGLYCERIDES BY DISTILLATION[a,b]

| | Carbon number distribution determined by GLC (wt %) | | | | | | | | | | | | | | | | | Yield (%) |
|---|---|---|---|---|---|---|---|---|---|---|---|---|---|---|---|---|---|---|
| | 20 | 22 | 24 | 26 | 28 | 30 | 32 | 34 | 36 | 38 | 40 | 42 | 44 | 46 | 48 | 50 | 52 | |
| Fraction I | 5.0 | 12.1 | 7.3 | 10.2 | 9.5 | 13.5 | 22.0 | 12.0 | 7.5 | 0.8 | — | — | — | — | — | — | — | 9.8 |
| Fraction II | — | — | — | — | 0.5 | 4.5 | 19.8 | 28.8 | 28.0 | 15.0 | 3.4 | — | — | — | — | — | — | 53.0 |
| Fraction III | — | — | — | — | — | — | 0.9 | 6.3 | 23.0 | 22.0 | 26.5 | 5.3 | 5.2 | 0.8 | — | — | — | 25.0 |
| Fraction IV | — | — | — | — | — | — | — | 3.0 | 7.7 | 19.6 | 27.5 | 23.8 | 12.7 | 5.2 | 0.2 | — | — | 2.3 |
| Residue | — | — | — | — | — | — | — | 6.7 | 1.2 | 6.8 | 17.1 | 25.8 | 20.3 | 13.2 | 9.0 | 4.4 | 1.3 | 10.0 |
| Original oil | tr | tr | 0.3 | 0.6 | 1.1 | 3.5 | 12.6 | 15.6 | 18.5 | 16.2 | 10.3 | 8.8 | 5.8 | 3.8 | tr | tr | tr | 100.0 |

[a] From Augustin (35).
[b] *Distillation conditions*: ASCO falling film molecular still with 15 mm distillation path; pressure, 5–8 μ Hg; fractions I, II, III, and IV were collected with distillation surface temperatures of 140°–150°, 160°–170°, 180°–190°, and 200°–220°, respectively.

# 9

## PARTIAL DEACYLATION REACTIONS

After triglyceride mixtures have been resolved as far as possible by chromatographic methods, further information on their composition can be obtained by using partial deacylation reactions. The derived mono- and diglycerides can then be utilized for three types of analyses:

(a) The diglycerides can be resolved by normal chromatographic methods to further identify the component triglycerides in the original sample.
(b) Fatty acid analysis of the mono- or diglycerides will determine the distribution of fatty acids between the $sn$-2- and the combined $sn$-1,3-positions.
(c) The diglycerides from deacylation or from (a) can be used for stereospecific analysis to distinguish between the fatty acid compositions at the $sn$-1- and $sn$-3-positions.

The first selective deacylation reagent, pancreatic lipase, was introduced in 1955 by Mattson and Beck (*615,616*) and by Savary and Desnuelle (*795,796*). Since that time, other enzymes producing selective deacylation have been found, and Yurkowski and Brockerhoff (*975*) have perfected chemical deacylation with Grignard reagents.

To be useful in analytical work, a deacylation technique must produce "representative" mono- and diglycerides, i.e., the exact relationship between the initial triglycerides and the resultant partial glycerides must be known. An *sn*-1,3-diglyceride product, for example, must exactly correspond to the *sn*-1- and *sn*-3-positions in the original triglycerides, both in fatty acid composition and in positional distribution of those acids. Similarly, an *sn*-2-monoglyceride product must have the same fatty acid composition as the *sn*-2-position in the original sample. Thus deacylation reagents for analytical work should not have undesirable specificities for certain fatty acids or triglycerides and should not promote acyl migration. A number of chemical and enzymatic reagents can meet these requirements if proper precautions are observed.

This chapter reviews the relative merits and limitations of the deacylation reactions currently used in triglyceride analysis work. Methods for stereospecific analysis of the diglycerides resulting from partial deacylation are discussed in Chapter 10.

## I. CHEMICAL DEACYLATION METHODS

### A. Grignard Reagents

The most useful method for producing representative diglycerides from a triglyceride sample is by deacylation with a Grignard reagent. This reaction was introduced in 1966 by Yurkowski and Brockerhoff (*975*) and is outlined in Fig. 9-1. The Grignard reagent reacts with one of the ester linkages in the triglyceride to produce, following hydrolysis of reaction intermediates, a diglyceride and a tertiary alcohol derived from the liberated acyl group. Deacylation continues further, so that all possible diglycerides and monoglycerides as well as free glycerol are found in the reaction products. The reaction is stopped by adding acetic acid at a point where the maximum yield of diglycerides is obtained. Ethyl magnesium bromide is the preferred reagent for this reaction since it produces a tertiary alcohol that is easily separated from the diglycerides during chromatographic isolation of the reaction products. Experience to date has not demonstrated any fatty acid or triglyceride specificity with $CH_3MgBr$, $CH_3MgI$, or $C_2H_5MgBr$; and Grignard deacylation is generally assumed (although not unequivocally proven) to be a random reaction. A typical semimicro Grignard deacylation procedure has been described by Christie and Moore (*170*):

> Forty milligrams of triglyceride is dissolved in 2 ml of dry diethyl ether, and 1 ml of $C_2H_5MgBr$ solution (0.5 $M$ in diethyl ether, freshly prepared) is added. The mixture is shaken for 60 seconds; then 0.05 ml of glacial acetic acid followed by 2 ml of water is added to stop the reaction. The lipid products are then extracted with diethyl ether. The extract is washed first with dilute

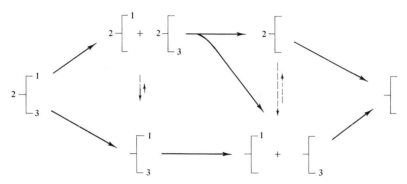

FIG. 9-1. Reaction sequence for triglyceride deacylation with a Grignard reagent. **1**, **2**, and **3** refer to the fatty acids at the $sn$-1-, $sn$-2-, and $sn$-3-positions in the original triglyceride.

aqueous $KHCO_3$, then with water, and finally dried over $MgSO_4$. After removal of the solvent, the diglycerides are quickly isolated by preparative TLC on silicic acid impregnated with 5% (w/w) boric acid to prevent acyl migration (*975*). A hexane/diethyl ether 50/50 solvent system resolves the $sn$-1,2(2,3)- and $sn$-1,3-diglycerides into separate bands. The yield of $sn$-1,2(2,3)-diglycerides is usually 6-7 mg (20–25%).

Similar macro and semimicro procedures have been published by Brockerhoff and co-workers (*119,975*), Åkesson (*12*), and Wood and Snyder (*964*).

## 1. Diglyceride Products

Grignard deacylation produces $sn$-1,2(2,3)- and $sn$-1,3-diglycerides in approximately a 2:1 ratio. Since the reagent is nonspecific in its attack on different triglyceride and fatty acid species, it is especially useful with triglyceride samples from which it is difficult to prepare representative $sn$-1,2,(2,3)-diglycerides by pancreatic lipase hydrolysis (i.e., fish oils containing 20:5ω3 and 22:6ω3, and samples having extremely different fatty acid compositions at the $sn$-1- and $sn$-3-positions).

Direct analysis of the $sn$-1,3-diglycerides gives the fatty acid composition of these combined positions, and the fatty acid composition of the $sn$-2-

position can be computed by difference:

$$\frac{\% \text{ X at}}{sn\text{-2-position}} = 3\left[\begin{array}{c}\% \text{ X in original} \\ \text{triglycerides}\end{array}\right] - 2\left[\begin{array}{c}\% \text{ X in } sn\text{-1,3-} \\ \text{diglycerides}\end{array}\right]$$

The fatty acid compositions of the $sn$-2- and $sn$-1,3-positions can also be calculated from the composition of the $sn$-1,2(2,3)-diglycerides:

$$\frac{\% \text{ X at}}{sn\text{-2-position}} = 4\left[\begin{array}{c}\% \text{ X in } sn\text{-1,2(2,3)-} \\ \text{diglycerides}\end{array}\right] - 3\left[\begin{array}{c}\% \text{ X in original} \\ \text{triglycerides}\end{array}\right]$$

$$\frac{\% \text{ X at}}{sn\text{-1,3-positions}} = 3\left[\begin{array}{c}\% \text{ X in original} \\ \text{triglycerides}\end{array}\right] - 2\left[\begin{array}{c}\% \text{ X in } sn\text{-1,2(2,3)-} \\ \text{diglycerides}\end{array}\right]$$

A small amount of undesirable acyl migration does occur during the Grignard reaction (dashed arrows in Fig. 9-1), but Yurkowski and Brockerhoff (975) and Christie and Moore (170) have demonstrated that isomerization can be limited to 1.5–2.0% with the $sn$-1,2(2,3)-diglycerides and 6–10% with the $sn$-1,3-diglycerides. However, the amount of isomerization can vary substantially from laboratory to laboratory, and one should always check that his reaction is producing representative diglycerides before proceeding to further analyses. This check is easily accomplished by comparing the fatty acid composition of the product diglycerides against that calculated from the fatty acid compositions of the original triglycerides and of the $sn$-2-monoglycerides derived from pancreatic lipase hydrolysis. With 16:0, for example, the proper calculations would be

$$\frac{\% \text{ 16:0 in}}{sn\text{-1,2(2,3)-diglycerides}} = \frac{3\left[\begin{array}{c}\% \text{ 16:0 in original} \\ \text{triglycerides}\end{array}\right] + \left[\begin{array}{c}\% \text{ 16:0 in 2-mono-} \\ \text{glycerides}\end{array}\right]}{4}$$

$$\frac{\% \text{ 16:0 in}}{sn\text{-1,3-diglycerides}} = \frac{3\left[\begin{array}{c}\% \text{ 16:0 in original} \\ \text{triglycerides}\end{array}\right] - \left[\begin{array}{c}\% \text{ 16:0 in 2-mono-} \\ \text{glycerides}\end{array}\right]}{2}$$

Lard triglycerides are an ideal test material since the great difference between the 16:0 content of the $sn$-2- and $sn$-1,3-positions makes any isomerization readily apparent (Table 9-1).

Christie and Moore (170) have presented evidence that diglyceride isomerization occurs only at the moment of hydrolysis and only in one direction without the establishment of a kinetic equilibrium. Thus any contamination of $sn$-1,2(2,3)-diglycerides by isomerization of $sn$-1,3-diglycerides would only change the composition of the 2-position (i.e., all acyl migration is 1 → 2 or 3 → 2). On the other hand, isomerization of $sn$-1,2(2,3)-diglycerides would contaminate the $sn$-1,3-diglycerides at both the $sn$-1- and the $sn$-3-positions (i.e., both 2 → 1 and 2 → 3 acyl migrations occur).

TABLE 9-1

FATTY ACID COMPOSITION OF THE PRODUCTS FROM GRIGNARD DEACYLATION OF LARD TRIGLYCERIDES[a]

| Method | Glyceride | | Fatty acids (mole %) | | | | | | | Relationship to original triglyceride structure |
|---|---|---|---|---|---|---|---|---|---|---|
| | | | 14:0 | 16:0 | 16:1 | 18:0 | 18:1 | 18:2 | 18:3 | |
| — | Original triglycerides | | 1.7 | 29.3 | 3.2 | 12.8 | 41.1 | 11.1 | 1.1 | — |
| Pancreatic lipase | sn-2-Monoglycerides | | 4.7 | 73.4 | 5.4 | 1.9 | 11.1 | 3.4 | 0.1 | Assumed correct |
| CH$_3$MgBr | sn-1,2(2,3)-Diglycerides | Found | 2.6 | 39.6 | 4.3 | 10.1 | 34.1 | 9.1 | 0.3 | } 1.5% isomerization |
| | | Calc. | 2.5 | 40.8 | 3.8 | 10.1 | 33.6 | 9.2 | 0.9 | |
| | sn-1,3-Diglycerides | Found | 0.6 | 9.2 | 2.4 | 17.7 | 55.7 | 13.8 | 0.6 | } 6% isomerization |
| | | Calc. | 0.2 | 7.3 | 2.1 | 18.3 | 56.1 | 15.0 | 1.6 | |
| | sn-2-Monoglycerides | Found | 4.6 | 58.5 | 3.6 | 4.6 | 23.3 | 5.2 | 0.2 | } Extensive isomerization |
| | | Lipase | 4.7 | 73.4 | 5.4 | 1.9 | 11.1 | 3.4 | 0.1 | |
| | sn-1(3)-Monoglycerides | Found | 1.2 | 13.6 | 2.2 | 15.7 | 52.2 | 13.8 | 1.3 | } Extensive isomerization |
| | | Calc. | 0.2 | 7.3 | 2.1 | 18.3 | 56.1 | 15.0 | 1.6 | |

[a] From Yurkowski and Brockerhoff (975).

Hence the sn-1,2(2,3)-diglycerides from Grignard deacylation reactions are more representative of the original triglycerides than the sn-1,3-diglycerides. Nevertheless, both types of diglycerides have been used for subsequent stereospecific analysis (Chapter 10) depending on the analytical error that can be tolerated. Brockerhoff (117) has proposed a reaction mechanism which might account for acyl migration during Grignard deacylation.

Further study is needed to determine if the amount of isomerization during Grignard deacylation can be further reduced by optimizing reaction conditions. Brockerhoff (117) has evaluated the effects of temperature, reaction time, and solvent on product composition but has found no suitable conditions for decreasing isomerization. It is also important to investigate whether samples containing short-chain acids ($C_4$–$C_8$) isomerize more readily than those containing only longer chain lengths.

## 2. Monoglyceride Products

The sn-1(3)- and sn-2-monoglycerides from the Grignard deacylation reaction cannot be used for analytical purposes, since their fatty acid compositions are not representative of the original triglyceride structure (Table 9-1).

## B. Other Reagents

It seems likely that other chemical reagents might also be used for the random deacylation of triglycerides. For the present, however, none has proven as satisfactory as the Grignard reagents. Preliminary experiments with NaOH, $NaOCH_3$, $HONH_2$, and $LiAlH_4$ have been unsuccessful (975), mainly because of excessive diglyceride isomerization. Thermal hydrolysis shows no positional specificity and very little fatty acid specificity (130,676), but unfortunately heat accelerates diglyceride isomerization (201).

Several workers (104,163,451,455) have noted that mildly basic conditions promote the selective hydrolysis of azelaic acid in oxidized triglycerides. Kartha (458,459) has suggested using this reaction to deacylate SSA into sn-1,2(2,3)- and sn-1,3-diglycerides, which can then be separated and analyzed to determine the structure of the original SSU. His procedure using $K_2CO_3$ hydrolyzes only 48–78% of the azelaic acid present in the sample. Thus any quantitative interpretation must assume that azelate hydrolysis is random, that no acyl migration occurs in the partial glycerides, and that no saturated acids are hydrolyzed. The quantitative validity of the procedure is rather doubtful, however, since the method has never been tested on triglyceride mixtures of known composition, and Kartha's results on kokum butter and malabar tallow (*458,459*) do not agree with pancreatic lipase analyses of the same fats (*189,623*).

## II. ENZYMATIC DEACYLATION METHODS

### A. Pancreatic Lipase

Mammalian pancreatic lipase catalyzes the hydrolysis of primary ester linkages in triglycerides to produce *sn*-1,2(2,3)-diglycerides, *sn*-2-monoglycerides, and free fatty acids* (Fig. 9-2). Its specificity for the *sn*-1- and *sn*-3-positions is near-absolute, making the enzyme a very useful reagent for the analytical deacylation of triglycerides. The *sn*-2-monoglycerides from lipase hydrolysis have been widely used for fatty acid analysis in studying the positional distribution of fatty acids in natural fats. The enzyme has also been employed in certain cases to prepare *sn*-1,2(2,3)-diglycerides for use in stereospecific analysis. Table 9-2 illustrates the different fatty acid compositions of the mono-, di-, and triglycerides produced by pancreatic lipolysis of illipe butter triglycerides.

The concept of primary ester specificity in the hydrolysis of triglycerides by pancreatic lipase was first proposed by Balls and Matlack (*43*) in 1938. The experimental proof for this idea was published in 1952–1956 by Mattson et al. (*615–617*), Borgström (*96,98*), Schønheyder and Volqvartz (*820,821*), and Savary and Desnuelle (*795,796*). Since then, the purification, specificity, optimum reaction conditions, and reaction products of

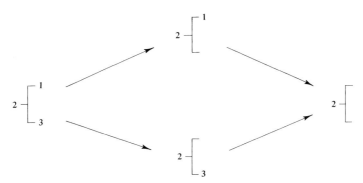

FIG. 9-2. Normal course of reaction during deacylation of a triglyceride by pancreatic lipase. **1**, **2**, and **3** refer to the fatty acids at the *sn*-1-, *sn*-2-, and *sn*-3-positions in the original triglyceride.

* Pancreatic lipase can also catalyze the resynthesis of primary ester linkages in partial glycerides (*99*), but all analytical deacylation reactions are carried out very rapidly and at an alkaline pH so that the amount of resynthesis is negligible.

TABLE 9-2
FATTY ACID COMPOSITION OF THE GLYCERIDES PRODUCED BY DEACYLATION OF
ILLIPE BUTTER TRIGLYCERIDES WITH PANCREATIC LIPASE[a]

| Glyceride | | Fatty acids (mole %) | | | | | |
|---|---|---|---|---|---|---|---|
| | | 16:0 | 16:1 | 18:0 | 18:1 | 18:2 | 18:3 |
| Triglycerides[b] | | 16.5 | 0.3 | 45.8 | 35.4 | 0.6 | 1.4 |
| sn-1,2(2,3)-Diglycerides | Found | 11.7 | — | 36.9 | 49.3 | 1.0 | 1.1 |
| | Calc. | 12.8 | 0.2 | 35.2 | 49.9 | 0.8 | 1.1 |
| sn-2-Monoglycerides | | 1.8 | — | 3.4 | 93.6 | 1.2 | — |

[a] *Reaction conditions:* 950 mg of illipe butter triglycerides, 1.2 $M$ $NH_4Cl/NH_4OH$ buffer solution at pH 8.5, 2 ml 22% $CaCl_2 \cdot 6H_2O$ solution, 0.1 ml of 25% bile salts solution, 100 mg of pancreatin, 37.5°. From Coleman (*187*).

[b] The fatty acid compositions of the initial and residual triglycerides were identical within the limits of experimental error.

pancreatic lipase have been widely studied, and it is more fully characterized than any other lipase.

## 1. Enzyme

The most widely used source of pancreatic lipase is hog pancreatin, a powder obtained by dehydrating and defatting pig pancreas with acetone and diethyl ether. The pancreatin (steapsin) produced in this manner is stable for a considerable time and can be purchased from most biochemical supply houses. Although the lipase activity of pancreatin is rather low, it has proven quite satisfactory for the analytical deacylation of triglycerides. All enzyme preparations should be tested with a blank run to be certain they will not contribute unwanted contaminants to the reaction products; if such artifacts are found, additional extraction with acetone and/or diethyl ether will usually remove the contaminants (*421,440,587,783*). Human (*99,182,184*), dog (*343*), rat (*99,626*), and skate (*118*) pancreas produce lipases with equivalent specificity.

In addition to pancreatic lipase, small amounts of two other carboxylic ester hydrolases are also found in the mammalian pancreas: an esterase which hydrolyzes water-soluble ester linkages (*274,275,626,788,790*) and a nonspecific lipase that can hydrolyze the water-insoluble esters of both primary and secondary alcohols including cholesterol esters (*626,657,870, 902*).The latter nonspecific lipase may produce a few percent of atypical 1,3-diglycerides and 1(3)-monoglycerides during triglyceride deacylation. Mattson and Volpenhein (*626,627*) report that the nonspecific lipase in rat pancreatic juice can be inactivated without losing appreciable pancreatic lipase activity by (i) hydrolysis in the absence of bile salts, (ii) self-digestion of pancreatic lipase at pH 9.0 and 40° for 1 hour, or (iii) treating pancreatic lipase with 0.0005 $M$ diethyl-$p$-nitrophenyl phosphate for 1 hour. However, only (i) has been tested and found effective with hog pancreatin (*870*).

If desired, pancreatic lipase of greater purity and higher specific activity can be prepared by one of the many published purification procedures (*252,300,303,749,822,929*). The detailed enzymology of pancreatic lipase is discussed in several recent reviews (*214,414,471,957*).

## 2. Reaction Conditions

The reaction conditions are chosen to achieve deacylation as rapidly as possible (<90 seconds) and to minimize undesirable acyl migration in the partial glycerides. Optimum reaction conditions for hog pancreatic lipase include a pH near 8 (*58,309,819*), a 0.5–1.5 $M$ electrolyte concentration (*615*), the presence of calcium ion (*66,162,181,216,819*), a high enzyme/sample ratio, vigorous agitation (*284,311*), and an emulsifier (bile salts,* gum arabic, polyvinyl alcohol, etc.) to maximize interfacial surface area (*788,789*). A reaction temperature of 37°–40° is employed to permit rapid deacylation without appreciable enzyme degradation (*289,309*) and to ensure that triglyceride substrates (except for SSS) are in the liquid state.

The semimicro procedure of Luddy *et al.* (*587*) has been widely used for the deacylation of triglycerides with pancreatic lipase:

> Fifty milligrams of triglyceride is weighed into a 5 ml screw-cap vial along with sufficient pancreatin (~9 mg) to achieve the desired amount of hydrolysis. Then 1.0 ml of 1 $M$ tris(hydroxymethyl)methylamine (adjusted to pH 8.0), 0.1 ml of 22% $CaCl_2$ solution, and 0.25 ml of 1% bile salts solution are added. The vial and its contents are first warmed in a water bath at 40° for 1 minute without shaking. The cap is then screwed on tightly, taped in place, and the vial is shaken for 45–90 seconds at 3000 strokes/minute in a dental amalgamator (Crescent Dental Manufacturing Co., Chicago, Ill.) modified for use with small glass vials (*588*). At the end of the reaction time, the mixture is *immediately* transferred to a separatory funnel and extracted with diethyl ether.† The extract is washed with water, dried over $Na_2SO_4$, filtered, and evaporated. The individual products are quickly isolated by preparative TLC on silicic acid containing 8% (w/w) boric acid to prevent acyl migration (*975*). Yields of ~30% diglyceride or monoglyceride are obtained under optimum reaction conditions.

* Although bile salts are obvious surface-active agents, their role in lipolysis is more complex than that of a simple emulsifier (*97,289,625,823*).

† Luddy *et al.* (*587*) originally recommended stopping lipolysis by adding 0.5 ml of 6 $N$ HCl, but Snyder and Piantadosi (*848*) have demonstrated that such acid conditions promote acyl migration in partial glycerides.

The amount of hydrolysis can be determined by quantitating the fatty acids released or by measuring the relative amounts of monoglyceride, diglyceride, triglyceride, and free glycerol (Table 2-3). Similar semimicro deacylation procedures have been described by Jurriens (*440*) and Privett and Nutter (*740*). Macro procedures have been published by Mattson and Volpenhein (*620*) and Brockerhoff and Yurkowski (*122*). Deacylations of *sn*-1,2(2,3)-diglycerides (*916,976*) or their acetate derivatives (*512,759*) are carried out in the same manner as with triglycerides.

Cocoa butter triglycerides are a good test material for lipolysis reactions since (i) they are readily available, (ii) the 1,3- and 2-positions have such widely different fatty acid compositions, and (iii) the composition of the expected monoglycerides is well known (*122,443,619*).

When oxygenated or short-chain acids interfere with proper separation of lipolysis products by TLC, gas chromatography of the silylated products may give a clearer picture of their composition (*877,878*). Pancreatic lipase is inhibited by cyclopropene fatty acids (*446*), benzene (*569*), and glyceryl phosphatides (*337,514*); and all reaction mixtures therefore should be free of such materials.

### 3. *Specificity*

Pancreatic lipase possesses near-absolute specificity for the hydrolysis of primary ester linkages in triglycerides. Recent studies (*99,250,566,627*) indicate >97% specificity for the *sn*-1,3-positions, and any release of *sn*-2-position fatty acids is generally attributed to acyl migration or to contamination with a nonspecific lipase. According to Tattrie *et al.* (*882*), Karnovsky and Wolff (*450*), and Jensen *et al.* (*418*), pancreatic lipase attacks both primary ester groups of a triglyceride at the same rate, i.e., there is no preferential hydrolysis of either the *sn*-1- or the *sn*-3-position when the same fatty acid is attached to both positions.

Pancreatic lipase often appears to hydrolyze unsaturated fatty acids more rapidly than saturated acids of the same chain length. During lipolysis of lard triglycerides, Coleman (*187*) found that both the triglyceride and diglyceride fractions became more saturated as the reaction proceeded. Luddy *et al.* (Table 9-3) noted that the hydrolysis of *rac*-1-oleo-2,3-distearin with pancreatic lipase produced distearin as the principal diglyceride. Other workers (*414,616,796*), however, have reported only a slight preferential hydrolysis of unsaturated acids; so apparently reaction conditions influence this effect. It has not been clearly established whether this preferential hydrolysis of unsaturated fatty acids by pancreatic lipase represents a true specificity for unsaturated fatty acids or is actually a specificity for unsaturated triglycerides and diglycerides; both explanations fit the

TABLE 9-3

Fatty Acid Compositions of the Products Formed by Pancreatic Lipase or Milk Lipase Deacylation of Triglycerides Containing Both Saturated and Unsaturated Acids at the 1,3-Positions

| Enzyme | Fatty acid composition of lipolysis products (mole %) | | | | | | | | |
|---|---|---|---|---|---|---|---|---|---|
| | Free fatty acids | | | Monoglycerides | | | Diglycerides | | |
| | 16:0 | 18:0 | 18:1 | 16:0 | 18:0 | 18:1 | 16:0 | 18:0 | 18:1 |
| Pancreatic lipase (587) | | | | | | | | | |
| rac-1-Oleo-2,3-distearin | | | | | | | | | |
| Found | 1 | 42 | 57 | 1 | 99 | — | 1 | 97 | 2 |
| Theory | — | 50 | 50 | — | 100 | — | — | 75 | 25 |
| Milk lipase (422) | | | | | | | | | |
| rac-1-Palmito-2,3-diolein | | | | | | | | | |
| Found | 52 | — | 48 | — | — | 100 | 30 | — | 70 |
| Theory | 50 | — | 50 | — | — | 100 | 25 | — | 75 |

available experimental facts. In any case, the effect can be reduced by adding a small amount of hexane to the lipolysis reaction (110,171,780).

Comparative studies on triglycerides of oleic and elaidic acids (179,423) indicate that pancreatic lipase does not differentiate between these two *cis* and *trans* isomers. Although pancreatic lipase has frequently been used to determine the positional distribution of oxygenated acids in natural fats (3,332,664,876,878), the precise specificity of the enzyme for molecules containing such acids has never been tested using triglyceride mixtures of known composition.

Pancreatic lipase hydrolyzes fatty acids having double bonds, alkyl branching, or other functional groups at the 2-, 3-, 4-, or 5-carbon atoms more slowly than their corresponding isomers having the functional groups further removed from the ester linkage. This effect is no doubt due to steric hindrance. Thus Kleiman and co-workers (509) reported that 16:1-3t, 18:1-3t, and 18:3-3t,9c,12c resist hydrolysis; and Bottino et al. (102) and Brockerhoff (110,975) encountered the same problem with 20:5-5c,8c,11c,14c,17c and 22:6-4c,7c,10c,13c,16c,19c in seal, whale, and cod liver triglycerides. In addition, Tryding et al. (71,895) and Garner and Smith (301) found that 2-alkyl and 3-alkyl fatty acids seriously inhibit the action of pancreatic lipase. Similar inhibitory effects have been noted by Brockerhoff (116) during lipolysis of the fatty acid esters of several monohydroxy alcohols.

It was originally thought that pancreatic lipase possessed a distinct specificity for short-chain acids such as 4:0, since the butyric acid content

TABLE 9-4
Varying Composition of Residual Triglycerides during Deacylation of an Equimolar Mixture of rac-PBB and OOO by Pancreatic Lipase[a,b]

| Elapsed time (minutes) | Residual triglyceride composition (mole %) | |
|---|---|---|
| | rac-PBB | OOO |
| 0.0 | 52.8 | 47.2 |
| 2.5 | 45.2 | 54.8 |
| 5.0 | 29.2 | 70.8 |
| 10.0 | 16.9 | 83.1 |

[a] From Sampugna et al. (785).
[b] Reaction conditions: 70 mg rac-PBB, 132 mg OOO, 8 ml 0.25 $M$ tris(hydroxymethyl)-methylamine buffer (pH 8.0) containing 10% (w/v) gum arabic, 0.5 ml of 4 $M$ $CaCl_2$, 0.2 ml of 1% (w/v) bile salts, and 25 mg of hog pancreatin were emulsified together and incubated at 37°.

of the unhydrolyzed triglycerides decreases during the progressive lipolysis of butterfat. However, Jensen and his co-workers (*424,785*) have proven that this effect is actually the result of preferential hydrolysis of lower molecular weight triglycerides. Hydrolysis of pure rac-PBB released 4:0 and 16:0 in approximately equimolar quantities, proving a lack of chain length specificity. When a mixture of rac-PBB and OOO was hydrolyzed, however, the OOO content of the residual triglycerides increased as the reaction proceeded (Table 9-4), indicating preferential hydrolysis of rac-PBB.

There is also some evidence for the preferential hydrolysis of long-chain over short-chain diglycerides by pancreatic lipase. When Sampugna et al. (*785*) hydrolyzed rac-PBB, the resultant diglycerides contained more than twice as much dibutyrin as palmitobutyrin. Similarly, lipolysis of rac-PPB produced much more palmitobutyrin than dipalmitin. Sampugna and his co-workers attributed these results to the partial solubility of butyrate diglycerides in water, causing preferential hydrolysis of the more hydrophobic diglycerides. This possibility needs further investigation.

Many workers (*44,788,818,956,957*) have demonstrated that pancreatic lipase acts only at an oil/water interface and only on liquid, water-insoluble substrates. Hence water-soluble molecules such as triacetin and high-melting (>40°) triglycerides such as tristearin resist hydrolysis under normal reaction conditions. High-melting SSS molecules can be hydrolyzed, however, if they are dissolved in a liquid carrier (triolein, methyl oleate, methyl decanoate, etc.) which will not interfere with subsequent analysis (*51,345,440,587*).

*4. Diglyceride Products*

The *sn*-1,2(2,3)-diglycerides obtained by deacylation with pancreatic lipase may or may not be representative of the original triglycerides. The various fatty acid and triglyceride specificities of the enzyme (see above) can cause nonrepresentative diglycerides to be produced from certain types of samples. Thus Brockerhoff and his co-workers (*110,111,121,122*) could prepare representative diglycerides from corn, olive, cocoa butter, peanut, dog, cat, horse, turkey, and frog triglycerides, but not with seal blubber or cod liver triglycerides, and only intermittantly with lard. Anderson et al. (*25*) compared several lipolysis procedures and concluded that the production of representative diglycerides depends not only on sample composition but also on the reaction conditions used. Unfortunately, the precise reaction conditions which favor random deacylation of the *sn*-1- and *sn*-3-positions are as yet undefined. Therefore, any worker using pancreatic lipase to prepare representative *sn*-1,2(2,3)-diglycerides should firmly establish that random 1,3-hydrolysis definitely does occur with the specific sample and with the specific reaction conditions being used in his laboratory. This is easily done (*111*) by checking the composition of the diglycerides against that calculated from the *sn*-2-monoglyceride and triglyceride compositions. For example, the proper calculation for 16:0 would be

$$\frac{\% \ 16:0 \ \text{in}}{sn\text{-}1,2(2,3)\text{-diglycerides}} = \frac{3\left[\begin{array}{c}\% \ 16:0 \ \text{in original} \\ \text{triglycerides}\end{array}\right] + \left[\begin{array}{c}\% \ 16:0 \ \text{in mono-} \\ \text{glycerides}\end{array}\right]}{4}$$

Acyl migration is usually not a problem in preparing representative diglycerides when rapid lipolysis reactions and neutral product extraction conditions (*848*) are employed with triglycerides containing $C_{10}$–$C_{22}$ acids. The isomerization of long-chain *sn*-1,2(2,3)-diglycerides to 1,3-diglycerides (Fig. 9-3) occurs very slowly in aqueous dispersion at pH 8.0 (*250,569,622*). Diglycerides containing $C_4$–$C_8$ acids at the 2-position may show significant isomerization, however (*67,251*).

*5. Monoglyceride Products*

The monoglycerides produced by pancreatic lipase are usually representative of the 2-position and have been widely used for determining the

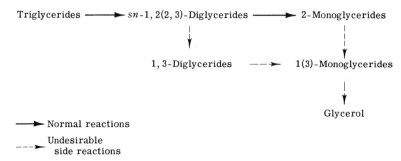

FIG. 9-3. Undesirable side reactions caused by acyl migration in the partial glycerides during deacylation of triglycerides with pancreatic lipase. Primary esters are much more stable than secondary esters in this regard; hence any acyl migration during *very short* lipolysis reactions is mostly unidirectional, i.e., 2-positions → 1(3)-positions.

positional distribution of fatty acids in triglycerides:

$$\frac{\% \text{ X at 2-position}} = \frac{\% \text{ X in monoglycerides produced by hydrolysis with pancreatic lipase}}$$

$$\frac{\% \text{ X at combined 1,3-positions}} = \frac{3\left[\begin{array}{c}\% \text{ X in original} \\ \text{triglycerides}\end{array}\right] - \left[\begin{array}{c}\% \text{ X at} \\ \text{2-position}\end{array}\right]}{2}$$

The 2-monoglycerides have a pronounced tendency to isomerize under the normal reaction conditions used for lipase hydrolysis (Fig. 9-3, *250,251,372,386,421,622*). As a result, the monoglycerides isolated after pancreatic lipase hydrolysis are usually a mixture of the 2- and 1(3)-isomers with the former predominating (Table 9-5). Since the 1(3)-monoglycerides arise almost entirely from isomerization of the 2-monoglycerides, analysis of the fatty acid composition of the total monoglycerides still gives an accurate analysis of the 2-position in the original triglycerides. If appreciable amounts of 1(3)-monoglycerides are hydrolyzed to glycerol, however, there may be some selectivity in their lipolysis causing the remaining monoglycerides to become unrepresentative of the 2-position. Fortunately, rapid lipolysis reactions generate very little (~2%) glycerol (*251,407*) except when $C_4$–$C_8$ acids are present in the sample (*251,821*).

Some workers have presumed that the various fatty acid and triglyceride specificities which prevent pancreatic lipase from producing representative diglycerides will also prevent the production of representative monoglycerides. Theoretically, this should be true; but it is apparently not the case with many natural triglyceride mixtures. Coleman (*187,188*), for example, has examined the fatty acid compositions of the mono-, di-, and triglycerides produced by lipolysis of lard triglycerides for varying lengths of time.

TABLE 9-5
MONOGLYCERIDE ISOMERS PRODUCED DURING DEACYLATION OF
SYNTHETIC AND NATURAL TRIGLYCERIDES WITH PANCREATIC
LIPASE[a,b]

| Triglyceride | % 2-Monoglyceride | % 1(3)-Monoglyceride |
|---|---|---|
| β-POP | 98 | 2 |
| β-PPO | 81 | 19 |
| OOO | 80 | 20 |
| LLL | 70 | 30 |
| LnLnLn | 59 | 41 |
| DDD | 81 | 19 |
| OcOcOc | 74 | 26 |
| HHH | 63 | 37 |
| Beef tallow | 99 | 1 |
| Lard | 91 | 9 |
| Coconut oil | 88 | 12 |
| Olive oil | 84 | 16 |
| Soybean oil | 72 | 28 |

[a] From Hirayama and Nakae (*372,376*).

[b] *Reaction conditions:* 500 mg of triglyceride was added to 6 ml of 0.5 $M$ tris(hydroxymethyl)methylamine (pH 7.5), 0.5 ml of 45% $CaCl_2$, 2 ml of 1% deoxycholate solution, and 100 mg of pancreatin suspended in 3 ml of 0.5 $M$ tris buffer (pH 8.0). This mixture was homogenized for 10 minutes at 40°, 5 ml of 1.0 $N$ HCl/95% $C_2H_5OH$ (3/5) was added, and the lipids were then extracted with diethyl ether or $CHCl_3$. Monoglyceride isomers were determined by periodic acid oxidation before and after $HClO_4$ isomerization.

He found that the tri- and diglyceride fractions became progressively more saturated with increasing hydrolysis, but the fatty acid composition of the monoglycerides was independent of the amount of hydrolysis. Coleman suggested that this apparent anomaly may be due to a random esterification of 2-position fatty acids among the triglycerides present in lard (i.e., as in a 1-random–2-random–3-random or a 1,3-random–2-random distribution). Such a possibility is illustrated by the calculations in Table 9-6, where lard triglycerides are divided into three classes (I, II, and III) according to the fatty acid composition of their 1- and 3-positions. Any amounts of I, II, and III may be chosen, but the β-XSX:β-XUX ratio in each class is fixed at 76:24 (i.e., a 2-random distribution prevails). Since U is more readily removed from the 1- and 3-positions than S, class III triglycerides will be hydrolyzed more rapidly than class II, and class II more rapidly than class I. Hence the $sn$-1,2(2,3)-diglycerides and residual triglycerides will become increasingly more saturated as hydrolysis proceeds. However, each triglyceride class will yield the same monoglycer-

TABLE 9-6

Triglyceride Composition of Lard Having a 2-Random Distribution[a] Causing the 2-Monoglyceride Composition To Be Independent of the Triglyceride Class Hydrolyzed[b]

| Triglyceride class[c] | Fatty acids at 1- and 3-positions | Triglycerides yielding saturated 2-monoglycerides | | Triglycerides yielding unsaturated 2-monoglycerides | | % S in 2-monoglycerides produced |
|---|---|---|---|---|---|---|
| | | Triglyceride | Mole % | Triglyceride | Mole % | |
| I | S + S | SSS | 10.7 | β-SUS | 3.3 | $\dfrac{10.7}{10.7 + 3.3}(100) = 76$ |
| II | S + U | β-SSU | 35.8 | β-SUU | 11.2 | $\dfrac{35.8}{35.8 + 11.2}(100) = 76$ |
| III | U + U | β-USU | 29.7 | UUU | 9.3 | $\dfrac{29.7}{29.7 + 9.3}(100) = 76$ |

[a] In a 2-random distribution, all the fatty acids esterified at the 2-position are randomly distributed among all the triglycerides present (see Chapter 12, Section II,A).
[b] From Coleman (187).
[c] The relative amounts of class I, II, and III triglycerides are unimportant; but the β-XSX:β-XUX ratio in each class is fixed at 76:24 (i.e., a 2-random distribution prevails).

ide composition (76% S), causing the monoglyceride composition to remain constant throughout the entire hydrolysis (assuming no appreciable hydrolysis of monoglycerides). Since there is evidence for a 2-random distribution in many natural fats (Chapter 12, Section II), it follows that representative monoglycerides can often be produced from natural triglyceride samples containing lipase-resistant acids. Yurkowski and Brockerhoff (975), for example, have shown this to be true with seal blubber triglycerides by comparing pancreatic lipase and Grignard reagent deacylation results.

### 6. Free Fatty Acid Products

The free fatty acids released by pancreatic lipase are generally not representative of the combined $sn$-1- and $sn$-3-positions of the original triglycerides. Two problems tend to make them unsuitable for analytical purposes:

(a) The fatty acid and triglyceride specificities of pancreatic lipase often cause the free fatty acid composition to vary with the extent of hydrolysis.
(b) If any monoglyceride is completely hydrolyzed to glycerol (1–2% is usual), some of the 2-position acids will contaminate the free fatty acids.

When the fatty acid composition of the combined 1,3-positions is required, it can be accurately calculated from the compositions of the monoglycerides and the original triglycerides (Section II,A,5).

## B. Milk Lipase

Cow's milk contains a lipase that resembles pancreatic lipase in its specific hydrolysis of the sn-1,3-positions of triglycerides. Experiments by Jensen et al. (298,299,415,416,420,422,425) have proved that milk lipase is an effective agent for the analytical deacylation of triglycerides to representative sn-1,2(2,3)-diglycerides and sn-2-monoglycerides.

There is extensive evidence (231,279,280,589,825,826,967) that milk lipase is not just a single enzyme but a group of four or more different proteins. Thus for analytical work, it is essential to use enzyme preparations of proven specificity. Jensen and his co-workers have found characteristic milk lipase specificity in three such preparations: (i) lyophilized raw skim milk (298); (ii) a concentrate obtained by adsorption of milk proteins on $Mg(OH)_2$ (656); and (iii) a concentrate prepared from milk clarifier slime by acetone extraction, $(NH_4)_2SO_4$ precipitation, and gel filtration (145). Specificity for primary ester groups has been proven with all three preparations (298,299,416,420,422,425), but the lack of preferential hydrolysis of unsaturated fatty acids and the selective hydrolysis of low molecular weight triglycerides (see below) have only been demonstrated with preparation (iii) (422,425). Other methods for the purification of milk lipase are available (231,278,292,293), but the specificity of these preparations has not been established.

Jensen et al. (422) have used the following general procedure for the deacylation of triglycerides with milk lipase:

> Two hundred milligrams of triglyceride is emulsified into 10 ml of 0.25 M of tris(hydroxymethyl)methylamine buffer (pH 8.0) containing 10% gum arabic. One milliliter of milk lipase concentrate is added, and the mixture is shaken for 5–15 minutes in a 38° water bath. Lipolysis is terminated by acidulation (see footnote, p. 175), and the lipids are extracted with diethyl ether. The resultant products are separated into triglycerides, diglycerides, monoglycerides, and free fatty acids by preparative thin-layer chromatography (Chapter 2, Section II,C).

As in all lipase deacylations, reaction time should be kept as short as possible to minimize acyl migration within the resultant partial glycerides. Milk lipase is sensitive to light (281,501,852) and to dissolved oxygen (526), and appropriate precautions should be taken when using the enzyme.

The positional specificity of milk lipase for the primary ester linkages of triglycerides is well established (299,420,422). Under optimum reaction conditions, its positional specificity is probably equivalent to that of pancreatic lipase. Studies to date have not revealed any marked fatty acid specificity with milk lipase, but it has only been tested with triglycerides containing the common fatty acids. The lack of preferential hydrolysis of unsaturated acids is particularly noteworthy (*422*); lipolysis of *rac*-POO releases nearly equal amounts of 16:0 and 18:1.

Milk lipase was originally thought to exhibit a specificity for short-chain acids such as 4:0, but this is now known to be a specificity for low molecular weight triglycerides. Jensen and co-workers (*420*) showed that milk lipase releases approximately equal amounts of 16:0 and 4:0 during lipolysis of *rac*-PBB. However, when an equimolar mixture of *rac*-PBB and OOO is hydrolyzed, both 16:0 and 4:0 are released more rapidly than 18:1 (*425*). Hence milk lipase is similar to pancreatic lipase in its preferential attack on low molecular weight triglycerides.

### 1. Diglyceride Products

The preparation of diglycerides using milk lipase involves the same general problems and precautions as previously mentioned for pancreatic lipase (Section II,A,4). Jensen et al. (*298,420,422*) have demonstrated that milk lipase can produce representative *sn*-1,2(2,3)-diglycerides from individual triglyceride substrates such as *rac*-POP, *rac*-OPO, *rac*-POO, and *rac*-PBB; but further testing on triglyceride mixtures is needed. Since milk lipase exhibits very little preferential hydrolysis of unsaturated fatty acids (Table 9-3), it is superior to pancreatic lipase for the preparation of representative *sn*-1,2(2,3)-diglycerides from β-SSU and β-SUU triglycerides.

### 2. Monoglyceride Products

Present evidence (*298,299,422*) indicates that deacylation with milk lipase produces representative *sn*-2-monoglycerides when the sample contains only common $C_{14}$–$C_{18}$ acids. Previous comments on pancreatic lipase monoglyceride products (Section II,A,5) also apply to the monoglycerides obtained with milk lipase.

### C. *Rhizopus arrhizus* Lipase

The mold *Rhizopus arrhizus* produces an extracellular lipase that hydrolyzes the primary esters of triglycerides in a manner similar to pancreatic lipase. The mold enzyme has been highly purified by Laboureur and Labrousse (*548,549*), and Semeriva et al. (*829,831*), and optimum reaction conditions have been established. Experiments with long-chain triglycerides (*548,680a,830*) indicated at least a 98% specificity for the *sn*-1- and *sn*-3-positions. Thus *R. arrhizus* lipase appears to be suitable for the analytical deacylation of triglycerides, but it offers no known advantages over pancreatic lipase.

## D. *Geotrichum candidum* Lipase

The mold *Geotrichum candidum* produces an unusual extracellular lipase that has a pronounced specificity for the hydrolysis of oleic and linoleic acids from triglycerides. Studies by Alford et al. (*17*) and Jensen and co-workers (*427,608*) have indicated that this lipase is highly specific for fatty acids with cis-9-unsaturation regardless of their position on the triglyceride substrate. Deacylation of triglyceride mixtures by *G. candidum* lipase can produce representative diglycerides in which the free hydroxyl group is known to have been occupied by a cis-9-unsaturated acid.

The preparation of *G. candidum* extracellular lipase has been described by Alford and Smith (*18*). The organism is grown in a salts–glucose–protein hydrolysate medium for 4 days at 25°. After filtering off the cells, the filtrate is dialyzed and lyophilized to produce a powder which can be stored at 0° for more than a year with little change in activity. No purification of *G. candidum* lipase has yet been published, although this might prove beneficial in minimizing side reactions during deacylation of triglycerides.

A suitable procedure for the deacylation of triglycerides with *Geotrichum candidum* lipase has been described by Sampugna and Jensen (*781*):

> Two milliliters of 1% gum arabic solution is added to 50 mg of triglyceride sample, and the mixture is emulsified by sonication. Then 7.5 ml of distilled water and 0.5 ml of 0.1 $M$ $CaCl_2$ solution are added, and the mixture is allowed to equilibrate at 37° with magnetic stirring in a recording pH-Stat set at 8.5 pH. Next 10–15 mg of *G. candidum* lipase is added, and the digestion is allowed to proceed for 15 minutes. During digestion the pH is kept constant at 8.5 by the automatic addition of 0.09 $N$ NaOH. At the end of 15 minutes, the reaction mixture is transferred to a separatory funnel containing 25 ml of distilled water, and the lipids are extracted with petroleum ether/diethyl ether 1/1. The interfacial fluff (Ca soaps of the fatty acids) is discarded with the water layer. The organic phase is dried with $Na_2SO_4$, filtered, and concentrated under reduced pressure. The sn-1,2(2,3)- and sn-1,3-diglycerides are then isolated by preparative TLC on silicic acid containing 5% (w/w) boric acid to prevent acyl migration. Yields up to 45% diglyceride (all isomers) can be obtained.

Sampugna and Jensen (781) emphasize the necessity for very short reaction times to prevent acyl migration within the partial glycerides and to avoid some hydrolysis of saturated fatty acids. Further study of both these problems is needed.

### TABLE 9-7
### Fatty Acid Composition of Diglycerides Produced by Deacylation of Pure Triglycerides with *Geotrichum candidum* Lipase[a,b]

| | | Calc.[c] | Found | | |
|---|---|---|---|---|---|
| Substrate | sn-Diglycerides | 16:0 | 16:0 | 18:0 | 18:1 |
| rac-StPO | 1,2 + 2,3 | 50.0 | 49.8 | 50.2 | tr |
| rac-POSt | 1,3 | 50.0 | 52.4 | 47.6 | tr |
| rac-PStO | 1,2 + 2,3 | 50.0 | 48.7 | 51.3 | tr |
| sn-PStO | 1,2 + 2,3 | 50.0 | 45.9 | 54.1 | tr |
| sn-OStP | 1,2 + 2,3 | 50.0 | 48.9 | 51.1 | tr |
| rac-PPO | 1,2 + 2,3 | 100.0 | 100.0 | — | — |
| sn-OOP | 1,3 | 50.0 | 51.6 | — | 48.4 |
| sn-OOP | 1,2 + 2,3 | 50.0 | 50.3 | — | 49.7 |

[a] From Sampugna and Jensen (*781*).
[b] Substrate digested with 15 mg of crude enzyme for 15 minutes at 37° and 8.5 pH.
[c] Assuming only oleic acid is hydrolyzed.

The specificity of *Geotrichum candidum* lipase for *cis*-9-unsaturated acids has been well characterized (*17,427,608,781*) using palmitoleoyl, oleoyl, and linoleoyl triglycerides. *Cis*-6-, *trans*-6-, *trans*-9-, and *cis*-11-octadecenoic acids definitely resist hydrolysis; but other positional isomers have not yet been tested. On the other hand, the slight hydrolysis of saturated acids by *G. candidum* lipase is not fully understood. It remains to be seen whether this represents a true lack of absolute specificity for *cis*-9-unsaturated acids or is due to the presence of a second nonspecific lipase. *Geotrichum candidum* lipase shows no preferential hydrolysis of the *sn*-1-, *sn*-2-, or *sn*-3-positions in glyceride substrates (Table 9-7; *608*). The triglyceride specificity of the enzyme has not been fully investigated. Work by Jensen *et al.* (*608,781*) indicated no marked triglyceride specificity between OOO and LLL or between the six stereoisomers of PStO. However, the relative hydrolysis rates of mono-, di-, and trioleotriglycerides *within the same system* have never been compared. Presumably the rate of hydrolysis would be approximately proportional to the amount of oleate present in the original triglyceride (*608*), but further studies are needed on this point.

### 1. Diglyceride Products

Sampugna and Jensen (*781*) have demonstrated that *G. candidum* lipase can deacylate pure mono- or pure dioleotriglycerides to produce representative diglycerides (Table 9-7). Similar representative diglycerides were obtained with a mixture of the six stereoisomers of PStO (*781*) and with the monoene triglycerides from cocoa butter (*782*). The quantitative application of *G. candidum* lipase deacylation to more complex natural triglyceride mixtures, while theoretically possible, remains undemonstrated; this would only be possible if the enzyme has no marked triglyceride specificity (see above).

## 2. Monoglyceride Products

The ability of *Geotrichum candidum* lipase to produce representative monoglycerides is uncertain. Jensen and co-workers (*608,781*) have repeatedly found that the crude enzyme produces some saturated monoglyceride from SSO substrates. It remains to be seen whether this undesirable side reaction can be eliminated by enzyme purification or by proper choice of reaction conditions.

### E. Other Lipases

Lipase activity is found in a great many plant and animal tissues (*72,217,957*). However, very few of these sources have been properly tested for specificity on mixed-acid triglycerides; and only pancreatic, milk, *Rhizopus arrhizus,* and *Geotrichum candidum* lipases have been sufficiently characterized for analytical applications. Many of the lipases examined so far exhibit a positional specificity similar to pancreatic lipase. Alford *et al.* (*17*), for example, surveyed 13 microbial lipases and found 10 of them showed preferential hydrolysis of the primary ester linkages in triglycerides.

The discovery of new lipases with unusual specificity characteristics would be a great aid to triglyceride analysis, and more exploratory studies are needed in this field. Two such possibilities deserve mention. Work by Kewson and his co-workers (*525*) suggested that *Vernonia anthelmintica* seeds might contain a lipase which hydrolyzes only the 2-position of triglycerides, but initial attempts by Olney *et al.* (*694*) to isolate such an enzyme were unsuccessful. It would also be extremely useful to have a lipase with no positional specificity, for it could be used to generate representative *sn*-1,3-diglycerides and possibly representative *sn*-1- and *sn*-3-monoglycerides. There are many reports of nonspecific lipases in the literature, but these must be interpreted with caution since excessive acyl migration during lipolysis can lead to apparently nonspecific hydrolysis. Recent work on the acid lipase of castor beans (*680*) indicates that it has no positional specificity and might possibly prove useful for analytical purposes.

# 10

# STEREOSPECIFIC ANALYSIS

The chemical and enzymatic deacylation techniques described in Chapter 9 can be used to determine the fatty acids esterified at the *sn*-2- and the combined *sn*-1,3-positions of a triglyceride sample. For a complete analysis of triglyceride positional isomers, however, it is necessary to distinguish between the *sn*-1- and *sn*-3-positions, which are stereochemically distinct (Chapter 1, Section I,A). Such a procedure, called "stereospecific analysis," was first introduced in 1965 by Brockerhoff (*111*). He demonstrated that the fatty acid compositions of the *sn*-1- and *sn*-3-positions could be separately analyzed based on the hydrolytic stereospecificity of phospholipase A for synthetic phospholipids derived from the original triglycerides. The following year, Lands *et al.* (*556*) introduced an alternative method for distinguishing the *sn*-1- and *sn*-3-fatty acids utilizing the stereospecific phosphorylation of derived diglycerides by diglyceride kinase. Stereospecific analysis techniques have subsequently been improved and applied to a wide variety of samples, contributing greatly to our understanding of the triglyceride composition of natural fats.

This chapter describes the various techniques now available for stereospecific analysis of triglycerides and reviews their application to the determination of the positional distribution of fatty acids and the analysis of component triglycerides.

## I. METHODS

Stereospecific analysis of triglycerides involves three basic reactions:

(a) Degradation of the triglycerides to representative diglycerides
(b) Phosphorylation of the diglycerides to produce phospholipids
(c) Hydrolysis of the phospholipids by phospholipase A

After each reaction, the products are separated by TLC on silicic acid and analyzed where necessary for fatty acid composition. Many variations and combinations of reactions (a), (b), and (c) are possible; but the sequences most frequently employed are the $sn$-1,2(2,3)-diglyceride method of Brockerhoff (*111*), the $sn$-1,3-diglyceride method of Brockerhoff (*114*), and the procedure of Lands et al. (*556*). These specific reaction sequences will be briefly outlined before the individual steps are discussed in detail.

## A. $sn$-1,2(2,3)-Diglyceride Method of Brockerhoff

The $sn$-1,2(2,3)-diglyceride method of Brockerhoff (*110,111,119,975*) is outlined in Fig. 10-1. The initial step is incubation of the triglycerides with pancreatic lipase to obtain $sn$-2-monoglycerides. Either simultaneously or subsequently, the triglycerides are deacylated to representative $sn$-1,2(2,3)-diglycerides with pancreatic lipase or a Grignard reagent. After isolation of the $sn$-1,2(2,3)-diglycerides by preparative TLC, they are reacted with phenyl dichlorophosphate to produce a mixture of $sn$-1,2-diacyl-3-phosphatidylphenol and $sn$-2,3-diacyl-1-phosphatidylphenol. Incubation of these compounds with phospholipase A liberates the fatty acids from the 2-position of the $sn$-3-phosphatide but leaves the $sn$-1-phosphatide unhydrolyzed. Separation and fatty acid analysis of the various reaction products allows computation of the composition of the $sn$-1-, $sn$-2-, and $sn$-3-positions in the original triglyceride as follows:

$sn$-1 = lysophosphatide from phospholipase A hydrolysis
$sn$-2 = monoglyceride from pancreatic lipase hydrolysis
$sn$-2 = free fatty acids from phospholipase A hydrolysis
$sn$-3 = 3(original triglyceride) — (lysophosphatide)
        — (monoglyceride)
$sn$-3 = 2(unhydrolyzed phosphatide) — (monoglyceride)

This method allows direct determination of the $sn$-1- and $sn$-2-positions, the latter by two independent methods. The $sn$-3-position is determined using two difference calculations, which should agree within 2% absolute for minor ($<10\%$) and 3–4% absolute for major ($>10\%$) components (*122*) except when labile five- and six-double-bond fatty acids are present (*120*).

Brockerhoff's original procedure (*111,119*) for 1–2 g triglyceride samples has since been adapted for 10–40 mg samples by Christie and Moore (*170*), Åkesson (*12*), and Wood and Snyder (*964*). Typical results from the $sn$-1,2(2,3)-diglyceride method of Brockerhoff are given in Table 10-1.

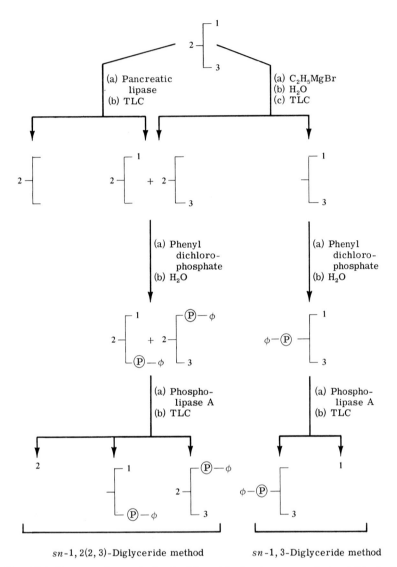

FIG. 10-1. Stereospecific analysis of triglycerides by the two methods of Brockerhoff (*111,114,115,119*). Ⓟ—Φ, phenyl phosphate group; **1**, **2**, and **3** refer to the fatty acids at the *sn*-1-, *sn*-2-, and *sn*-3-positions in the original triglyceride.

## B. *sn*-1,3-Diglyceride Method of Brockerhoff

Brockerhoff's *sn*-1,3-diglyceride method (*114*) is also outlined in Fig. 10-1. Although originally developed using 1000 mg triglyceride samples, the procedure is easily adapted to 20–50 mg samples in the same manner

TABLE 10-1

STEREOSPECIFIC ANALYSIS OF POLAR BEAR BLUBBER TRIGLYCERIDES BY THE $sn$-1,2(2,3)-DIGLYCERIDE METHOD OF BROCKERHOFF[a]

| | $sn$-Position | Fatty acid composition (mole %) | | | | | | | | | |
|---|---|---|---|---|---|---|---|---|---|---|---|
| | | 14:0 | 16:0 | 16:1 | 18:1 | 20:1 | 22:1 | 20:5 | 22:5 | 22:6 | Others |
| Lysophosphatide | 1 | 1.7 | 5.0 | 7.0 | 26.8 | 30.7 | 7.0 | 3.6 | 5.8 | 6.6 | 5.8 |
| Free fatty acid from phospholipase A hydrolysis | 2 | 5.0 | 8.5 | 25.7 | 47.0 | 4.7 | 0.7 | 0.3 | 1.2 | 0.1 | 6.8 |
| Monoglyceride from pancreatic lipase hydrolysis | 2 | 5.4 | 6.7 | 24.3 | 45.9 | 3.9 | 0.8 | 0.7 | 2.2 | 1.5 | 8.6 |
| 3(Triglyceride) − (lysophosphatide) − (monoglyceride) | 3 | 0.7 | 2.4 | 8.6 | 30.8 | 25.1 | 1.1 | 3.2 | 14.2 | 13.5 | 0.3 |
| 2(Unhydrolyzed phosphatide) − (monoglyceride) | 3 | 0.8 | 2.9 | 7.7 | 29.5 | 24.3 | 2.8 | 5.7 | 12.4 | 11.9 | 2.4 |
| Triglyceride | 1,2,3 | 2.6 | 4.7 | 13.3 | 34.5 | 19.9 | 3.0 | 2.5 | 7.4 | 7.2 | 4.9 |
| Unhydrolyzed phosphatide | 2,3 | 3.1 | 4.8 | 16.0 | 37.8 | 14.1 | 1.5 | 3.2 | 7.3 | 6.7 | 5.5 |

[a] From Brockerhoff et al. (119).

TABLE 10-2

COMPARISON OF BROCKERHOFF'S TWO METHODS FOR THE STEREOSPECIFIC ANALYSIS OF CORN OIL TRIGLYCERIDES[a]

| Method | Position | Fatty acid composition (mole %) | | | | | |
|---|---|---|---|---|---|---|---|
| | | 16:0 | 16:1 | 18:0 | 18:1 | 18:2 | 18:3 |
| sn-1,2(2,3)-Diglycerides | 1 | 17.9 | 0.3 | 3.2 | 27.5 | 49.8 | 1.2 |
| | 2 | 2.3 | 0.1 | 0.2 | 26.5 | 70.3 | 0.7 |
| | 3 | 13.5 | 0.1 | 2.8 | 30.6 | 51.6 | 1.0 |
| sn-1,3-Diglycerides | 1 | 18.5 | 0.4 | 3.5 | 28.1 | 48.5 | 1.0 |
| | 2 | 1.8 | 0.1 | 0.2 | 25.8 | 71.2 | 0.9 |
| | 3 | 12.6 | 0.5 | 2.2 | 31.0 | 52.6 | 1.1 |

[a] From Brockerhoff (114).

as the sn-1,2(2,3)-digyceride method (12,170,964). The sn-2-position is again characterized by analysis of the 2-monoglyceride from pancreatic lipolysis. Then the sn-1,3-diglyceride products from Grignard deacylation are isolated and converted to sn-1,3-diacyl-2-phosphatidylphenol using phenyl dichlorophosphate. Treatment with phospholipase A will hydrolyze only the sn-1-acyl group from an sn-2-phosphatide (209,210), leaving a lysophosphatide containing the sn-3-acyl chain. Separation and fatty acid analysis of the various reaction products permits direct determination of all three positions of the original triglyceride:

sn-1 = free fatty acids from phospholipase A hydrolysis
sn-2 = monoglyceride from pancreatic lipase hydrolysis
sn-3 = lysophosphatide from phospholipase A hydrolysis

Problems with isomerization of the sn-1,3-diglycerides during Grignard deacylation (Chapter 9, Section I,A,1) makes this technique slightly less accurate than the sn-1,2(2,3)-diglyceride method of Brockerhoff, but this increased error is somewhat balanced by the fact that the sn-3-position can be determined directly rather than by difference. Table 10-2 compares stereospecific analyses of corn oil triglycerides by both the sn-1,2(2,3)- and sn-1,3-diglyceride methods of Brockerhoff.

## C. Method of Lands

Lands and his co-workers (556) have used a somewhat different approach to the stereospecific analysis of triglycerides, based on the stereo-

# I. METHODS

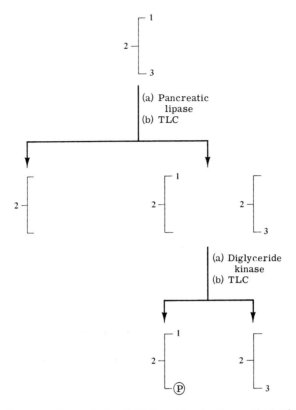

FIG. 10-2. Stereospecific analysis of triglycerides by the method of Lands et al. (556). Ⓟ, phosphate group; 1, 2, and 3 refer to the fatty acids at the sn-1-, sn-2-, and sn-3-positions in the original triglyceride.

specificity of diglyceride kinase (Fig. 10-2). The initial step is a partial hydrolysis of 2–5 mg triglyceride samples with pancreatic lipase followed by isolation of the sn-2-monoglycerides and the sn-1,2(2,3)-diglycerides from the reaction products. Incubation of the diglycerides with diglyceride kinase from *Escherichia coli* stereospecifically converts the sn-1,2-diglyceride to sn-1,2-diacyl-3-phosphatidate, while the sn-2,3-diglyceride remains unphosphorylated. Isolation and fatty acid analysis of the phosphatidate as well as the monoglycerides from lipolysis allows computation of the sn-1-, sn-2-, and sn-3-positions in the original triglycerides:

sn-1 = 2(phosphatidate) − (monoglyceride)
sn-2 = monoglyceride from pancreatic lipase hydrolysis
sn-3 = 3(original triglyceride) − 2(phosphatidate)

TABLE 10-3
STEREOSPECIFIC ANALYSIS OF RAT LIVER TRIGLYCERIDES BY THE METHOD OF LANDS[a]

|  | sn-Position | \multicolumn{7}{c}{Fatty acid composition (mole %)} | | | | | | |
|---|---|---|---|---|---|---|---|---|
|  | sn-Position | 14:0 | 16:0 | 16:1 | 18:0 | 18:1 | 18:2 | Others |
| 2(Phosphatidate) −(monoglyceride) | 1 | 1.7 | 62.2 | −0.8 | 5.2 | 18.2 | 13.5 | 0.0 |
| Monoglyceride from pancreatic lipase hydrolysis | 2 | 1.1 | 8.6 | 4.0 | 1.0 | 39.6 | 42.1 | 3.6 |
| 3(Triglyceride) −2(phosphatidate) | 3 | 1.4 | 10.5 | 5.8 | 0.4 | 32.2 | 19.7 | 30.0 |
| Triglyceride | 1,2,3 | 1.4 | 27.1 | 3.0 | 2.2 | 30.0 | 25.1 | 11.2 |
| Phosphatidate | 1,2 | 1.4 | 35.4 | 1.6 | 3.1 | 28.9 | 27.8 | 1.8 |

[a] From Slakey and Lands (843).

Only the sn-2-position is determined directly in the Lands procedure, for the sn-1- and sn-3-positions must be calculated by difference. Direct determination of the sn-1-position would be possible, however, by hydrolysis of the phosphatidate with phospholipase A (95,555). The phosphorylation of sn-1,2-diglycerides by diglyceride kinase does not go to completion under the conditions described by Lands et al. (556), so analysis of the unphosphorylated diglycerides cannot provide a check on the above calculations. Data from a stereospecific analysis of rat liver triglycerides by the Lands method are given in Table 10-3.

## D. Choice of Method

It is evident from the above discussion that numerous alternatives are available in selecting a suitable procedure for stereospecific analysis of triglycerides. The specific choice usually depends on whether accuracy, speed, or the determination of minor fatty acids is the principal goal.

For maximum accuracy, the method of choice is clearly the sn-1,2(2,3)-diglyceride method of Brockerhoff (Fig. 10-1). First of all, the sn-1,2(2,3)-diglycerides prepared by Grignard deacylation (or by pancreatic lipase with some samples) are more representative of the original triglyceride structure than sn-1,3-diglycerides (Chapter 9, Section I,A,1). Second, the direct determination of the fatty acid composition of the sn-1-position and the dual calculation procedures for the sn-3-position should yield more accurate data than the Lands method in which the sn-1- and *sn*-3-positions are both determined by difference* and there is no alternative calculation for checking results. Brockerhoff and Yurkowski (*122*) have estimated that the positional analyses of fatty acid composition by the *sn*-1,2(2,3)-diglyceride method are accurate to 2% absolute for major (>10%) and within 1% absolute for minor (<10%) acids. With marine animal triglycerides containing the labile 20:5, 22:5, and 22:6 acids, however, the estimated error increases to 8% relative for major components (*120*). The accuracy of the Lands method is somewhat less: 11% absolute for major and 5% absolute for minor acids in the *sn*-1- and *sn*-3-positions (*556*).

For speed of analysis, the Lands method is obviously the most rapid procedure. Both Brockerhoff methods include deacylation, phosphorylation, and phospholipase A hydrolysis steps, requiring a total of 3–4 days of work for each stereospecific analysis. The Lands technique can be completed in about two-thirds this time since no phospholipase A hydrolysis is necessary.

If the amount or the radioactivity of minor fatty acids in the *sn*-3-position must be determined, then the *sn*-1,3-diglyceride method of Brockerhoff is often preferable. The error from difference calculations for minor acids is usually greater than the error from slightly nonrepresentative *sn*-1,3-diglycerides, making direct measurement of the *sn*-3-position acids the more accurate procedure (*114*).

### E. Deacylation of Triglycerides to Representative Diglycerides

The first step in all stereospecific analysis procedures is deacylation of the triglyceride sample to produce representative diglycerides. Suitable *sn*-1,2(2,3)-diglycerides can be prepared with Grignard reagent, pancreatic lipase, milk lipase, or *Geotrichum candidum* lipase. Representative *sn*-1,3-diglycerides can be obtained with Grignard reagent or *Geotrichum candidum* lipase. The specific deacylation procedures and their relative merits and limitations for producing representative diglycerides are discussed in detail in Chapter 9.

* Experimental errors can be magnified severalfold when difference calculations are used to determine the positional distribution of fatty acids. The Lands method for computing the *sn*-3-position illustrates this point: *sn-3* = 3(*triglyceride*) − 2(*phosphatidate*). Suppose that a given triglyceride mixture contains 35% 18:1 at all three positions, but 36% 18:1 is found in the triglycerides and 34% in the phosphatidate through normal GLC error. By difference calculations, *sn-3* = 3(*36*) − 2(*34*) = 40%, which is 5% absolute too high. On the other hand, it is also possible for the level of error to remain unchanged, provided both experimental values err in the same direction. For example, if 36% 18:1 were found in the phosphatidate, then *sn-3* = 3(*36*) − 2(*36*) = 36%. In practice, difference calculation errors probably fall somewhere in between these two extremes.

Since specificity and isomerization problems are sometimes encountered in all the available deacylation methods, it is essential to establish that truly representative diglycerides have been obtained before proceeding further with stereospecific analysis. This is easily done by checking the fatty acid composition of the diglycerides against that calculated from the compositions of the original triglycerides and the 2-monoglycerides from pancreatic lipase hydrolysis (25,170,975). The calculation procedure for each fatty acid is

$$\frac{\% \text{ X in } sn\text{-}1,2(2,3)\text{-diglycerides}}{} = \frac{3\left[\% \text{ X in triglycerides}\right] + \left[\% \text{ X in 2-monoglycerides}\right]}{4}$$

$$\frac{\% \text{ X in } sn\text{-}1,3\text{-diglycerides}}{} = \frac{3\left[\% \text{ X in triglycerides}\right] - \left[\% \text{ X in 2-monoglycerides}\right]}{2}$$

The experimental and calculated fatty acid compositions for $sn$-1,2(2,3)-diglycerides should agree within 5% relative for major acids (>10%) and 1% absolute for minor acids (<10%) to be considered representative and suitable for stereospecific analysis (Tables 9-1 and 9-2). Tolerance levels for $sn$-1,3-diglycerides from Grignard deacylation are necessarily slightly higher than these limits (Table 9-1), making the $sn$-1,3-isomer less suitable for accurate stereospecific analysis. The above calculations cannot be applied to diglycerides prepared with *Geotrichum candidum* lipase because of the fatty acid specificity of the enzyme.

## F. Phosphorylation of Diglycerides

The next step in stereospecific analysis of triglycerides is conversion of the diglycerides into phospholipids. Both chemical and enzymatic methods have been used for this purpose.

### 1. Chemical Synthesis

The original stereospecific analysis procedure of Brockerhoff (*111*) employed phenyl dichlorophosphate ($C_6H_5OPOCl_2$, also called phenyl phosphorodichloridate) for the conversion of diglycerides to phosphatidylphenol. The reaction is essentially quantitative for both $sn$-1,2(2,3)- and $sn$-1,3-isomers (*111,114*) and has been widely adopted. A typical procedure has been described by Brockerhoff (*119*):

> The diglycerides (93 mg) are dissolved in 1 ml of anhydrous diethyl ether and added dropwise with stirring to a mixture of 1 ml of dry pyridine, 1 ml of diethyl ether, and 0.5 ml of freshly distilled phenyl dichlorophosphate.

After 60 minutes at room temperature, 5 ml of pyridine, 3 ml of diethyl ether, and several drops of water are added with cooling. The reaction mixture is then added to a separatory funnel containing 30 ml of methanol, 25 ml of water, 30 ml of chloroform, and 1 ml of triethylamine. After shaking, the lower $CHCl_3$ layer is recovered and evaporated to obtain the phosphatidylphenol.

Similar semimicro procedures have been described by Åkesson (*12*) and Sampugna and Jensen (*781*).

For accurate results from stereospecific analyses, it is essential that preparation, isolation, and phosphorylation of the diglyceride be carried out in as short a time as possible so that acyl migration is held to an absolute minimum. Positional purity of the phospholipid can be checked by TLC on silicic acid, since *sn*-2- and *rac*-1,3-phosphatidylphenols can be separated with a chloroform/methanol/3.5 $N$ $NH_4OH$ 80/17.5/2.5 solvent system (*12*). As a further precaution, Åkesson (*12*) recommends preparative TLC for removing all traces of the undesired isomer. If extensive acyl migration has occurred, however, even such purified phosphatidylphenols may not be representative, since some fatty acid chains may undergo acyl migration more rapidly than others (*251*).

## 2. Enzymatic Synthesis

Diglyceride kinase has been used by Lands and his co-workers (*556*) for the stereospecific conversion of *sn*-1,2-diglycerides to *sn*-1,2-diacyl-3-phosphatidate. This enzyme is prepared from *Escherichia coli* cells using the methods of Pieringer *et al.* (*397,719*) including heating the enzyme for 10 minutes at 100° to minimize phosphatase activity. Lands *et al.* (*556*) have demonstrated that diglyceride kinase is stereospecific for the conversion of *sn*-1,2-dipalmitin to *sn*-1,2-dipalmito-3-phosphatidate, while *sn*-2,3-dipalmitin remains unphosphorylated (Fig. 10-3). An initial experiment with mixed *sn*-1,2(2,3)-diglycerides indicated that diglyceride kinase has no marked specificity for diglycerides bearing certain acyl groups (*556*), but further testing is needed to establish how widely this applies. The enzyme is inhibited by 2′,7′-dichlorofluorescein (*556*); hence this compound should not be used to visualize TLC bands during diglyceride isolation. Lands *et al.* (*556*) have published the following procedure for phosphorylation of *sn*-1,2-diglycerides using diglyceride kinase:

One to three milligrams of *sn*-1,2(2,3)-diglycerides is placed in a test tube and the following reagents are added: 10 $\mu$l of 200 mg/ml mixed bile salts; 0.10 ml of 0.05 $M$ adenosine triphosphate; 0.05 ml of 1.0 $M$ $MgCl_2$; 0.05 ml of 0.50 $M$ sodium phosphate buffer (pH 7.95); and 0.8 mg of crude diglyceride kinase in 0.10 ml of cysteine phosphate buffer. The reaction is incubated at 37° with constant shaking. After 1 hour, 0.20 ml of 1 $N$ HCl is added, and the lipids are extracted with 2.0 ml of $CHCl_3/CH_3OH$ 2/1

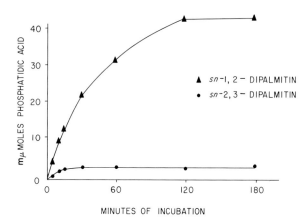

FIG. 10-3. Stereospecificity of diglyceride kinase from *Escherichia coli* for *sn*-1,2- and *sn*-2,3-dipalmitin. *Reaction conditions:* 0.47 mmole of diglyceride; 2.4 moles of $^{32}$P-ATP; 93 mmoles of $MgCl_2$; 0.93% (v/v) Cutscum (Fisher Scientific Co., Fairlawn, N.J.); 46.5 mmoles of sodium phosphate buffer, pH 7.5; 0.58 mg of protein from heat-treated particulate preparation of *E. coli* suspended in 0.05 ml of 0.1% cysteine hydrochloride 0.01 *M* sodium phosphate, pH 7.0; final volume, 0.215 ml; incubation temperature, 37°. From Lands *et al.* (556).

followed by 1.3 ml of $CHCl_3$. A drop of triethylamine is added to the combined extracts, and the solvent is evaporated. The *sn*-1,2-diacyl-3-phosphatidate is isolated by preparative TLC on silicic acid using chloroform/methanol/water 65/36/8 as the developing solvent (128).

## G. Hydrolysis of Phospholipid with Phospholipase A

The final step in stereospecific analysis is hydrolysis of the phospholipid using phospholipase A followed by separation of the reaction products and determination of their fatty acid compositions (Fig. 10-1). Phospholipase A is stereospecific in its hydrolysis of the 2-position fatty acid from an *sn*-1,2-diacyl-3-phosphatide, but the enzyme will not attack the enantiomorphic *sn*-2,3-diacyl-1-phosphatide (210,343,881,903). When presented with an *sn*-1,3-diacyl-2-phospholipid substrate, phospholipase A hydrolyzes the fatty acid in the *sn*-1-position (209,210).

The usual source of the phospholipase A used in analytical reactions is lyophilized snake venoms (Ross Allen Reptile Institute, Silver Springs, Fla.; Miami Serpentarium Laboratories, Miami, Fla.) which are not fractionated to concentrate the enzyme. Available evidence indicates that snake venom phospholipase A is absolute in its specificity for the *sn*-2-position of *sn*-3-phospholipids and the *sn*-1-position of *sn*-2-phospholipids, but the enzyme has a marked specificity for substrates bearing certain acyl groups.

Thus van Deenen and co-workers (*79,903*) have shown that *sn*-3-lecithins bearing different acyl groups (1,2-dioleo; 1,2-distearo; 1,2-dipentadecano; 1,2-didecano; 1,2-dibutyro; 1-oleo-2-butyro; and 1-butyro-2-oleo) are hydrolyzed at different rates by phospholipase A from *Crotalus adamanteus* venom. Nutter and Privett (*685*) followed the composition of the fatty acids liberated during lecithin hydrolysis by venoms of seven different species of snakes. Only *Ophiophagus hannah* venom did not show a marked preferential hydrolysis of saturated fatty acids during the early part of the reaction, and Nutter and Privett recommend that this venom be adopted for analytical work.

For stereospecific triglyceride analyses, it is desirable that the phospholipase A hydrolysis be carried to completion. This assures that no preferential hydrolysis of certain fatty acids has taken place and that the unhydrolyzed phosphoglyceride is all *sn*-2,3-diacyl-1-phosphatidylphenol.

A typical semimicro procedure for the hydrolysis of phosphatidylphenol with phospholipase A has been described by Christie and Moore (*170,172*):

> The phosphatidylphenol (7–10 mg) is dissolved in 3 ml of diethyl ether. Then 0.1 ml of Tris buffer containing $CaCl_2$ [0.5 $M$ tris(hydroxymethyl)methylamine, 0.002 $M$ $CaCl_2$, pH 7.5] and 0.5 mg of *Ophiophagus hannah* venom are added, and the mixture is shaken overnight. The following morning, 5 ml of isobutanol and 0.02 ml of acetic acid are added and the mixture is taken to dryness on a rotary-film evaporator. The residue is dissolved in a small amount of chloroform/methanol 2/1 and applied in a band on a silicic acid TLC plate, which is then developed in hexane/diethyl ether/formic acid 50/50/1. The top third of the plate is sprayed with 2′,7′-dichlorofluorescein solution, and the free fatty acid band is located and recovered. The TLC plate is then redeveloped in chloroform/methanol/14 $M$ $NH_4OH$ 90/8/2 up to the level of the previous spray. After spraying with Rhodamine 6G solution, the lysophosphatidylphenol and unhydrolyzed phosphatidylphenol bands are located and recovered. All three products are then converted to methyl esters for GLC analysis of fatty acid composition.

Similar phospholipase A hydrolysis procedures have been reported by Brockerhoff *et al.* (*119*), Sampugna and Jensen (*781*), and Åkesson *et al.* (*12,15*).

## II. APPLICATIONS

### A. Positional Distribution of Fatty Acids

Stereospecific analysis procedures have been widely used to determine the positional distribution of fatty acids between the *sn*-1-, *sn*-2-, and *sn*-3-

positions of the triglycerides in natural fats. The fatty acid compositions obtained are normally reported on the basis of 100 mole percent at each of the three positions analyzed. A typical stereospecific analysis of polar bear blubber triglycerides by the $sn$-1,2(2,3)-diglyceride method of Brockerhoff is given in Table 10-1. These data illustrate the type of agreement obtained between the multiple determinations of the $sn$-2- and $sn$-3-positions by this technique. A comparison of stereospecific analyses of corn oil triglycerides by both the $sn$-1,2(2,3)- and the $sn$-1,3-diglyceride methods of Brockerhoff (Table 10-2) shows close agreement between the two results, even though $sn$-1,3-diglycerides are slightly less representative of the original sample (Chapter 9, Section I,A,1). Table 10-3 presents data from a stereospecific analysis of rat liver triglycerides by the method of Lands.

Difficulties other than nonrepresentative diglycerides and acyl migration are encountered with samples containing certain types of fatty acids. The extensive sample manipulation required during stereospecific analyses exposes polyunsaturated fatty acids to atmospheric oxidation and consequent loss. Note in Table 10-1, for example, that the content of 20:5, 22:5, and 22:6 in the free fatty acid from phospholipase A hydrolysis (which underwent extensive sample handling) is considerably lower than in the monoglyceride from pancreatic lipase hydrolysis (which was subjected to less sample manipulation). Precautions such as nitrogen blanketing and the use of antioxidants (Table 2-1) can help minimize this problem, but some exposure to atmospheric oxidation is inevitable during the lengthy analytical procedure.

Stereospecific analysis of triglycerides containing short-chain or oxygenated acids presents serious problems since the presence of these unusual acids changes $R_f$ values during TLC (Chapter 8, Sections I,B and I,E) resulting in multiple bands for a single lipid class. For example, triglycerides containing both ricinoleic (R) and normal (X) fatty acids could produce three separable diglyceride bands (RR–, RX–, XX–) as well as multiple phosphoglyceride bands. Such problems can sometimes be solved by using other chromatographic isolation techniques or by acetylating the hydroxy groups so that overlap of lipid classes does not occur during separation.

## B. Composition of Triglyceride Mixtures

Stereospecific analysis results can also be used to determine the individual components in simple triglyceride mixtures separated from natural fats. The effectiveness of this procedure depends on the complexity of the isomeric mixture, however.

## II. APPLICATIONS

A single diacid triglyceride has three possible isomers:

sn-PLL     sn-LPL     sn-LLP

The isomeric composition of such a mixture is easily determined by straightforward stereospecific analysis. The amount of 16:0 found at the sn-1-, sn-2-, and sn-3-positions directly corresponds to the amounts of sn-PLL, sn-LPL, and sn-LLP present. A mixture of two different diacid triglycerides

sn-PLL     sn-LPL     sn-LLP<br>
sn-OOL     sn-OLO     sn-LOO

is analyzed similarly using both direct correlation or difference calculations:

$$sn\text{-PLL} = P_1$$
$$sn\text{-LPL} = P_2$$
$$sn\text{-LLP} = P_3$$
$$sn\text{-OOL} = L_3 - P_1 - P_2$$
$$sn\text{-OLO} = L_2 - P_1 - P_3$$
$$sn\text{-LOO} = L_1 - P_2 - P_3$$

where $P_1$ is % 16:0 at sn-1-position; $L_1$, % 18:2 at sn-1-position; $P_2$, % 16:0 at sn-2-position; etc. When the same two fatty acids appear in two different diacid triglycerides,

sn-PLL     sn-LPL     sn-LLP<br>
sn-PPL     sn-PLP     sn-LPP

then the isomeric composition is not determinable by a single stereospecific analysis. If, however, this mixture is first resolved into PLL and PPL bands by Ag$^+$ TLC or liquid–liquid partition chromatography, then stereospecific analysis of each fraction quantitates all six components.

A triacid triglyceride can occur in six different isomeric forms:

sn-MPO     sn-OMP     sn-POM<br>
sn-OPM     sn-PMO     sn-MOP

A single stereospecific analysis of such a mixture cannot determine its isomeric composition (113,803). For example, if there is 30% M at the sn-1-position, then sn-MPO + sn-MOP = 30%, but the relative proportion of these two triglycerides is unknown. In sn-MPO, the sn-2-position contains P, and the sn-3-position contains O. However, the P at the sn-2-position is shared by sn-MPO and sn-OPM; the O at the sn-3-position is shared by sn-MPO and sn-PMO; and again these ratios remain unknown. It follows that the proportions of isomers in one of these pairs

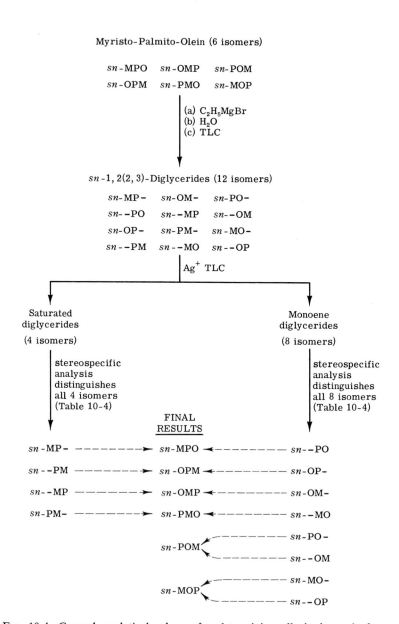

FIG. 10-4. General analytical scheme for determining all six isomeric forms of a triacid triglyceride by (i) deacylation with Grignard reagent or pancreatic lipase, (ii) separation of the resultant diglycerides by unsaturation using Ag⁺ TLC, and (iii) stereospecific analysis of the two diglyceride fractions. Equivalent results are obtained if the chromatographic separation is carried out at the phospholipid level.

## II. APPLICATIONS

must be determined before the other components can be calculated. Jensen *et al.* (*426*) and Lands and Slakey (*557*) have pointed out that this problem can be solved by chromatographic separation at the diglyceride or phosphoglyceride level followed by stereospecific analysis of each of the two fractions obtained.

A general analytical scheme for determining all six isomers of myristo-palmitoolein is shown in Fig. 10-4. The triglycerides are first deacylated with $C_2H_5$ MgBr to produce twelve $sn$-1,2(2,3)-diglyceride species, which are then separated into saturated and monoene fractions by $Ag^+$ TLC. Stereospecific analysis of each fraction distinguishes all the diglyceride isomers in each (Table 10-4), and these 12 results are combined to give the amounts of the six isomeric triglycerides in the original sample. Equivalent results can be obtained by numerous variations of the same basic approach, as long as the diglycerides or phosphoglycerides are resolved into two fractions for separate stereospecific analyses. Possible alternatives include: other deacylation agents, use of $sn$-1,3-diglycerides, diglyceride frac-

TABLE 10-4
Calculation of Component $sn$-1,2(2,3)-Diglycerides after Stereospecific Analysis by the Method of Brockerhoff

**Nomenclature**
  $M_1$ = % 14:0 in $sn$-1-acyl-3-phosphatidylphenol from phospholipase A hydrolysis
  $M_2$ = % 14:0 in 2-monoglycerides from pancreatic lipase hydrolysis
  $M_2^*$ = % 14:0 in free fatty acids from phospholipase A hydrolysis
  $M_{23}$ = % 14:0 in $sn$-2,3-diacyl-1-phosphatidylphenol after phospholipase A hydrolysis
  Nomenclature for 16:0 (P) is analogous
    $a$ = mole fraction of $sn$-1-acyl-3-phosphatidylphenol in total phosphoglycerides after phospholipase A hydrolysis

**Saturated $sn$-1,2(2,3)-diglycerides from MPO (Fig. 10-4)**
  $sn$-MP- = $M_1 a$
  $sn$--PM = $P_2 - M_1 a$
  $sn$--MP = $M_2 - P_1 a$
  $sn$-PM- = $P_1 a$

**Monoene $sn$-1,2(2,3)-diglycerides from MPO (Fig. 10-4)**
  $sn$--PO = $P_2 - P_2^* a$
  $sn$-OP- = $P_2^* a$
  $sn$-OM- = $M_2^* a$
  $sn$--MO = $M_2 - M_2^* a$
  $sn$-PO- = $P_1 a$
  $sn$--OM = $2M_{23}(1 - a) - M_2 + M_2^* a$
  $sn$-MO- = $M_1 a$
  $sn$--OP = $2P_{23}(1 - a) - P_2 + P_2^* a$

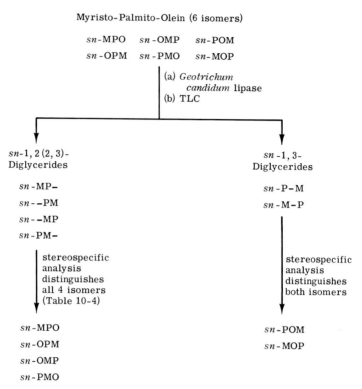

FIG. 10-5. Analytical scheme for determining all six isomeric forms of a triacid triglyceride containing oleic acid by (i) deacylation of oleoyl groups with *Geotrichum candidum* lipase, (ii) chromatographic sepation of sn-1,2(2,3)- and sn-1,3-diglycerides by TLC on silicic acid, and (iii) stereospecific analysis of the two diglyceride fractions. From Jensen et al. (*426,781*).

tionation by liquid–liquid partition chromatography, or separation of phosphoglycerides rather than diglycerides. One of the more novel variations is that of Jensen and his co-workers (*426,781*), which is based on the specific deacylation of a *cis*-9-acid with *Geotrichum candidum* lipase (Fig. 10-5). The resultant sn-1,2(2,3)- and sn-1,3-diglycerides are then separated by TLC on silicic acid, and each diglyceride fraction is subjected to stereospecific analysis to distinguish the six original isomeric triglycerides.

When the distribution of isomers in a triacid triglyceride is such that one fatty acid is almost absent at one position on the glycerol, the approximate amounts of the major isomers can be estimated from a single stereospecific analysis. This short-cut procedure was used by Slakey and Lands (*843*) who isolated **012** from rat liver triglycerides and found 3.7 mole percent saturated acid at the $sn$-2-position and 33.0% diene acid at the $sn$-3-position. Thus $sn$-**102** + $sn$-**201** = 3.7%, and the amount of $sn$-**102** present must lie in the 0.0–3.7% range. Since $sn$-**102** + $sn$-**012** = 33.0%, it follows that $sn$-**012** must fall between 29.3% and 33.0%. Such range calculations sacrifice accuracy for speed of analysis and are only applicable to special samples.

Mixtures of two triacid triglycerides or of a triacid and a diacid triglyceride usually require a combination of chromatographic separation and stereospecific analysis techniques for complete isomer determination (assuming there is a common fatty acid in all components present). Analytical schemes for resolving such complex mixtures are discussed in Chapter 13.

### C. Composition of Derived Diglycerides

The use of stereospecific analysis to determine the composition of mixtures of derived diglycerides follows the same approach as for triglycerides but without the initial deacylation step. Either the $sn$-1,2(2,3)- or $sn$-1,3-diglyceride methods of Brockerhoff (Fig. 10-1) or the method of Lands (Fig. 10-2) may be used, provided the appropriate diglyceride isomers are present or are isolated by TLC. The $sn$-1,2(2,3)-diglycerides require a separate hydrolysis with pancreatic lipase to determine the fatty acid composition of the 2-position.

Stereospecific analysis of $sn$-1,2(2,3)-diglyceride mixtures cannot assume (as is done with triglyceride deacylation products) that equal quantities of the $sn$-1,2- and $sn$-2,3-isomers are present in the sample. The proportions of these two isomeric forms can be determined by measuring the relative amounts of $sn$-1-acyl-3-phosphatidylphenol and $sn$-2,3-diacyl-1-phosphatidylphenol after hydrolysis with phospholipase A. This can be accomplished by (i) addition of an internal standard during GLC analysis of the derived methyl esters (*781,782*), (ii) analysis of the phosphorus content of each product, or (iii) determination of the glycerol content of each product (Table 2-3). Suitable calculation procedures for the determination of component $sn$-1,2(2,3)-diglycerides are outlined in Table 10-4.

# 11

# PHYSICAL PROPERTIES

The classical methods of qualitative organic analysis stress the identification of compounds by their physical properties (melting point, refractive index, density, spectra, etc.). Such techniques can only be effective when a specific compound has been isolated in pure form, a very difficult task with most natural fat triglycerides. Chromatographic and enzymatic methods for triglyceride analysis have now become so effective in their ability to characterize specific *groups* of triglycerides ($C_{52}$, monounsaturated, $\beta$-SUS, etc.) that traditional physical measurements are now mainly employed in a confirmatory role. There are occasional instances, however, in which a physical property plays a primary role in the identification of triglyceride positional isomers. An example is found in the recent work of Kleiman et al. (511,512), who determined the configuration of sn-1,2-di-($C_{16}/C_{18}$)-3-acetin in *Euonymus verrucosus* seed fat by measuring its rotation of polarized light.

This chapter describes how individual molecular species of triglycerides can be distinguished by measuring certain physical properties. Since these methods are infrequently used in triglyceride analysis, discussions are brief and detailed experimental procedures are not included.

## I. MASS SPECTROMETRY

The application of mass spectrometry to the analysis of triglycerides is still in the initial stages of development. The technique has a number of positive advantages: very little sample (1–10 $\mu$g) is required, molecular

I. MASS SPECTROMETRY

weights up through $C_{66}$ can be examined, and pure triglycerides can be identified from their degradation patterns in only a few minutes. On the other hand, mass spectrometers are very expensive to purchase and costly to maintain, and this prevents their being available for use in most laboratories. Their application to the quantitative analysis of triglyceride mixtures has only been explored briefly in one laboratory. Despite these limitations, mass spectrometry holds considerable promise as a useful technique for rapid triglyceride analysis and will probably be used more widely as suitable new methodology is developed.

For a detailed discussion of the equipment and the experimental techniques employed for mass spectrometry of lipids, the reader is referred to one of the excellent recent reviews on the subject (*151,776,860,866*).

## A. Pure Triglycerides

The mass spectra of tristearin and *rac*-1-myristo-2-stearo-3-palmitin are shown in Fig. 11-1, and the major fragments for tristearin are identified

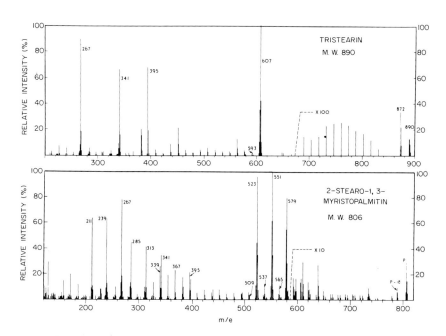

FIG. 11-1. Mass spectra of tristearin and β-1-myristo-2-stearo-3-palmitin. Probable identifications of the major fragments for tristearin are listed in Table 11-1. *Operating conditions:* A.E.I. MS-9 mass spectrometer, direct introduction of samples into the ionization chamber at 170°–200°. From Barber *et al.* (*48*).

TABLE 11-1
Major Fragments in the Mass Spectrum of Tristearin[a]

| $m/e$ | Origin | Fragment | | Relative abundance |
|---|---|---|---|---|
| 267 | RCO | Acyl ion | | 90 |
| 341 | RCO + 74 | RCOO—◇—$\overset{+}{O}$H | | 66 |
| 382 | RCO + 115 | RCOO—[structure with $\overset{+}{O}\cdot$, C=CH$_2$] | | 20 |
| 395 | RCO + 128 + 14$n$ | | $n = 1$ | 68 |
| 409 | | | $n = 2$ | 4 |
| 423 | | RCOO—[structure with $\overset{+}{O}\cdot$, C=CH–(CH$_2$)$_n$] | $n = 3$ | 2 |
| 437 | | | $n = 4$ | 8 |
| 451 | | | $n = 5$ | 21 |
| etc. | | | | |
| 593 | M − R$_\alpha$COOCH$_2$ | Loss of acyloxymethylene group from $sn$-1- or $sn$-3-position | | 3 |
| 606 | M − RCOOH | Loss of RCOOH | | 36 |
| 607 | M − RCOO | Loss of acyloxy group | | 100 |
| 872 | M − 18 | Loss of H$_2$O | | 0.3 |
| 890 | M | Molecular ion | | 0.2 |

[a] Based on the findings of Barber et al. (*48*) and Holman and co-workers (*1, 558*).

in Table 11-1. The fatty acid composition of a triglyceride is easily deduced from the large [M − RCOO]$^+$ and [RCO]$^+$ peaks. In the mass spectrum of β-MStP, for example, there are three peaks at 523, 551, and 579 $m/e$ corresponding to the loss of three different acyloxy groups from the molecular ion. The three [RCO]$^+$ peaks at 211, 239, and 267 confirm this identification. The fatty acid esterified at the 2-position can be determined since it produces little or no [M − RCOOCH$_2$]$^+$ fragmentation compared with the acids at the 1- and 3-positions (*47,558*). Thus comparison of the 509, 537, and 565 peaks from β-MStP shows 509 to be much smaller than 537 and 565, indicating 18:0 is esterified at the secondary hydroxyl. Mass spectrometry does not differentiate between the $sn$-1- and $sn$-3-positions of triglycerides.

Lauer et al. (*558*) report that acyl chain length has a marked influence on ion yield. Monoacid triglycerides of very-short-chain acids exhibit no molecular ions in their mass spectra. Experiments with mixed-acid triglycerides show that the amount of [M − RCOO]$^+$ ion produced increases with the chain length of the acyloxy group lost. Thus the ratio of $[M - C_6H_{11}O_2]^+$ to $[M - C_{18}H_{35}O_2]^+$ ion intensities was 13:87 for hexanodistearin, while the corresponding $[M - C_{14}H_{27}O_2]^+$ to $[M - C_{18}H_{35}O_2]^+$ ion ratio for myristodistearin was 28:72. This effect is independent of the position of the shorter-chain acid in the triglyceride molecule.

Unsaturated triglycerides are easily recognized since each double bond diminishes the mass of a fragment of two units. $[RCO - 1]^+$ ions are also prominent if a monounsaturated acyl group is present (558). The homology of $[RCO + 128 + 14\ n]^+$ fragments (Table 11-1) would obviously be interrupted by the presence of any double bonds in the hydrocarbon chain that is ruptured. Double-bond mobility during fragmentation prevents the direct determination of double-bond location by this means (558). However, Lauer et al. (558) have demonstrated that if an unsaturated triglyceride is deuterated first, the location of the double bonds in a monoacid triglyceride can be deduced from its mass spectrum (Fig. 11-2). Deuterated mixed-acid triglycerides yield overlapping $[RCO + 128 + 14\ n]^+$ series, making double-bond location much more difficult.

A detailed investigation of fragmentation mechanisms in the mass spectrometry of triglycerides has been published by Holman and his co-workers (1,558). Other investigators have published mass spectra for sn-StOSt and β-StStO (48), β-2-lauro-1,3-didecanoin (776), PPP and OOO (515), various diacetotriglycerides (430), β-1-(5-hydroxydodecano)-2,3-dipalmitin (227), and an estolide triglyceride from *Sapium sebiferum* seeds (851).

## B. Natural Triglyceride Mixtures

Although individual molecular species of triglycerides can be readily identified from their mass spectra, quantitative analysis of triglyceride mixtures is considerably more difficult. The influence of acyl chain length on ion yield and the overlapping of fragmentation patterns from different molecular species are two of the major problems to be overcome.

Hites (382) has recently proposed the quantitative analysis of triglyceride mixtures based on the combined intensities of the $[M]^+$ and $[M - 18]^+$ ions. Using a single-focusing mass spectrometer, this technique can distinguish between triglycerides which differ by one or more mass units. Thus molecules are classified according to their carbon number *and* their degree of unsaturation. Hites' procedure involves the following steps:

(a) The 700–1000 *m/e* spectrum of the triglyceride mixture is recorded several times.

(b) The [M]⁺ and [M-18]⁺ ion intensities for each type of triglyceride are summed for each run, and replicate runs are averaged.
(c) Appropriate correction factors are applied for heavy isotopes, variation of vapor pressure with molecular weight [assuming $\log p = k$ (carbon number)], and loss of unsaturated molecules (by empirical calibration).
(d) The percentage of each component is then calculated.

Table 11-2 gives an analysis of cocoa butter triglycerides by this method. Hites' procedure for the rapid analysis of triglyceride mixtures by mass spectrometry shows considerable promise, but further calibration studies

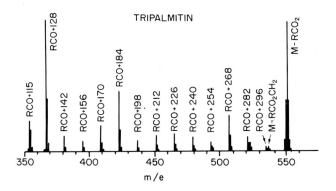

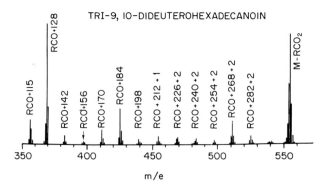

Fig. 11-2. Mass spectra of tripalmitin and tri-9,10-dideuterohexadecanoin showing the homologous series of [RCO + 128 + 14 n]⁺ fragments. The presence of deuterium at the 9- and 10-positions on the acyl chain raises the mass number of the [RCO + 212]⁺ fragment by one and all the higher homologs by two. Thus the location of the double bonds in an unsaturated monoacid triglyceride can be determined from the mass spectrum of its deuterated derivative. *Operating conditions:* Hitachi RMU-6D mass spectrometer operated at 70 eV, direct introduction of sample into the ionization chamber. From Lauer et al. (558).

TABLE 11-2

TRIGLYCERIDE COMPOSITION OF COCOA BUTTER
DETERMINED BY DIRECT MASS SPECTROMETRY[a]

| Triglyceride | Mole % | Triglyceride | Mole % |
|---|---|---|---|
| 50:1[b] | 13.4 | 52:4 | 1.1 |
| 50:2 | 2.3 | 54:1 | 20.4 |
| 50:3 | 0.8 | 54:2 | 11.2 |
| 52:1 | 30.7 | 54:3 | 4.9 |
| 52:2 | 10.1 | 54:4 | 1.5 |
| 52:3 | 3.0 | 54:5 | 0.7 |

[a] From Hites (382).
[b] (Carbon number):(number of double bonds).

are necessary to determine its quantitative accuracy. The use of the newer field ionization and chemical ionization techniques, which yield more prominent parent ions and simpler spectra (258a,772a), might also prove advantageous for such quantitative work.

Another approach to the use of mass spectrometry for triglyceride analysis would be tandem operation of a gas chromatograph and a mass spectrometer. The extreme sensitivity of mass spectrometry would make it ideal for identifying the fatty acid composition of each triglyceride peak as it was eluted from the GLC column. However, all triglyceride mass spectrometry to date has been by direct introduction of the sample into the heated ionization chamber. The coupling of a gas chromatograph with a mass spectrometer necessitates an effective carrier gas separator which will work with high molecular weight triglycerides without causing them to condense and produce undesirable "memory effects" (776). Tandem gas chromatography/mass spectrometry has been successfully used for $C_{32}$–$C_{36}$ diglyceride derivatives (47,394), so presumably triglycerides up to $C_{36}$ could be similarly analyzed. Further investigation is necessary, however, to find effective carrier gas separators for the usual $C_{48}$–$C_{54}$ triglycerides.

## C. Derived Diglycerides

The mass spectra of diglycerides have been studied in a number of laboratories (47,394,667,709). Best results have been obtained using the trimethylsilyl ether (TMS) derivatives, which yield different fragmentation patterns for 1,2- and 1,3-diglycerides as shown in Fig. 11-3. [RCO]⁺, [RCO + 74]⁺, [RCOO + 74]⁺, [M − RCOOH]⁺, [M − RCOO]⁺, and [M]⁺ appear in the spectra of both isomers, analogous to triglyceride frag-

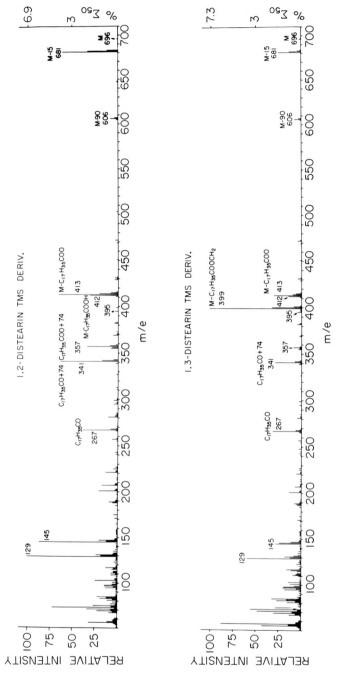

Fig. 11-3. Mass spectra of 1,2-distearin and 1,3-distearin as their trimethylsilyl ether (TMS) derivatives. *Operating conditions:* LKB Model 9000 mass spectrometer operated at 70 eV, direct introduction of sample into the ionization chamber. From Horning *et al.* (394).

mentation. $[M-15]^+$ and $[M-90]^+$ ions are also present due to the —$OSi(CH_3)_3$ group. The 1,3-diglyceride isomer is readily distinguished, however, by the appearance of a prominent $[M-RCOOCH_2]^+$ fragment in its spectrum. Diacid diglycerides produce a corresponding mixture of ions due to the two different acyl groups present. Positional isomers of 1,2-diglycerides are characterized by the preferential loss of the acyloxy group from the 2-position. With $\beta$-StP–, for example, the $[M-C_{15}H_{31}CO]^+$ ion is twice as numerous as $[M-C_{17}H_{35}CO]^+$; while the reverse is true for $\beta$-PSt– (47).

The analysis of individual molecular species in diglyceride mixtures using tandem gas chromatography/mass spectrometry has been investigated by Barber et al. (47) and Horning and co-workers (394); but the technique has not proven as useful as had been expected. GLC peaks containing one component are easily identified and quantitated, and two diglyceride species in the same peak can be identified if they are major components (394). However, the quantitative interpretation of mass spectra from mixtures of diglycerides has not been attempted. Another problem in tandem GLC/MS operation is the tendency of spectra from isomeric mixed-acid 1,2-diglycerides (i.e., $\beta$-StP– and $\beta$-PSt–) to become indistinguishable, due to some unknown isomerization effect (47).

## II. MELTING POINT

The identification of individual triglycerides by melting point determination presents a number of formidable difficulties:

(a) Triglycerides solidify in several different crystalline forms (polymorphism), giving rise to multiple melting points (Table 11-3). An individual triglyceride may exhibit one or more of these melting points depending on its thermal history.

(b) The highest melting points of many common triglycerides are very similar. For instance the most stable ($\beta$) polymorphs of $\beta$-POP and $\beta$-POSt melt at 35.1° and 35.5°, respectively (560).

(c) A melting point close to that of the pure compound does not always ensure high purity. The common $C_{48}$–$C_{54}$ triglycerides often have such similar structures that they can form mixed crystals. For example, adding 10% sn-StPSt to pure StStSt lowers the melting point only 1° (593).

(d) Accurate melting point data must be available before an unknown can be compared. This often requires synthesis of a suitable triglyceride standard.

TABLE 11-3
PHYSICAL PROPERTIES OF THE THREE MAIN CRYSTAL FORMS FOR SATURATED, MONOACID TRIGLYCERIDES[a]

| Physical property | Crystal form | | |
|---|---|---|---|
| | α | β' | β |
| Melting point | Lowest, nonalternating[b] | Intermediate, non-alternating | Highest, alternating |
| X-Ray diffraction (short spacings) | 4.1 Å | 3.8 and 4.2 Å | 4.6 Å (strongest) |
| Chain packing | Hexagonal | Orthorhombic | Triclinic |
| Density | Least dense | Intermediate | Most dense |
| Infrared spectrum | Characteristic spectrum, single band at 720 cm$^{-1}$ | Characteristic spectrum, doublet at 727 and 719 cm$^{-1}$ | Characteristic spectrum, single band at 717 cm$^{-1}$ |
| NMR spectrum | Considerable proton mobility | Slight proton mobility | Slight proton mobility |

[a] Adapted from Chapman (*147*).
[b] Melting points "alternate" when a plot of melting point vs. carbon number shows separate curves for the even and odd members of an homologous series.

(e) Individual pure triglycerides suitable for melting point determinations are exceedingly difficult to isolate from natural fats, even with modern chromatographic techniques.

These problems make melting point determination far less useful than chromatographic and enzymatic procedures for the identification of triglycerides.

Triglyceride melting points can be accurately determined using only a few milligrams of sample. Suitable techniques include the American Oil Chemists' Society standard capillary melting point (*23*), the "thrust-in" capillary melting point (*263*), a microscope having a Kofler hot stage (*803*), variable temperature X-ray diffraction equipment (*770,859*), and differential thermal analysis/differential scanning calorimetry (*65,712*). If measurements are to be made only on the most stable polymorphic form (the highest melting point), then the sample is crystallized from acetone. Other crystal forms can be produced by quenching the melt, followed by prolonged tempering a few degrees below the melting point of a particular polymorph (*592,595*).

Several compilations of triglyceride melting point data are available in the literature (*41,147,601,691*).

## II. MELTING POINT

A number of early workers (*356,361,590,595,636–640*) used melting point data to identify symmetrical and unsymmetrical isomers in mixed-acid triglycerides. Schlenk (*803*) has suggested that the careful determination of mixed melting points might even be used to identify the configuration of an asymmetric triglyceride (Fig. 11-4). While both these techniques are theoretically possible, their application presents two major difficulties:

(a) For accurate identification, samples must contain only one isomeric form. Triglycerides isolated from natural fats, however, may well be an unequal mixture of two or more isomeric forms.
(b) Interpretation of results is based on temperature differences of only a few degrees, necessitating extremely accurate measurement of melting points and very pure samples.

In view of these problems, the technique cannot be recommended now that pancreatic lipase and stereospecific analysis procedures are available for characterizing such isomers.

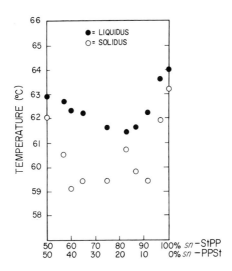

FIG. 11-4. Melting point diagram for mixtures of *sn*-1-stearo-2,3-dipalmitin and *sn*-1,2-dipalmito-3-stearin. *Solidus point* is the temperature at which liquid is first observed; the *liquidus point* is the temperature at which all solid disappears. To permit critical evaluation, no curves are drawn. The presence of a eutectic point between the racemate and an enantiomorph suggests that the two could be distinguished by a mixed melting point procedure if very pure samples were available. Melting points were measured on acetone-crystallized triglyceride using a microscope equipped with a Kofler hot stage heated at 0.2°/minute. From Schlenk (*803*).

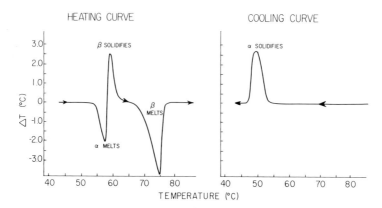

Fig. 11-5. Heating and cooling curves for differential thermal analysis of tristearin. *Operating conditions:* 0.6°/minute heating rate, 0.6°/minute cooling rate, DTA apparatus constructed by authors. Redrawn from Perron et al. *(713).*

## III. DIFFERENTIAL COOLING CURVES

Conventional differential thermal analysis (DTA) and differential scanning calorimetry (DSC) heating curves for triglycerides show multiple peaks due to their polymorphic behavior (Fig. 11-5, left). Attempts to correlate the composition of triglyceride mixtures with the peaks in their DTA/DSC curves have generally been unsuccessful *(306).* Melting and

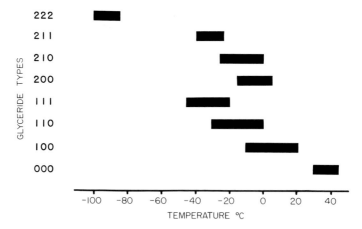

Fig. 11-6. Temperature ranges for the crystallization of several types of triglycerides using differential thermal analysis with rapid cooling. *Operating conditions:* sample size, ~40 mg; cooling rate, 6°/minute; DTA apparatus constructed by authors. From Berger and Akehurst *(69).*

## III. DIFFERENTIAL COOLING CURVES

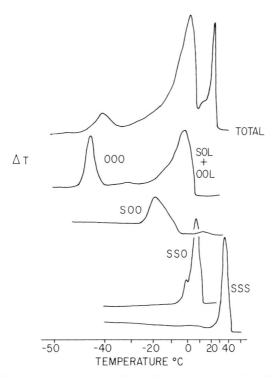

FIG. 11-7. Differential thermal analysis cooling curves for palm oil and its various triglyceride fractions separated by silver ion thin-layer chromatography. *Operating conditions:* sample size, ~40 mg; cooling rate, 6°/minute; DTA apparatus constructed by authors. From Berger and Akehurst (69).

solidification peaks for the various components overlap and cannot be readily differentiated.

On the other hand, DTA/DSC cooling curves are quite simple (Fig. 11-5, right). Rapidly cooled triglycerides crystallize solely in the α-phase, producing only a single exothermic peak. A novel procedure for using DTA/DSC cooling curves for the analysis of triglyceride mixtures has been developed by Berger and Akehurst (69). They report that the peaks in a DTA cooling curve correspond to specific types of triglycerides (Fig. 11-6). This was confirmed by fractionating palm oil triglycerides by silver ion TLC and showing that each fraction had a cooling curve peak corresponding to a peak in the cooling curve of the total mixture (Fig. 11-7). The precise crystallization temperature for each triglyceride type depends on the cooling rate, but the relative positions of the peaks in the curve are constant. Thus it is possible to obtain a qualitative idea of the triglycerides in a mixture by running a simple DTA/DSC cooling curve. This promising technique merits further development.

## IV. INFRARED SPECTROSCOPY

Triglycerides are transparent to visible and ultraviolet wavelengths but possess many infrared absorption bands that are characteristic of their general structure (Fig. 11-8; Table 11-4). Most long-chain triglycerides have very similar infrared spectra, although the presence of functional groups in the fatty acid chains will add additional bands. Hence individual triglyceride species are more easily identified by chromatographic and enzymatic techniques than from their infrared spectra.

The inability of infrared spectroscopy to differentiate between individual triglycerides makes it a good method for quantitating total triglyceride. The strong band at 1742 cm$^{-1}$ due to carbonyl stretching can be used to

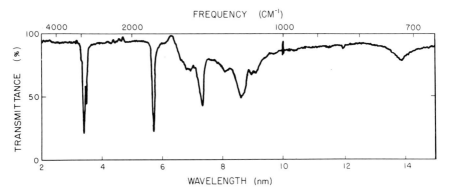

FIG. 11-8. Infrared spectrum of the conidial triglycerides produced by the fungus *Glomerella cingulata*. See Table 11-4 for band assignments. *Operating conditions:* 10 mg of triglyceride/ml CS$_2$, 1 mm path NaCl cells, Perkin-Elmer 221 double-beam infrared spectrophotometer. From Jack (*408*).

TABLE 11-4
ASSIGNMENTS OF BANDS IN INFRARED SPECTRA OF TRIGLYCERIDES[a]

| | |
|---|---|
| C—H stretching (CH$_2$ and CH$_3$) | 3030–2967 cm$^{-1}$ |
| C=O stretching (COOR) | 1751–1733 cm$^{-1}$ |
| C—H bending (CH$_2$/CH$_3$ groups) | 1464–1453 cm$^{-1}$ |
| C—H bend (symmetrical deform of CH$_3$) | 1383–1361 cm$^{-1}$ |
| C—H in plane wagging or rocking of CH$_2$ groups | 1261–1250 cm$^{-1}$ |
| C—O stretching (COOR) | 1179–1166 cm$^{-1}$ |
| CH$_2$ rocking mode | 730–717 cm$^{-1}$ |

[a] From Chapman (*150*).

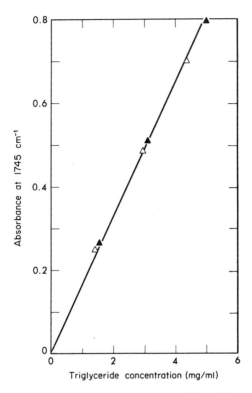

FIG. 11-9. Typical calibration curve measuring triglyceride absorption at 1745 cm$^{-1}$. *Operating conditions:* triolein in CCl$_4$ solution, 1.0 mm path NaCl cells, Perkin-Elmer 1421 dual-grating spectrophotometer at 0.25 mm slit width. From Freeman (*285*).

measure total moles of triglyceride (Fig. 11-9), provided no other ester compounds are present. Suitable spectra can be obtained with as little as 1–10 mg of sample dissolved in CS$_2$, CHCl$_3$, or CCl$_4$. Freeman (*285,286*), and Krell and Hashim (*523*) have published procedures for quantitating total triglycerides in this manner. Neither fatty acid chain length nor degree of unsaturation significantly affects absorptivity at 1742 cm$^{-1}$ (*523,774*), and the amount of absorption is close to a linear function of triglyceride concentration (*285,286,523*). Other frequencies (1100, 1160, 1470, and 2990 cm$^{-1}$) can also be used to measure total triglycerides (*682*), but they are either less sensitive than 1742 cm$^{-1}$ or are influenced by acyl chain length and degree of unsaturation.

Although most triglycerides have similar infrared spectra in solution, there are a few cases in which specific types of triglyceride molecules can be recognized. Acetotriglycerides have a prominent band at 1235 cm$^{-1}$ due

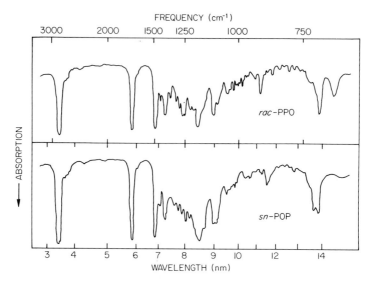

FIG. 11-10. Infrared spectra of the most stable crystalline forms of *rac*-1-oleo-2,3-dipalmitin and *sn*-2-oleo-1,3-dipalmitin. Note differences in 730–715 cm$^{-1}$ region. *Operating conditions:* capillary thickness of sample crystallized between NaCl flats, Grubb-Parsons S3 double-beam prism spectrometer. From Chapman (*146,150*).

to the acetate group (*512*). Saturated triglycerides of different chain length can be distinguished by measuring the ratio of the C—H (2926 cm$^{-1}$) and the ester carbonyl (1742 cm$^{-1}$) absorption bands after suitable calibration (*502*). Positional isomers such as β-PPA and β-PAP (*338*) and β-StStAc and β-StAcSt (*385*) can be differentiated by their spectra in the 1050–1120 cm$^{-1}$ region.

The infrared spectra of solid triglycerides can be used to distinguish between their various polymorphic forms (Table 11-3), and this is occasionally a help in their characterization. Studies by Chapman (*146*) and by Golikova *et al.* (*310*) have correlated the number and frequencies of the 1185–1350 cm$^{-1}$ bands with the structure of various solidified pure triglycerides containing 14:0, 16:0, 18:0, 18:1, 18:2, and 18:3. Infrared spectroscopy of solid samples can also differentiate between certain positional isomers. Since the most stable polymorphic form of β-SOS is β and the most stable polymorph of β-SSO is β′, these two isomers can be distinguished by their infrared spectra after samples have been crystallized from acetone (Fig. 11-10; *146*). This technique has been used to identify 2-palmitooleostearin in lard (*154*), 2-oleopalmitostearin in cocoa butter (*154*), β-PPO in olive oil (*132*), and to distinguish between β-PPSt and β-PStP (*711*).

More detailed discussions of the infrared spectroscopy of triglycerides are to be found in various reviews (*149,150,692*).

## V. X-RAY DIFFRACTION

The different crystalline forms of triglycerides can also be identified by their characteristic X-ray diffraction patterns. If polymorphic behavior is known to be a function of triglyceride structure, then individual molecular species can be identified by this technique.

The experimental techniques for producing X-ray diffraction patterns with triglycerides and compilations of experimental data have been presented in several review articles (*40,147,153,594,601,691*). Only a few milligrams of powdered crystal is required for analysis. The diffraction patterns exhibit two groups of lines corresponding to long and short spacings. The long spacings correspond to the planes formed by the —$CH_3$ or ester groups and are related to the chain length of the fatty acid moieties. The short spacings reflect the cross-sectional packing of the hydrocarbon chains (Table 11-3). Both long and short spacings are useful for identification purposes.

Before the advent of modern enzymatic methods for the positional analysis of triglycerides, positional isomers were sometimes differentiated by their X-ray diffraction patterns. This technique has assisted in the identification of $\beta$-SOS in cocoa butter (*154,591*), 2-palmitooleostearin in lard (*154*), $\beta$-StOSt in kokum butter (*590*), $\beta$-POP in *Sapium sebiferum* fruit coat fat (stillingia tallow) (*595*), and $\beta$-POP in *Caryocar villosum* (piquia) seed fat (*595*). Such triglyceride isomers are more easily identified by pancreatic lipase today, however.

Since the crystal structure of a pure enantiomorphic triglyceride differs from that of its corresponding racemate, pure asymmetric and racemic triglycerides can be distinguished by their X-ray diffraction patterns. Schlenk (*801–803*) has investigated the X-ray diffraction patterns of *sn*-1-lauro-2,3-dipalmitin, *sn*-1,2-dipalmito-3-laurin, and *rac*-1,2-dipalmito-3-laurin crystals in their most stable crystalline form (Fig. 11-11). The *sn*-LaPP and *sn*-PPLa have identical diffraction patterns; while the corresponding racemate, whether synthesized *de novo* or obtained by mixing equal quantities of the two enantiomers, has a markedly different X-ray diffraction pattern. Thus the exact configuration of a triglyceride having two different fatty acids at the primary hydroxyls could be determined using the X-ray diffraction technique. The first step is to synthesize the corresponding racemate and one of the enantiomers. Then the X-ray diffraction patterns of the three compounds are compared, and the unknown triglyceride is

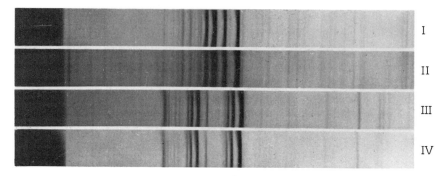

Fig. 11-11. X-Ray diffraction patterns (short spacings) of the most stable crystalline forms of (I) sn-1-lauro-2,3-dipalmitin; (II) sn-1,2-dipalmito-3-laurin; (III) rac-1,2-dipalmito-3-laurin prepared by direct synthesis; and (IV) rac-1,2-dipalmito-3-laurin prepared by mixing equal quantities of I and II. From Schlenk (802).

identified as an enantiomer or a racemate. If an enantiomer is found, the natural and synthetic enantiomers are mixed in equal quantities, and the diffraction pattern of the mixture is determined. If the mixture still has the pattern of an enantiomer, then the natural and synthetic enantiomers have the same configuration. If the mixture has the diffraction pattern of a racemate, however, then the natural enantiomer has the opposite configuration of the synthetic compound. Schlenk has identified the 2-oleopalmitostearin in cocoa butter as a racemate by this method (801–803).

Schlenk's technique for configuration identification presents three difficulties which make it less useful than stereospecific analysis procedures. (i) The isolation of a single triglyceride species from a natural fat is difficult to achieve with present chromatographic techniques. (ii) Two triglyceride standards must be synthesized for each determination. (iii) The behavior of unequal mixtures of enantiomers is unknown. Perhaps a systematic study of the X-ray diffraction patterns of enantiomorphic triglycerides might reveal predictable differences between asymmetric and racemic triglycerides. If this should prove true, then difficulty (ii) would be partially eliminated, and a triglyceride might be classified as asymmetric or racemic solely on the basis of its own diffraction pattern.

## VI. NUCLEAR MAGNETIC RESONANCE

The nuclear magnetic resonance (NMR) spectra of triglycerides exhibit characteristic peaks, each produced by protons of different character. The position of each peak indicates the chemical environment of those protons, and the area under the peak reflects the number of protons with that environment (Fig. 11-12). Differences between the NMR spectra of individual triglycerides are strictly a function of fatty acid compositon (double bonds, chain length, hydroxyl groups, etc.). Although triglycerides of different fatty acid composition could be distinguished by their specific NMR spectra, the technique has been little used for this purpose since triglyceride fatty acid composition is more easily determined by GLC. One exception, however, is molecules containing acetic acid; acetotriglycerides are clearly distinguished by a sharp signal from the —OOCCH$_3$ protons at 7.95$\tau$ (Fig. 11-12; *152,391,512,884*). NMR can also distinguish between the various crystalline forms of triglycerides (Table 11-3; *54,152,155,670*) but not between triglyceride positional isomers under normal conditions (*148*). However, the addition of a chemical shift reagent such as tris-(1,1,1,2,2,3,3-heptafluoro-7,7-dimethyl-4,6-octanedionato)europeum (III) permits unsaturated positional isomers such as β-POP and β-PPO to be distinguished by their different α-methylene signals at 0.8–1.2$\tau$ (Fig. 11-13; *714a*).

It is possible to determine the total triglyceride content of a sample by comparing the area of the 5.8$\tau$ NMR signals (glyceride —CH$_2$O— protons) with suitable calibration standards (*434*).

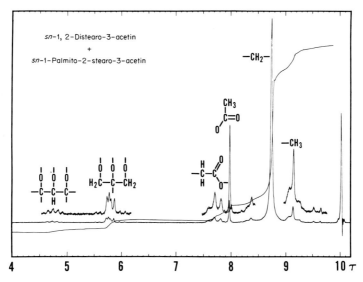

FIG. 11-12. Nuclear magnetic resonance spectra of a mixture of *sn*-1,2-distearo-3-acetin and *sn*-1-palmito-2-stearo-3-acetin from hydrogenated *Euonymus verrucosus* seed triglycerides. *Operating conditions:* sample dissolved in CDCl$_3$ containing tetramethylsilane as an internal standard, Varian 60 MHz NMR spectrometer. From Kleiman *et al.* (*512*).

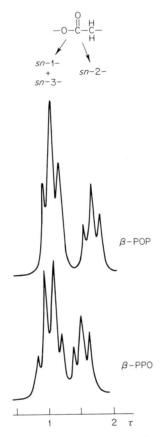

Fig. 11-13. Nuclear magnetic resonance spectra of sn-2-oleo-1,3-dipalmitin and rac-1,2-dipalmito-3-olein in the presence of a chemical shift reagent. *Operating conditions:* $2.85 \times 10^{-5}$ moles of triglyceride and $6.62 \times 10^{-5}$ moles of tris-(1,1,1,2,2,3,3-heptafluoro-7,7-dimethyl-4,6-octanedionato)europeum(III) in 360 µl of $CCl_4$ containing tetramethylsilane as an internal standard. From Pfeffer and Rothbart (*714a*).

## VII. ROTATION OF POLARIZED LIGHT

### A. Triglycerides

When the sn-1- and sn-3-positions of a triglyceride are occupied by different fatty acids, the central carbon atom of the glycerol is asymmetrically substituted, and the resultant molecule should rotate plane-polarized light. In 1939, Baer and Fischer (*36*) synthesized sn-1,2-dilauro-3-palmitin, sn-1,2-distearo-3-laurin, and sn-1,2-dipalmito-3-stearin and found that none of these compounds had a measurable optical rotation at 589 nm (the D line of sodium). It seems certain that the triglycerides did, indeed, contain asymmetric carbon atoms and that no appreciable racemization occurred during synthesis, since (i) they were prepared from optically active monoglycerides, and (ii) aromatic triglycerides made by the same procedure (*849*) *did* show appreciable optical activity. On the other hand, Schlenk (*803*) has synthesized sn-1,2-dibutyro-3-myristin by the same technique and found an $(\alpha)_{589\,nm} = +0.85°$. Thus the optical rotation of asymmetric triglycerides is only measurable when there is a very marked difference between the structures of the acids at the sn-1- and sn-3-positions.

The rotatory dispersion (specific rotation at various wavelengths) of several optically active triglycerides reported in the literature are compiled in Table 11-5. These data lead to the following conclusions about asymmetric, trialiphatic triglycerides:

(a) Optical activity is enhanced when the acids esterified at the 1- and 3-positions are as different as possible. This can be a difference in chain length or the presence of double bonds or branching near the carboxyl group. The difference between 16:0 and 18:0 or 18:1 and 18:0 is insignificant as far as optical activity is concerned, even in the ultraviolet.

(b) Optical activity increases substantially with decreasing wavelength. The largest specific rotations are at ultraviolet frequencies.

(c) Specific rotation varies with the solvent in which the triglyceride is dissolved. The limited evidence available in Table 11-5 does not point to any one solvent as being superior for specific rotation measurements. Solution concentration also affects specific rotation values as Baer and Mahadevan (*38*) have shown for diglycerides.

(d) With *straight chain, saturated* fatty acids, the specific rotation of an optically active triglyceride is positive when the acid in the sn-1-position is shorter than the acid in the sn-3-position and negative when the sn-3-acid is shorter than the sn-1-acid.

It is apparent that the asymmetry of optically active triglycerides containing only normal $C_{16}$ and $C_{18}$ fatty acids is so small that their rotation of polarized light cannot be measured by present techniques, even in the ultraviolet.* Therefore, the use of optical rotation measurements for the

---

* Such asymmetric triglycerides containing expected but unmeasurable optical activity have been termed *cryptoactive* triglycerides by Schlenk (*803*).

TABLE 11-5
Optical Rotatory Dispersion of Asymmetric Triglycerides

| Triglyceride | Solution | Specific rotation (°) at nm | | | | | | | | | | | References |
|---|---|---|---|---|---|---|---|---|---|---|---|---|---|
| | | 589 | 578 | 545–546 | 450 | 436 | 405–408 | 366 | 350 | 334 | 313 | 300–302 | 297 | 250 | |
| sn-1,2-Divalero-3-pivalin | Neat | — | +1.35 | +1.60 | — | +2.70 | +3.20 | +4.12 | — | +4.9 | +5.6 | +6.0 | — | — | 803 |
| | 10% in pyridine | — | +0.20 | +0.22 | — | +0.42 | +0.48 | +0.55 | — | — | — | — | — | — | 803 |
| | 10% in benzene | — | +0.30 | +0.25 | — | +0.15 | +0.05 | +0.05 | — | −0.20 | −0.85 | — | — | — | 803 |
| | 10% in methanol | — | +1.90 | +2.20 | — | +3.20 | +3.70 | +4.70 | — | +6.00 | +7.00 | +7.65 | +8.00 | — | 803 |
| | 10% in dioxane | — | +1.04 | +1.08 | — | +2.00 | +2.56 | +3.48 | — | — | — | — | — | — | 803 |
| | 10% in acetone | — | +1.50 | +1.65 | — | +2.82 | +3.40 | +4.42 | — | — | — | — | — | — | 803 |
| | 10% in ethyl acetate | — | +1.22 | +1.42 | — | +2.40 | +2.85 | +3.80 | — | — | — | — | — | — | 803 |
| | 10% in cyclohexane | — | +1.80 | +2.06 | — | +3.46 | +4.20 | +5.60 | — | +6.8 | +8.0 | +8.8 | — | — | 803 |
| sn-1,2-Distearo-3-acetin | 2% in hexane | −0.6 | — | — | −1.1 | — | — | — | −2.0 | — | — | −2.9 | — | −9.1 | 39, 512 |
| sn-1,2-Dibutyro-3-laurin | Neat | — | +0.05 | +0.05 | — | +0.11 | +0.17 | — | — | — | — | — | — | — | 803 |
| | 10% in benzene | — | +1.10 | +1.40 | — | +2.00 | +2.20 | +2.65 | — | — | — | — | — | — | 803 |
| | 10% in methanol | — | +0.02 | +0.10 | — | +0.16 | +0.12 | +0.14 | — | — | — | — | — | — | 803 |
| sn-1,2-Dibutyro-3-myristin | Neat | — | +0.09 | +0.10 | — | +0.16 | +0.18 | — | — | — | — | — | — | — | 803 |
| | 30% in benzene | — | +0.88 | +0.94 | — | +1.40 | +1.66 | +2.20 | — | — | — | — | — | — | 803 |
| | 34% in methanol | — | +0.13 | +0.15 | — | +0.18 | +0.32 | — | — | — | — | — | — | — | 803 |
| sn-1-Myristo-2,3-dibutyrin | Neat | — | −0.09 | −0.13 | — | −0.17 | — | — | — | — | — | — | — | — | 803 |
| | 30% in pyridine | — | −0.99 | −1.12 | — | −1.96 | −2.36 | −3.08 | — | — | — | — | — | — | 803 |
| | 30% in benzene | — | −0.79 | −0.91 | — | −1.62 | −1.91 | −2.57 | — | −2.74 | — | — | — | — | 803 |
| | 31% in methanol | — | −0.18 | −0.17 | — | −0.24 | −0.24 | −0.29 | — | — | — | — | — | — | 803 |
| sn-1,2-Diisovalero-3-myristin | Neat | — | −0.55 | −0.64 | — | −1.25 | −1.39 | −1.72 | — | — | — | — | — | — | 803 |
| | 34% in pyridine | — | −0.33 | −0.42 | — | −0.67 | −0.90 | −1.16 | — | — | — | — | — | — | 803 |
| | 33% in benzene | — | −0.41 | −0.48 | — | −0.92 | −1.11 | −1.53 | — | −1.95 | — | — | — | — | 803 |
| | 33% in methanol | — | −0.79 | −0.86 | — | −1.51 | −1.88 | −2.50 | — | −3.12 | — | — | — | — | 803 |
| sn-1-Isovalero-2,3-dimyristin | 30% in benzene | — | — | — | — | −0.06 | −0.12 | −0.25 | — | −0.43 | −0.43 | — | — | — | 803 |
| sn-1,2-Disorbo-3-palmitin | 7% in benzene | — | −6.85 | −8.00 | — | — | — | — | — | — | — | — | — | — | 803 |
| sn-1-Sorbo-2,3-dipalmitin | 16% in benzene | — | −5.57 | −9.3 | — | −17.9 | −21.1 | −30.95 | — | — | — | — | — | — | 803 |
| sn-1,2-Dipalmito-3-sorbin | Chloroform | — | — | +3.66 | — | — | — | — | — | — | — | — | — | — | 661 |
| sn-1-Lauro-2,3-dipalmitin | 32% in benzene | — | +0.10 | +0.10 | — | +0.11 | +0.10 | +0.16 | — | +0.16 | +0.20 | +0.23 | +0.28 | — | 803 |
| sn-1,2-Dipalmito-3-laurin | 21% in benzene | 0.0 | −0.06 | −0.07 | — | −0.08 | −0.11 | −0.18 | — | — | — | — | — | — | 36 |
| sn-1,2-Dipalmito-3-stearin | 8% in chloroform | 0.0 | — | — | — | — | — | — | — | — | — | — | — | — | 36 |
| | 5% in pyridine | — | — | — | — | — | — | — | — | — | — | — | — | — | |
| sn-1-Stearo-2-oleo-3-palmitin | 10% in benzene | — | 0.0 | — | — | — | 0.0 | — | — | — | — | 0.0 | — | — | 803 |

direct positional analysis of triglycerides is limited to those species containing a short-chain acid at either the sn-1- or sn-3-position. Such analyses have been used by Kleiman et al. (*511,512*) and Bagby and Smith (*39*) to prove the presence of sn-1,2-($C_{16}$/$C_{18}$)-3-acetotriglycerides in *Euonymus verrucosus* and *Impatiens edgeworthii* seed fats.

## B. Derived Diglycerides

Since most long-chain, asymmetric triglycerides have no measurable optical rotation, methods have been proposed for converting them to diglyceride derivatives which will rotate polarized light. The asymmetry of the original triglyceride can thus be determined indirectly by measuring the specific rotation of the derived diglycerides. Two such methods are outlined in Fig. 11-14.

In the technique of Morris (*661,662*), ~500 mg of a triglyceride such as sn-1-palmito-2,3-diolein is reacted with pancreatic lipase or $C_2H_5MgBr$. The resulting $\alpha,\beta$-diglycerides, a mixture of sn-1-palmito-2-olein and sn-2,3-diolein (~50 mg each), are isolated and converted to their trimethylsilyl ether derivatives to enhance their specific rotations and to prevent any subsequent isomerization. Asymmetry cannot be measured at this point, since the two diglycerides are virtual enantiomers having equal and opposite rotations. But since the two differ in unsaturation, they are easily separated by $Ag^+$ adsorption chromatography. If the original triglyceride were cryptoactive, then the two individual diglyceride ethers will be optically active; and the directions of rotation will indicate the positioning of acids in the initial sample. If the original triglyceride were racemic, then the derived diglyceride ethers will also be racemic. Morris' technique is only applicable to triglycerides which produce easily separated diglyceride ethers. Silver ion adsorption chromatography, silicic acid adsorption chromatography, and liquid–liquid partition chromatography can be used to separate diglyceride ethers that differ in unsaturation, functional group, or chain length. Morris (*662*) has used his technique to demonstrate the presence of asymmetric triglycerides in lard, palm oil, malabar tallow, and cocoa butter.

Coleman (*190*) has proposed a similar technique in which the diglyceride derivatives need not be separated. His method (Fig. 11-14) also begins with pancreatic lipase or $C_2H_5MgBr$ hydrolysis of the original triglyceride and isolation of the resultant $\alpha,\beta$-diglycerides. The diglycerides are then acetylated to stabilize them against any subsequent isomerization. The diglyceride acetates are subjected to von Rudloff oxidation (*937*) with $KMnO_4$/$KIO_4$ to convert the oleic acid to azelaic acid. Coleman assumes that the two products, sn-1-palmito-2-azelao-3-acetin and sn-1-aceto-2,3-

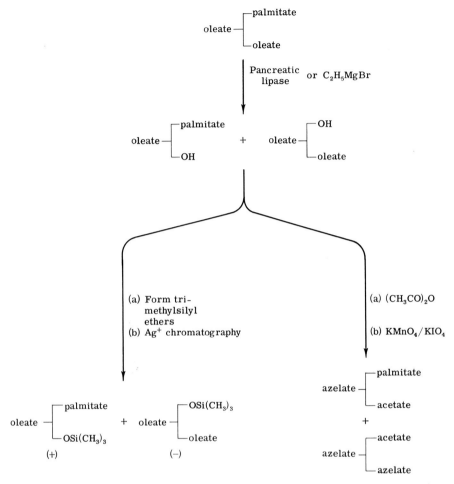

FIG. 11-14. Schematic outline of lipolysis/Ag⁺ chromatography method of Morris (*661,662*) and lipolysis/oxidation technique of Coleman (*190*) for positional analysis of an optically active triglyceride.

diazelain, will have significantly different specific rotations. Thus any rotation of polarized light by the products would indicate that the original triglyceride was cryptoactive, and the direction of rotation would indicate which enantiomorph predominated. The optical rotation of the products could probably be enhanced by converting the free carboxyl groups of the azelaic acid to metal salts. Coleman's method is only applicable to triglycerides having the *sn*-1- and *sn*-3-positions esterified to one saturated and one unsaturated fatty acid, since the resultant diglyceride acetates must differ in the number of azelaic acid residues they contain. To date, Coleman has not published any applications of his technique to the analysis of synthetic or natural triglycerides.

The choice of diglyceride derivative used in the techniques of Morris and Coleman is important since optical rotation varies widely with the derivative selected. Table 11-6 lists the specific rotations of several derivatives of sn-1,2- and sn-2,3-dipalmitin. Of those listed, the sorbate ester and the palmityl, benzyl, trimethylsilyl, and trityl ethers have the largest rotations; and specific rotation measurements would be more accurate with these compounds. Morris (662) used trimethylsilyl ethers in his work since this derivative is so easy to prepare. The derivative used in the Coleman lipolysis/oxidation technique must obviously be stable to $KMnO_4/KIO_4$ oxidation.

TABLE 11-6
Specific Rotations of Various Derivatives of sn-1,2(2,3)-Diglycerides

| Compound | Solvent | Concentration (g/100 ml) | Specific rotation (°) at nm | | | | Reference |
|---|---|---|---|---|---|---|---|
| | | | 589 | 545–546 | 450 | 350 | |
| 1,2-Dipalmitoyl-sn-glycerol | Chloroform | 0.75 | −2.9 | — | −4.6 | −9.0 | 976a |
| 1,2-Distearoyl-3-acetoyl-sn-glycerol | Hexane | 2.28 | −0.6 | — | −1.1 | −2.0 | 39, 512 |
| 1,2-Dipalmitoyl-3-sorboyl-sn-glycerol | Chloroform | — | — | +3.7 | — | — | 661 |
| 1-Sorboyl-2,3-dipalmitoyl-sn-glycerol | Benzene | 16.1 | — | −9.3 | — | −31.0[a] | 803 |
| 1,2-Dipalmitoyl-3-p-nitrobenzoyl-sn-glycerol | Chloroform | 11.9 | −1.6 | — | — | — | 849 |
| 1-Hexadecyl-2,3-dipalmitoyl-sn-glycerol | Chloroform | 10.0 | −4.0 | — | — | — | 561 |
| 1,2-Dipalmitoyl-3-benzyl-sn-glycerol | Chloroform | 8.53 | +6.3 | — | — | — | 849 |
| 1,2-Dipalmitoyl-3-trimethylsilyl-sn-glycerol | Chloroform | 3.4 | — | +4.7 | — | — | 662 |
| 1,2-Dipalmitoyl-3-(tetrahydropyran-2-yl)-sn-glycerol | Dioxane | 1.12 | +2.5 | — | +4.3 | +7.5 | 976a |
| 1-Trityl-2,3-dipalmitoyl-sn-glycerol | Chloroform | 2.0 | −6.9 | — | −13.0 | −24.5 | 976b |

[a] Measured at 366 nm.

## VIII. PIEZOELECTRIC EFFECT

Crystals having no center of symmetry become electrically charged when subjected to mechanical pressure in a direction of structural polarity. This *piezoelectric effect* is exhibited by any crystal composed of a single enantiomorph of an asymmetric compound. The phenomenon can be used in the analysis of cryptoactive triglycerides.

Schlenk (*801–803*) has demonstrated that the enantiomorphic or racemic nature of a triglyceride can be determined in the piezoelectrometer developed by Bergmann (*70*). With this instrument, several small crystals ($<2$ mg) of a pure triglyceride are placed between two metal plates which vibrate so as to apply and release pressure on the sample. The same plates also serve as electrodes which are connected to a voltmeter to measure the piezo charge on the crystals. The triglyceride powder obtained from acetone crystallization is used for the determination, since single triglyceride crystals are very difficult to obtain. Measurements are made at a temperature more than 20° below the melting point of the sample.

About 10 to 20 measurements are made on each sample, and the position of the material is changed each time so that a new portion of the sample is exposed to the electrodes. The piezoelectric effect is not apparent in all tests since the crystals are placed without orientation, and only a small portion of them may be in the optimum position having the polar axes in the direction of pressure. In addition, the effect may be weakened or cancelled when several crystals are pressed at the same time while their polar axes are in opposite directions. Therefore, numerous measurements are necessary to conclusively demonstrate the presence or absence of a piezoelectric effect in a triglyceride powder. Since single measurements are not reproducible, the test is only qualitative in nature. The piezoelectric effects of the two enantiomers of a triglyceride are identical.

Schlenk (*801–803*) has measured the piezoelectric effect of racemic and enantiomorphic forms of $\beta$-LaPP, $\beta$-PPSt, and $\beta$-POSt. In each case the pure enantiomer exhibited a positive piezoelectric effect, while the racemate showed none. Thus, piezoelectricity measurements can effectively demonstrate the presence or absence of cryptoactivity in a long-chain triglyceride.

## IX. OTHER PHYSICAL CONSTANTS

Other physical constants of liquid triglycerides are occasionally used as confirmatory identification techniques. Such measurements include refrac-

## IX. OTHER PHYSICAL CONSTANTS

### TABLE 11-7
### Physical Properties for Some Pure Triglycerides in the Liquid State

| | Refractive index (40°, $N_{aD}$) | Density (g/ml, 40°) | Critical solution temperature[a] (against $CH_3CN$) | Dielectric constant (40°) | Ultrasonic sound velocity (m/sec, 40°, 2 MHz) |
|---|---|---|---|---|---|
| Triacetin | 1.42332 | 1.1378 | — | 6.904 | 1319.4 |
| Tributyrin | 1.42758 | 1.0138 | — | 5.069 | 1269.9 |
| Trihexanoin | 1.43511 | 0.9648 | — | 4.280 | 1288.9 |
| Trioctanoin | 1.44062 | 0.9373 | — | 3.799 | 1315.0 |
| Tridecanoin | 1.44466 | 0.9203 | 53.0 | 3.480 | 1339.4 |
| Trilaurin | 1.44792 | 0.9085 | 91.5 | 3.281 | 1357.0 |
| Trimyristin | — | — | 120.0 | — | — |
| Tripalmitin | — | — | 145.0 | — | — |
| Tristearin | — | — | 170.0 | — | — |
| Triolein | 1.46220 | 0.8991 | 140.5 | 3.028 | 1397.2 |
| Trilinolein | 1.47120 | 0.9130 | 113.5 | 3.353 | 1407.7 |
| Trilinolenin | — | — | 88.0 | — | — |
| Triolein | 1.46220 | 0.8991 | 140.5 | 3.028 | 1397.2 |
| Trielaidin | 1.46124 | 0.8961 | 145.0 | 2.980 | 1388.4 |
| Tripetroselinin | — | — | 142.0 | — | — |
| rac-1-Palmito-2-oleo-3-stearin | 1.45612 | 0.8932 | 152.0 | 2.975 | 1392.4 |
| rac-1-Palmito-2-stearo-3-olein | 1.45619 | 0.8932 | 152.0 | 2.970 | 1393.1 |
| rac-1-Oleo-2-palmito-3-stearin | 1.45632 | 0.8941 | — | 2.978 | 1393.9 |
| Relative sensitivity to $-CH_2-$ group[b] | 11 | 4 | 13 | 7 | 6 |
| Relative sensitivity to $-CH=CH-$ group[c] | 60 | 9 | 18 | 22 | 7 |
| Sample size (mg) | 10–200 | 100–500 | 0.5–2 | | 5000 |
| References | *315, 807* | *315* | *807, 808* | *317* | *316, 404* |

[a] Critical solution temperature measurements have not been widely used in organic chemistry and may not be familiar to some readers. The mutual solubility of two liquids which are not miscible in all proportions is essentially a function of temperature. Raising the temperature usually increases the solubility, often reaching a point at which the two components become miscible in all proportions. This point, known as the upper *critical solution temperature*, can be used to characterize triglycerides. Fischer (*271,272*) and Schmid et al. (*806,808*) have described a convenient microtechnique in which 0.5–2.0 mg of triglyceride and a like amount of solvent are sealed together in a capillary tube and heated on a Kofler hot stage under a microscope. The temperature at which the meniscus between the sample and the test solvent disappears on heating or reappears on cooling is recorded as the critical solution temperature.

[b] Relative sensitivity to $-CH_2-$ group $= \dfrac{X_{\text{trilaurin}} - X_{\text{tridecanoin}}}{6 \left[ \begin{array}{c} \text{reproducibility in the} \\ \text{measurement of } X \end{array} \right]}$

Reproducibility of the measurement is assumed to be 5 units in the last decimal place.

$(X_{\text{trilaurin}} - X_{\text{tridecanoin}})$ rather than the more logical $(X_{\text{tristearin}} - X_{\text{tripalmitin}})$ is used to measure relative sensitivity to $-CH_2-$ group differences, since comparable data on StStSt and PPP are not available in all cases.

[c] Relative sensitivity to $-CH=CH-$ group $= \dfrac{X_{\text{trilinolein}} - X_{\text{triolein}}}{3 \left[ \begin{array}{c} \text{reproducibility in the} \\ \text{measurement of } X \end{array} \right]}$

Reproducibility of the measurement is assumed to be 5 units in the last decimal place.

tive index, density, critical solution temperature, dielectric constant, and ultrasonic sound velocity. It is convenient to consider these tests as a group, since they are very similar in their ability to distinguish between triglycerides of different molecular weight or unsaturation.

Selected physical constants for a number of pure triglycerides are listed in Table 11-7. The analytical usefulness of the various constants can be compared by evaluating the relative sensitivity of each measurement to changes in triglyceride molecular weight or unsaturation (see footnotes *a* and *b* of Table 11-7). Refractive index and critical solution temperature are the measurements most sensitive to differences in triglyceride molecular weight, while the refractive index is the constant most sensitive to differences in triglyceride unsaturation. Obviously the most generally useful physical constant for triglyceride characterization is refractive index. It can clearly distinguish between molecules such as β-POP and β-POSt *(315)* or between OOO and OOL. Thus refractomers are widely used to follow the course of hydrogenation with vegetable oils. The configuration of a double bond has some influence on the value of physical constants, but triglyceride positional isomers can not be distinguished by any of the physical properties listed in Table 11-7.

The values for the physical constants of liquid triglycerides generally follow a smooth curve when any homologous series is considered. Hence values for unknown triglycerides can be predicted by interpolation and extrapolation. Several authors have even developed empirical equations and nomographs for predicting the refractive index *(315,883)*, density *(315)*, dielectric constant *(317)*, and ultrasonic sound velocity *(316)* of triglycerides containing straight-chain $C_{10}$–$C_{22}$ saturated acids and oleic/linoleic/linolenic unsaturation.

# 12

# DISTRIBUTION OF FATTY ACIDS IN NATURAL TRIGLYCERIDE MIXTURES

Since the detailed analysis of component triglycerides in natural fats requires a lengthy experimental procedure, many workers have discussed the possibility of a theoretical approach to estimating triglyceride composition. This approach assumes that the compositions of natural triglyceride mixtures are the net result of (i) the fatty acids available for triglyceride biosynthesis and (ii) the substrate specificities of the enzymes that construct the triglycerides. Of course, the fatty acid composition of natural triglycerides varies widely with species, tissue, diet, and environment; but enzyme specificities are often identical for large groups of organisms (pancreatic lipase in mammals, for example). If the enzymatic specificities governing triglyceride construction could be defined, then it might be possible to estimate the triglyceride composition of a natural fat from its fatty acid composition alone, and lengthy analytical procedures could be minimized or avoided.

This problem has been approached in two different ways. On the one hand, analytical chemists have accumulated considerable data on the triglyceride composition of natural fats and have tried to discern common patterns of acyl group distribution. On the other hand, biochemists have developed a tentative and, as yet, incomplete idea of the biosynthesis of triglycerides and the specificities of the enzymes involved. The analytical chemists have achieved considerable success in developing empirical formulas for approximating the positional distribution of fatty acids and the component triglycerides of seed fats; and this accomplishment apparently demonstrates the validity of their approach. However, the fatty acid distribution hypotheses have not proven as applicable to triglyceride mixtures of animal origin where fatty acids originate from both exogenous and endogenous sources.

Spirited controversies and extravagant claims by the proponents of various fatty acid distribution hypotheses have caused many lipid chemists to regard all distribution hypotheses with suspicion. This is unfortunate, since the hypotheses definitely serve a useful purpose when a *quick estimate* of the triglyceride composition of a natural fat is required. Experimental analytical procedures will always remain the most accurate way for determining triglyceride composition; but fatty acid distribution hypotheses provide handy guidelines and rules-of-thumb which are useful in planning experiments and in minimizing experimental work.

This chapter discusses three approaches to estimating the distribution of fatty acids in natural triglyceride mixtures: (i) positional distribution patterns, (ii) triglyceride composition patterns, and (iii) biosyntheses of triglycerides.

## I. POSITIONAL DISTRIBUTION PATTERNS

Analyses of natural fat triglyceride mixtures by pancreatic lipase and by various stereospecific analysis procedures have revealed a number of regular patterns in the distribution of fatty acids between the *sn*-1-, *sn*-2-, and *sn*-3-positions of the total triglycerides. These positional distribution patterns are distinctly different in plant and animal triglycerides.

### A. Plant Triglycerides

#### 1. Palmitic, Stearic, $C_{20}$, $C_{22}$, and $C_{24}$ Acids

Palmitic acid is esterified almost exclusively at the combined 1,3-positions of seed triglycerides. Mattson and Volpenhein (*618,619,623*) first demonstrated this by examining several dozen seed fats by pancreatic lipase hydrolysis, and other workers (*324*) have confirmed these findings. When the Mattson-Volpenhein data are used to plot mole percent 16:0 at the 1,3- or 2-positions versus mole percent 16:0 in the total triglycerides, the graph shown in Fig. 12-1 is obtained. It follows that the palmitic acid content of the combined 1,3-positions can be estimated by the empirical formula:

$$\% \ 16{:}0 \text{ in } 1{,}3\text{-positions} = 1.47 \ x \qquad \textit{Seed Fats}$$

## I. POSITIONAL DISTRIBUTION PATTERNS

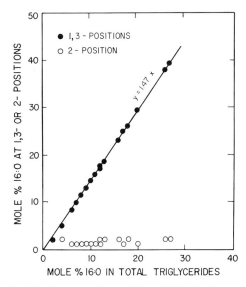

Fig. 12-1. Relationship between the positional distribution of palmitic acid and the total 16:0 content of seed triglycerides. Plotted from the data of Mattson and Volpenhein (623).

where $x$ is mole percent 16:0 in the total triglycerides and $0 < x < 30$. A small amount (1–2%) of 16:0 is found at the 2-position of some seed fats (causing the constant in the above equation to be 1.47 instead of 1.50); but since this amount is close to the experimental error in the pancreatic lipase technique (Chapter 9, Section II,A,3), the precise amount of 16:0 in the 2-position is still uncertain. It is not yet clear how 16:0 is divided between the sn-1- and sn-3-positions in seed triglycerides. Stereospecific analyses by Brockerhoff and Yurkowski (122) have shown that the 16:0 contents of the sn-1- and sn-3-positions are approximately equal in rapeseed oil, soybean oil, and cocoa butter, but slightly different in peanut, linseed, corn, and olive oils. Stereospecific analysis of maize seed triglycerides from 12 genotypes of varying fatty acid composition (212) has indicated a regular positional distribution for the saturated acids (16:0 + 18:0) between the three positions of the glycerol for this one species (see Fig. 12-5).

Stearic, arachidic, behenic, and lignoceric acids have not been studied as extensively, but available data (619,623) indicate that they are esterified almost totally at the combined 1,3-positions in the same manner as palmitic acid.

Erucic acid is also esterified almost exclusively at the combined 1,3-positions of seed triglycerides (32,320,573,619). Typical pancreatic lipase

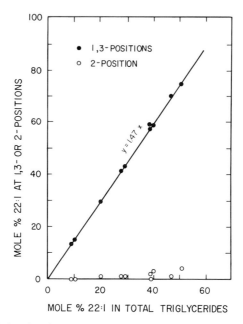

FIG. 12-2. Relationship between the positional distribution of erucic acid and the total 22:1 content of seed triglycerides. Plotted from the data of Mattson and Volpenhein (*619*).

analyses by Mattson and Volpenheim (*619*) on various erucic acid seed fats are plotted in Fig. 12-2. The same empirical relationship found for 16:0 also fits the positional distribution data for erucic acid:

$$\% \text{ 22:1 in 1,3-positions} = 1.47\,x \qquad \textit{Seed Fats}$$

where $x$ is mole % 22:1 in the total triglycerides and $0 < x < 50$. A small amount of 22:1 is found in the 2-position, but this may or may not be significant (see above). A single stereospecific analysis of rapeseed triglycerides (*122*) showed 35% 22:1 at the *sn*-1-position and 51% 22:1 at the *sn*-3-position; no other erucic acid fats have been examined by this technique.

Eicosenoic and tetracosenoic acids have the same positional distribution as erucic acid according to the limited evidence now available (*32,320,571,573,619*).

## 2. *Oleic, Linoleic, and Linolenic Acids*

The positional distribution of oleic, linoleic, and linolenic acids in seed triglycerides is the subject of continuing research. Trends have been recognized, but clearcut patterns applicable to all species have not yet been defined. Gunstone (*323*) and Mattson and Volpenhein (*623*) first suggested that 18:1, 18:2, and 18:3 are randomly distributed among the free hydroxy groups remaining after 16:0, 18:0, and $C_{20}$, $C_{22}$, and $C_{24}$ acids are esterified at the 1,3-positions. Further study of lipolysis results by Gunstone *et al.* (*324,325*) and Mattson and Volpenhein (*623*), however, has indicated that among the $C_{18}$ unsaturated acids, oleic and linolenic show a slight preference for the 1,3-positions while linoleic shows a slight preference for the 2-position. This was taken into account by Evans *et al.* (*257*) when they proposed the following three rules for estimating the positional distribution of 18:1, 18:2, and 18:3 in seed triglycerides:

(a) Saturated acids and those with chain lengths greater than 18 carbons are first distributed equally at the 1- and 3-positions.
(b) Oleic and linolenic acids are then distributed equally and randomly on the unfilled 1-, 2-, and 3-positions, with any excess from the 1- and 3-positions being added to the 2-position.
(c) All remaining positions are filled by linoleic acid.

Table 12-1 compares the positional distributions of fatty acids in seven seed fats as estimated by the Evans hypothesis with that determined experimentally by lipolysis. Agreement between the two values is fairly good, except for Cruciferae seed fats such as rapeseed oil, where large differences between the calculated and experimental values for linoleic acid occur (*257;* Table 12-1).

A different type of approach has been taken by Litchfield (*573*) using graphical correlation studies on lipolysis data from 24 species of Cruciferae seed triglycerides. When the ratio

$$\frac{\%\ 18:1\ \text{in 2-position}}{\%\ 18:1\ \text{in total triglycerides}}$$

for each species is plotted versus the content of Category I acids (16:0, 18:0, plus all $C_{20}$, $C_{22}$, and $C_{24}$ acids) in the total triglycerides, a smooth curve is obtained (Fig. 12-3). Application of suitable statistical procedures yields a best-fitting curve, from which an equation expressing the % 18:1 in the 2-position as a function of the fatty acid composition of the total triglycerides can be derived:

$$\%\ 18:1\ \text{in 2-position} = \frac{x[113 - C_\mathrm{I}]}{108 - 1.39\, C_\mathrm{I}} \quad \begin{array}{l}\textit{Cruciferae}\\ \textit{Species}\\ \textit{Only}\end{array}$$

where $x$ is mole percent 18:1 in the total triglycerides and $C_\mathrm{I}$ is mole percent Category I acids (16:0, 18:0, plus all $C_{20}$, $C_{22}$, and $C_{24}$ acids) in the total triglycerides. The percent 18:1 in the 1,3-positions is then

TABLE 12-1
COMPARISON OF THE POSITIONAL DISTRIBUTION OF FATTY ACIDS IN SEED
TRIGLYCERIDES AS ESTIMATED BY THE EVANS HYPOTHESIS WITH
EXPERIMENTAL RESULTS OBTAINED BY LIPASE HYDROLYSIS

| Source | Fatty acid | Composition (mole %) | | | | |
|---|---|---|---|---|---|---|
| | | Total triglycerides | 2-Position | | 1,3-Positions | |
| | | Found | Found | Evans hypothesis | Found | Evans hypothesis |
| Sesame seed | 16:0 | 10 | 1 | — | 14 | 15 |
| (623) | 18:0 | 6 | 1 | — | 8 | 9 |
| | 18:1 | 40 | 43 | 40 | 39 | 40 |
| | 18:2 | 44 | 55 | 60 | 39 | 36 |
| Cashew nut | 16:0 | 13 | 2 | — | 19 | 20 |
| (623) | 18:0 | 9 | — | — | 13 | 13 |
| | 18:1 | 60 | 56 | 60 | 62 | 60 |
| | 18:2 | 18 | 42 | 40 | 6 | 7 |
| Cottonseed | 16:0 | 26 | 2 | — | 38 | 39 |
| (623) | 18:0 | 2 | 1 | — | 2 | 3 |
| | 18:1 | 17 | 20 | 17 | 16 | 17 |
| | 18:2 | 55 | 77 | 83 | 44 | 41 |
| Linseed | 16:0 | 6 | 1 | — | 9 | 9 |
| (623) | 18:0 | 4 | — | — | 6 | 6 |
| | 18:1 | 22 | 27 | 22 | 20 | 22 |
| | 18:2 | 15 | 23 | 25 | 11 | 10 |
| | 18:3 | 52 | 48 | 52 | 54 | 52 |
| Wild rose seed | 16:0 | 4 | — | — | 6 | 6 |
| (328) | 18:0 | 1 | — | — | 1 | 1 |
| | 18:1 | 10 | 11 | 10 | 10 | 10 |
| | 18:2 | 49 | 57 | 53 | 45 | 47 |
| | 18:3 | 36 | 32 | 36 | 38 | 36 |
| Soybean | 16:0 | 12 | 1 | — | 17 | 18 |
| (623) | 18:0 | 4 | — | — | 6 | 6 |
| | 18:1 | 25 | 22 | 25 | 27 | 25 |
| | 18:2 | 51 | 69 | 67 | 42 | 43 |
| | 18:3 | 8 | 8 | 8 | 8 | 8 |
| Rapeseed | 16:0 | 4 | 1 | — | 5 | 6 |
| (619) | 18:0 | 1 | 1 | — | 1 | 1 |
| | 18:1 | 17 | 32 | 29 | 10 | 11 |
| | 18:2 | 17 | 41 | 51 | 5 | — |
| | 18:3 | 11 | 22 | 19 | 5 | 7 |
| | 20:1 | 10 | — | — | 15 | 15 |
| | 22:1 | 40 | 3 | — | 59 | 60 |

I. POSITIONAL DISTRIBUTION PATTERNS 239

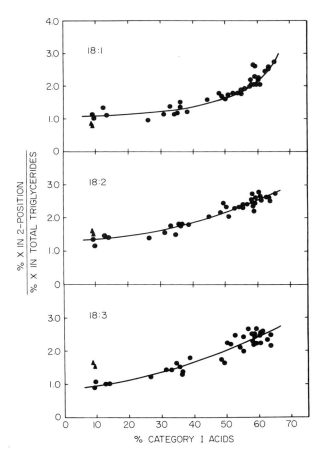

FIG. 12-3. Correlation of the lipolysis results for oleic, linoleic, and linolenic acids [expressed as (% X in 2-position)/(% X in total triglycerides)] with the content of Category I acids (i.e., 16:0, 18:0, and all $C_{20}$, $C_{22}$, and $C_{24}$ acids) in Cruciferae seed triglycerides. ●, results for individual species; ▲, zero erucic variety of *Brassica napus*. From Litchfield (573).

readily calculated by difference (Chapter 9, Section II,A,5). Similar treatment of lipolysis data for linoleic and linolenic acids in Cruciferae seed triglycerides (Fig. 12-3) produces analogous relationships for estimating the positional distribution of 18:2 and 18:3:

% 18:2 in 2-position =            *Cruciferae*
  $y [1.37 - 0.000467\ C_I + 0.000337\ C_I^2]$    *Species*
% 18:3 in 2-position =            *Only*
  $z [0.854 + 0.00941\ C_I + 0.000288\ C_I^2]$

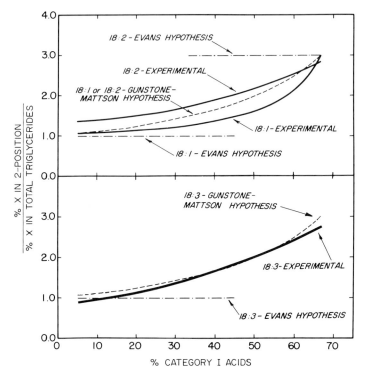

FIG. 12-4. Graphical comparison of various hypotheses for describing the positional distribution of oleic, linoleic, and linolenic acids in Cruciferae seed triglycerides. Compare with Fig. 12-3. From Litchfield (573).

where $y$ is mole percent 18:2 in the total triglycerides and $z$ is mole percent 18:3 in the total triglycerides. Comparison of calculated and experimental values for the percent 18:1, 18:2, or 18:3 at the 2-position of Cruciferae seed triglycerides showed a standard error of 2.8% absolute or less in all three cases (573). Graphical comparison of the Litchfield correlation formulas with the Gunstone-Mattson and Evans hypotheses (Fig. 12-4) indicates that the correlation relationships are more accurate for estimating the positional distribution of 18:1, 18:2, and 18:3 in Cruciferae seed triglycerides. Attempts to apply these correlation formulas to non-Cruciferae seed fats have shown good agreement in many cases, but also some large discrepancies, especially in fats containing a high level of a single unsaturated acid where a Gunstone-Mattson distribution pattern is most likely. It seems probable, therefore, that the positional distributions of 18:1, 18:2, and 18:3 in seed triglycerides are somewhat interdependent, and the Cruciferae-based formulas need further refinement before they can be applied to other plant families.

Thus the best available methods for estimating the $\alpha,\beta$-positional distributions of oleic, linoleic, and linolenic acids are the Evans hypothesis for non-Cruciferae species and the Litchfield correlation formulas for the Cruciferae. Both methods, however, provide only estimates which may deviate slightly from the actual fatty acid distribution.

Too few seed fats have been examined by stereospecific analysis (*122,168,212,718,945*) to establish any general patterns for the distribution of oleic, linoleic, and linolenic acids between the *sn*-1- and *sn*-3-positions. De la Roche *et al.* (*212,945*) have analyzed maize seed triglyceride from 12 genotypes of widely varying fatty acid composition and found regular distribution patterns for 18:1 and 18:2 at all three positions (Fig. 12-5). However, these relationships are only valid for a single species.

### 3. Other Acids

Many other acids occur in a limited number of seed fats, but only in a few cases are sufficient data available to establish definite patterns of positional distribution.

Litchfield (*570,572*) has examined the positional distribution of decanoic, lauric, and myristic acids by pancreatic lipase hydrolysis of seed triglycerides from 24 species. His results, summarized in Table 12-2, indi-

TABLE 12-2

POSITIONAL DISTRIBUTION PATTERNS FOR DECANOIC, LAURIC, AND MYRISTIC ACIDS IN SEED TRIGLYCERIDES[a]

| Taxonomic group<br>Class<br>  Family<br>    Subfamily | Species examined | Preferred position in triglycerides | | |
|---|---|---|---|---|
| | | 10:0 | 12:0 | 14:0 |
| Dicotyledons | | | | |
|   Lauraceae | 8 | 2 | 1, 3 | 1, 3 |
|   Myristicaceae<br>  Salvadoraceae<br>  Ixonanthaceae | 4 | — | 1, 3 | 2 |
| Monocotyledons | | | | |
|   Palmae | | | | |
|     Cocoideae | 4 | 1, 3 | 2 | Varies |
|     Coryphoideae<br>    Phoenicoideae<br>    Aricoideae | 8 | — | 1, 3 | 1, 3 |

[a] From Litchfield (*572*).

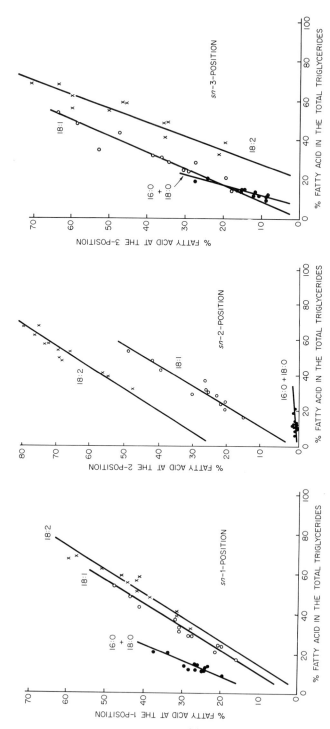

FIG. 12-5. Linear relationships between the positional distributions of saturated (16:0 + 18:0), oleic, and linoleic acids and the amount of that acid found in the total triglycerides of maize seed. Correlations are based on samples from 12 genotypes of widely varying fatty acid composition. Redrawn from de la Roche et al. (212).

cate that the positional distributions of 10:0, 12:0, and 14:0 vary between major taxonomic groups but are consistent within each individual group. The patterns indicated represent only trends, however, and it has not been possible to correlate these results on any mathematical basis.

Lipolysis studies (*194,717,875,878*) have revealed that vernolic acid (*cis*-12,13-epoxy-*cis*-9-octadecenoic acid) is preferentially esterified at the 2-position in seed triglycerides. Two very unusual acids, acetic in *Euonymus verrucosus* (*512*) and *Impatiens edgeworthii* (*39*) and an allenic estolide acid in *Sapium sebiferum* (*168*), are found exclusively at the *sn*-3-position in seed triglycerides. For further information on the positional distribution of unusual acids occurring in only a few isolated species, the reader is referred to the discussions of Gunstone *et al.* (*194,324*).

## B. Animal Triglycerides

Three factors have made it difficult to recognize positional distribution patterns in animal triglycerides: (i) dietary fat can alter the triglyceride composition of animal depot fats, (ii) positional distribution patterns are not the same for all types of animals, and (iii) positional distribution patterns probably vary between different body tissues in the same animal. Because of these difficulties, the positional distribution of fatty acids in animal triglycerides is poorly understood at the present time. Only a few definite patterns have been recognized.

### 1. $C_{20}$ and $C_{22}$ Acids

Docosahexaenoic acid (22:6-4c,7c,10c,13c,16c,19c) is regularly distributed in the triglycerides of certain aquatic animals. Litchfield (*568*) has examined the positional distribution of 22:6 in 29 fish, invertebrate, and turtle fats and found that the following empirical formulas define the experimental results:

$$\% \ 22{:}6 \text{ in } sn\text{-1-position} = 0.28\,x \qquad Fish$$
$$\% \ 22{:}6 \text{ in } sn\text{-2-position} = 2.06\,x \qquad Invertebrates$$
$$\% \ 22{:}6 \text{ in } sn\text{-3-position} = 0.66\,x \qquad Turtle$$

where $x$ is mole percent 22:6 in the total triglycerides and $0 < x < 30$. The correlation between predicted and found values for the *sn*-2-position is shown graphically in Fig. 12-6. Litchfield (*568*) has also examined the positional distribution of 22:6 in marine mammal blubber fats and found

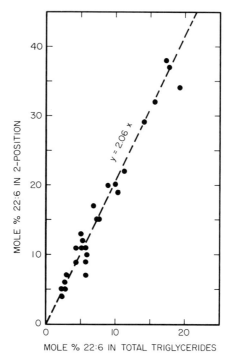

FIG. 12-6. Relationship between the mole percent 22:6 at the *sn*-2-position and the total 22:6 present in fish, invertebrate, and turtle triglycerides. From Litchfield (568).

a regular pattern that can be described by a different set of empirical formulas:

% 22:6 in *sn*-1-position = 0.94 $x$
% 22:6 in *sn*-2-position = 0.22 $x$     *Blubber of*
% 22:6 in *sn*-3-position = 1.84 $x$     *Marine Mammals*

where $x$ is mole percent 22:6 in the total triglycerides and $0 < x < 15$. A comparison of predicted and found values for the *sn*-3-position of marine mammal blubber fats is presented in Fig. 12-7. The positional distribution of 22:6 in nonaquatic animals apparently differs from the two groups mentioned above (568).

Docosapentaenoic acid (22:5-7*c*,10*c*,13*c*,16*c*,19*c*) appears to have the same positional distribution as 22:6 in fish, invertebrate, and turtle triglycerides and in marine mammal blubber triglycerides (568). This remains a tentative conclusion, however, since the 22:5 content of the aquatic animal fats examined never exceeded 7%.

I. POSITIONAL DISTRIBUTION PATTERNS 245

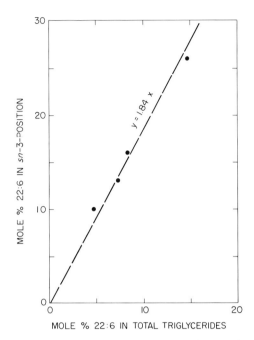

FIG. 12-7. Relationship between the mole percent 22:6 at the *sn*-3-position and the total 22:6 present in the blubber triglycerides of marine mammals. From Litchfield (*574*).

Eicosapentaenoic acid (20:5-*5c,8c,11c,14c,17c*) is preferentially esterified at the 2-position in fish and invertebrate triglycerides but not with the consistent pattern found with 22:6. A plot of mole percent 20:5 in the 2-position versus the amount of 20:5 found in the total triglycerides (Fig. 12-8) shows that all results lie on or above the line $y = x$, but the data points are too scattered to be defined by a single curve (*574*). One can generalize, however, by saying that

$$\% \ 20:5 \text{ in } sn\text{-2-position} \geqq x \qquad \textit{Fish and Invertebrates}$$

where $x$ is mole percent 20:5 in the total triglycerides and $0 < x < 30$. The same $y \geqq x$ relationship also applies to the amount of 20:5 found at the *sn*-3-position in marine mammal blubber triglycerides (*119,120*).

Docosenoic acid (22:1) follows a regular distribution pattern at the *sn*-3-position of all aquatic animal triglycerides (fish, mammals, invertebrates, turtle) examined to date. Litchfield (*574*) has plotted the mole percent 22:1 at the *sn*-3-position versus the amount of 22:1 found in the total triglycerides and obtained a smooth curve (Fig. 12-9) which can be

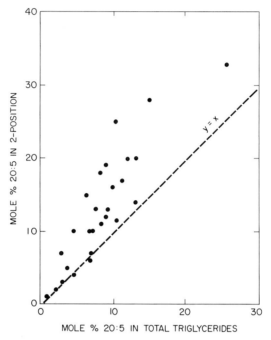

Fig. 12-8. Relationship between the mole percent 20:5 at the *sn*-2-position and the total 20:5 present in fish and invertebrate triglycerides. From Litchfield (*574*).

defined by the equation:

$$\% \ 22:1 \ \text{in} \ sn\text{-3-position} = 0.901 \ x + 0.0525 \ x^2 \qquad \begin{array}{l} Aquatic \\ Animals \end{array}$$

where $x$ is mole percent 22:1 in the total triglycerides and $0 < x < 25$. Curiously, there is no pronounced correlation of the amount of 22:1 found at the *sn*-1- and *sn*-2-positions with the 22:1 content of the triglycerides of aquatic animals.

### 2. Other Acids

Regular positional distribution patterns for other fatty acids in animal fats have not been recognized, but certain general tendencies have been noted. Each acid that is present in the total triglycerides can be found at all three positions. The saturated acid content of the *sn*-2-position is generally higher in animal than in plant triglycerides. Most animal fats are more asymmetric than plant fats, but some bird fats have almost equivalent fatty acid compositions at the *sn*-1- and *sn*-3-positions (*120,121*). Myristic and linoleic acids tend to concentrate in the *sn*-2-posi-

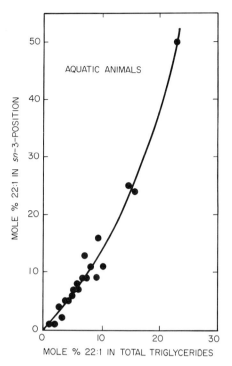

FIG. 12-9. Relationship between the mole percent 22:1 at the $sn$-3-position and the total 22:1 present in aquatic animal triglycerides. From Litchfield (574).

tion, palmitic acid is generally highest at the $sn$-1-position, while stearic acid content is usually lowest at the $sn$-2-position (120,121,628). Pig and peccary fats are unusual, for they have most of the palmitic acid concentrated at the $sn$-2-position (628) and very little 16:0 at the $sn$-3-position (121,170). Butyric and hexanoic acids in bovine milk fat are primarily esterified at the $sn$-3-position (108,724). For further discussion of the positional distribution of fatty acids in animal triglycerides, the reader is referred to the papers of Brockerhoff et al. (112,117,120,121).

Christie and Moore (174) have made a useful approach towards understanding the positional distribution of common fatty acids in animal triglycerides by examining the variations within a single species, the pig. Using stereospecific analysis data on 45 pig adipose tissue fats having a wide range of fatty acid compositions, they were able to demonstrate a good correlation between the fatty acid composition of the total triglycerides and the positional distribution of the four major acids (16:0, 18:0, 18:1, 18:2) within the triglyceride molecules (Fig. 12-10). Similar studies on the same type of tissue from several closely related genera might well

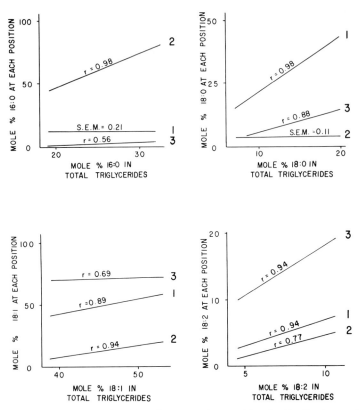

FIG. 12-10. Linear relationships between the positional distributions of 16:0, 18:0, 18:1, and 18:2 and the amount of that acid found in the total triglycerides of pig adipose tissue. Correlations are based on 45 samples from pigs raised under a variety of nutritional conditions. $r$, correlation coefficient; S.E.M., standard error of mean. Redrawn from Christie and Moore (*174*).

reveal common positional distribution patterns for specific groups of animals.

## II. TRIGLYCERIDE COMPOSITION PATTERNS

As quantitative data on the component triglycerides of natural fats have accumulated in the literature, there has been a constant effort to devise empirical mathematical formulas relating triglyceride composition to fatty acid composition. These "fatty acid distribution hypotheses" generally assume that chance plays a key role in the enzymatic construction of triglycerides from fatty acids and glycerol within the framework of certain re-

## II. TRIGLYCERIDE COMPOSITION PATTERNS

stricting conditions. This section describes the many fatty acid distribution hypotheses that have been proposed and discusses their validity in the light of current experimental results.

### A. 1-Random–2-Random–3-Random Hypothesis

This hypothesis, first proposed by Tsuda (*896*) in 1962, assumes that three different pools of fatty acids are separately distributed to the *sn*-1-, *sn*-2-, and *sn*-3-positions of all the glycerol molecules in a fat. Within its respective position, each pool of acids is distributed at random. The amount of each component triglyceride can be calculated from the general equation:

$$\% \, sn\text{-XYZ} = \begin{bmatrix} \text{mole } \% \text{ X at} \\ sn\text{-1-position} \end{bmatrix} \begin{bmatrix} \text{mole } \% \text{ Y at} \\ sn\text{-2-position} \end{bmatrix} \begin{bmatrix} \text{mole } \% \text{ Z at} \\ sn\text{-3-position} \end{bmatrix} (10^{-4})$$

Consider, for example, a fat such as illipe butter (*187*) which is composed of palmitic, stearic, and oleic acids; and let the subscripts 1, 2, and 3 represent the mole percent of an acid found at the *sn*-1-, *sn*-2-, and *sn*-3-positions of the total triglycerides. The amount of monoacid triglyceride PPP predicted by a 1-random–2-random–3-random distribution hypothesis would be calculated by the equation:

$$\% \, \text{PPP} = (P_1)(P_2)(P_3)(10^{-4})$$

The three positional isomers of the diacid triglyceride PPO would be calculated by the equations:

$$\% \, sn\text{-PPO} = (P_1)(P_2)(O_3)(10^{-4})$$
$$\% \, sn\text{-OPP} = (O_1)(P_2)(P_3)(10^{-4})$$
$$\% \, sn\text{-POP} = (P_1)(O_2)(P_3)(10^{-4})$$

If $P_1 \neq P_3$ or $O_1 \neq O_3$, then the enantiomers *sn*-PPO and *sn*-OPP will be present in unequal amounts in a 1-random–2-random–3-random distribution. The six positional isomers of the triacid triglyceride PStO would be calculated with the equations:

$$\% \, sn\text{-PStO} = (P_1)(St_2)(O_3)(10^{-4})$$
$$\% \, sn\text{-OStP} = (O_1)(St_2)(P_3)(10^{-4})$$
$$\% \, sn\text{-OPSt} = (O_1)(P_2)(St_3)(10^{-4})$$
$$\% \, sn\text{-StPO} = (St_1)(P_2)(O_3)(10^{-4})$$
$$\% \, sn\text{-POSt} = (P_1)(O_2)(St_3)(10^{-4})$$
$$\% \, sn\text{-StOP} = (St_1)(O_2)(P_3)(10^{-4})$$

Application of such procedures to all possible permutations and combinations of three acids yields the calculated triglyceride composition for the total fat.

To make a 1-random–2-random–3-random calculation, it is necessary to have stereospecific analysis data on the fatty acid compositions (*expressed in mole percent*) at the *sn*-1-, *sn*-2-, and *sn*-3-positions of the natural fat triglycerides. Alternatively (and less accurately), the positional distribution of fatty acids can sometimes be estimated as outlined in Section I.

## B. 1,3-Random–2-Random Hypothesis

The 1,3-random–2-random distribution hypothesis was proposed independently by Vander Wal (*910*) and Coleman and Fulton (*191*) in 1960–1961. This hypothesis assumes that two different pools of fatty acids are separately and randomly distributed to the 1,3- and 2-positions of all glycerol molecules in a fat. Since the *sn*-1- and *sn*-3-positions are both randomly esterified from the same pool, their fatty acid compositions will be equivalent. In mathematical terms, the 1,3-random–2-random distribution is a special case of the 1-random–2-random–3-random hypothesis where $X_1 = X_3$, $Y_1 = Y_3$, $Z_1 = Z_3$, etc. The amount of each component triglyceride can be calculated from the equation:

$$\% \ sn\text{-XYZ} = \begin{bmatrix} \text{mole \% X at} \\ \text{1,3-positions} \end{bmatrix} \begin{bmatrix} \text{mole \% Y at} \\ \text{2-position} \end{bmatrix} \begin{bmatrix} \text{mole \% Z at} \\ \text{1,3-positions} \end{bmatrix} (10^{-4})$$

Illipe butter containing P, St, and O again serves as a convenient example for the calculation procedure. The amount of monoacid triglyceride PPP predicted by the 1,3-random–2-random hypothesis is calculated from the equation:

$$\% \ \text{PPP} = (P_{1,3})^2 (P_2)(10^{-4})$$

where $P_{1,3} = P_1 = P_3$. The three positional isomers of the diacid triglyceride PPO can be calculated by equations:

$$\% \ \beta\text{-PPO} = sn\text{-PPO} + sn\text{-OPP} = 2(P_{1,3})(P_2)(O_{1,3})(10^{-4})$$
$$\% \ \beta\text{-POP} = (P_{1,3})^2 (O_2)(10^{-4})$$

Since $P_1 = P_3$ and $O_1 = O_3$, enantiomers are present in equal amounts; and it is customary to calculate only the total content of any two enantiomers. The six positional isomers of the triacid triglyceride PStO are calculated from the equations:

$$\% \ \beta\text{-PStO} = sn\text{-PStO} + sn\text{-OStP} = 2(P_{1,3})(St_2)(O_{1,3})(10^{-4})$$
$$\% \ \beta\text{-OPSt} = sn\text{-OPSt} + sn\text{-StPO} = 2(O_{1,3})(P_2)(St_{1,3})(10^{-4})$$
$$\% \ \beta\text{-POSt} = sn\text{-POSt} + sn\text{-StOP} = 2(P_{1,3})(O_2)(St_{1,3})(10^{-4})$$

Calculation of all possible permutations and combinations of the three acids gives the predicted triglyceride composition of the fat.

A 1,3-random–2-random distribution is usually calculated from pancreatic lipase hydrolysis data or, less accurately, from an estimated positional distribution of fatty acids (Section I). Perkins and Hanson (*708*) have published a FORTRAN computer program for perfoming such calculations.

## C. Other Fatty Acid Distribution Hypotheses

In addition to the 1-random–2-random–3-random and 1,3-random–2-random hypotheses, a number of other fatty acid distribution hypotheses for natural fats have been proposed and abandoned over the years. Most of these have been discarded because their description of natural fat triglyceride structure has been shown to be incorrect. Some are seldom used because they only differentiate between saturated and unsaturated acids or because they do not take triglyceride positional isomers into account. The 1,2,3-random, restricted random, and even or widest distribution hypotheses are briefly described here since they are occasionally referred to in the current literature. For a more detailed discussion of these and other older hypotheses, the reader is referred to the 1963 review by Coleman (*186*) or to the original publications.

### 1. 1,2,3-Random Hypothesis

The 1,2,3-random hypotheses was proposed by Longenecker (*584*) in 1941 and elaborated by Mattil and Norris (*614*) in 1947. It assumes that one pool of fatty acids is randomly distributed to all three positions of the glycerol molecules in a fat. Thus the fatty acid compositions of the *sn*-1-, *sn*-2-, and *sn*-3-positions will be equivalent. In mathematical terms, a 1,2,3-random hypothesis is a special case of the 1-random–2-random–3-random hypothesis where $X_1 = X_2 = X_3$, $Y_1 = Y_2 = Y_3$, $Z_1 = Z_2 = Z_3$, etc. The amount of each component triglyceride can be calculated from the equation

$$\% \, sn\text{-XYZ} = \left[\frac{\text{mole } \% \text{ X in}}{\text{total fat}}\right]\left[\frac{\text{mole } \% \text{ Y in}}{\text{total fat}}\right]\left[\frac{\text{mole } \% \text{ Z in}}{\text{total fat}}\right] (10^{-4})$$

Modern techniques for triglyceride positional analysis have shown that the 1,3- and 2-positions always have different fatty acid compositions in natural fats (Section I), thus disproving the 1,2,3-random hypothesis. However, the 1,2,3-random formulas are useful for predicting the composition of synthetic triglyceride mixtures having a completely random esterification of the acyl groups. A FORTRAN computer program for making such calculations has been published by Perkins und Hanson (708).

## 2. Restricted Random Distribution

Kartha (452–454) has proposed that the S and U acids in natural fat triglycerides are distributed according to a 1,2,3-random distribution pattern with the restriction that the SSS content must remain low enough for the fat to remain fluid *in vivo*. Under his restricted random distribution hypothesis, the SSS content of a fat must first be determined experimentally, and then the remaining S and U are assumed to be randomly distributed to form *only* SSU, SUU, and UUU. Kartha (452) has described a proportioning procedure for approximating such a distribution [see Coleman (186) for a clearer explanation], but a more accurate set of computation equations has been derived by Hammond and Jones* (341):

% SSS = determined experimentally
% SSU = $1.5\{100 + S - 2 \,(SSS)$
$\qquad - [10^4 - 3\,S^2 + 200\,S + 4\,S\,(SSS) - 400\,(SSS)]^{1/2}\}$
% SUU = $3\,S - 3\,(SSS) - 2\,(SSU)$
% UUU = $100 + 2\,(SSS) + SSU - 3\,S$

However, differences between results calculated by the two procedures are usually very small.

As first elaborated by Kartha (454), the restricted random distribution hypothesis assumed that isomeric triglycerides were present in equal amounts (i.e., *sn*-SSU = *sn*-SUS = *sn*-USS). More recently, Kartha (456) has proposed two biosynthetic mechanisms that would permit unequal proportions of positional isomers to be formed, but he presents no rules for predicting their relative amounts. The restricted random distribution hypothesis is seldom used today because it does not differentiate between individual fatty acids and because there is no way of calculating positional isomers.

## 3. Even or Widest Distribution

The scheme of even or widest distribution was proposed by Hilditch and his co-workers (77,192,363) to describe the pronounced tendency of fatty acids to form mixed-acid triglycerides in natural fats. This hypothesis holds that individual fatty acids are distributed as widely as possible among the triglyceride molecules of a fat. In its original form, this scheme was intended as a convenient rule-of-thumb without strict mathematical inter-

---

* The original Hammond-Jones equations were expressed on a mole fraction basis, but they have been converted to a mole percent basis for presentation here to achieve uniform nomenclature throughout this chapter.

pretation; and an "even distribution" of S has been defined as follows (*363*):

(a) If S $< 15\%$, S occurs only as SXX.
(b) If S $> 35\%$, S appears at least once in every triglyceride molecule.
(c) If $35\% < S < 65\%$, SSX will be present.
(d) If S $> 70\%$, only SSX and SSS will occur.

Later Hilditch and Meara (*357*) and Dutton *et al.* (*239*) attempted rigorous mathematical interpretation of the hypothesis, and this idea has been termed "widest distribution" (*323*):

(a) If $0\% < S < 33.3\%$, only XXX and SXX are present.
(b) If $33.3\% < S < 66.7\%$, only SXX and SSX are present.
(c) If $66.7\% < S < 100.0\%$, only SSX and SSS are present.

Such widest distribution calculations can only be applied to two-component systems [unless further assumptions are made (*239,256*)] and make no estimate of positional isomers. Hence they are rarely used today.

*4. Minor Fatty Acid Distribution Hypotheses*

A number of other theoretical descriptions of the fatty acid distribution in natural fat triglycerides have appeared in the literature: Youngs' hypothesis (*970*), the monoacid hypothesis (*186*), Bernstein's hypothesis (*73*), partial random distribution (*207,226*), 1-nonselective distribution (*897*), the 1-random–2,3-random hypothesis (*897*), and Tsuda's all-embracing ordered distribution (*896,897*). None of these hypotheses has proven widely accurate, however, when compared with modern experimental results on the component triglycerides found in natural fats.

## D. Validity of Distribution Hypotheses

*1. Nonhomogeneous Origin of Many Natural Fats*

Before the validity of the various fatty acid distribution hypotheses can be fairly evaluated, one must realize that two assumptions are implicit in this approach. Since each hypothesis states that triglyceride composition is a regular function of fatty acid composition, then the hypothesis can only be applied to natural triglyceride mixtures when

(a) the composition of the fatty acids supplied for triglyceride biosynthesis is the same in all cells producing the natural fat in question;
(b) the biosynthetic process for triglyceride synthesis is the same in all cells producing the fat.

It follows, therefore, that *fatty acid distribution hypotheses can only be applied to natural fats of homogeneous origin.*

There is accumulating evidence that many natural triglyceride mixtures are of nonhomogeneous origin. Kartha (*457*) and Galoppini and Lotti (*297*) have described regional variations in fatty acid composition *within* the endosperm of many seeds. The fatty acid composition of the germ has been found to be distinctly different from that of the rest of the seed in corn (*243*) and peanuts (*297*). Different genetic strains of corn (*945*) and rapeseed (*230*), and even individual seeds on the same soybean plant (*953*) have been shown to have different fatty acid compositions.

The same phenomenon also occurs in animal tissues; anatomical variations in depot fat composition have been well documented in rats (*180*), pigs (*50,139,172,840,861*), beef cattle (*139*), and sheep (*139,175,235,368*). This is sometimes due to the metabolic mixing of triglycerides from two different tissue sources. The liver, for example, can incorporate plasma triglycerides without prior hydrolysis (*613,757*). Triglyceride molecules synthesized in the liver are thus mixed with molecules from a completely different source, so that the total triglyceride composition of liver tissue may be unrepresentative of the molecules originating therein. Another similar example is plasma triglycerides which are known to originate from several different sources (*613*). Lack of equilibration between dietary fatty acids and depot fat triglycerides is another cause of nonhomogeneity. The turnover rate for triglycerides in adipose tissue is quite slow (*314,388*), and it takes up to 6 months before adipose tissue triglycerides can equilibrate with a new source of dietary fat (*180,736,889a*). If an animal has not received a diet of constant fatty acid composition for an extended period, its depot triglycerides will have been synthesized from fatty acid pools of different composition at different times.

The effect of this nonhomogeneous origin of natural triglyceride mixtures on the application of fatty acid distribution hypotheses has been examined by Litchfield and Reiser (*581*) and Gunstone and Padley (*328*). If there are minor variations in the relative amounts but not the types of fatty acids, this does not appreciably affect the use of statistical distribution formulas. Consider, for example, two individual linseeds found to contain 54% and 68% linolenic acid in their respective triglyceride mixtures. Comparison of the 1,3-random–2-random distributions for (i) linseed oil containing 54% 18:3, (ii) linseed oil containing 68% 18:3, and (iii) linseed oil having an intermediate compositon showed that there was no appreciable difference between (iii) and the mean of (i) and (ii) (*328*). On the other hand, if different types of fatty acids exist in different areas of the tissue or if the mechanism of triglyceride biosynthesis changes, then these differences must be taken into account to avoid large errors. Such errors would result if all the triglycerides of the palm fruit (*Elaeis guineensis*) were extracted together rather than analyzed separately as the fruit coat fat and the kernel oil (*581*).

It seems reasonable, therefore, to propose the following criteria which should be met before the validity of a fatty acid distribution hypothesis can be fairly evaluated:

(a) The triglyceride mixture should originate from a single, anatomically distinct tissue.
(b) All triglycerides present should have been synthesized in that tissue.
(c) The fatty acid composition of the triglycerides in the tissue should remain constant over the period in which the triglycerides are synthesized.

If these conditions are not met, then a natural triglyceride mixture may be too nonhomogeneous for valid comparison with fatty acid distribution hypotheses.

## 2. Comparison of Experimental and Predicted Triglyceride Compositions

The 1,3-random–2-random distribution hypothesis has proven rather accurate in predicting the triglyceride composition of seed fats containing only common fatty acids (16:0, 18:0, 18:1, 18:2, and 18:3) provided that optical isomers are not distinguished. Table 12-3 lists typical comparisons of experimental and predicted compositions for sunflower seed oil, candlenut oil, and cottonseed oil. Other reports (*83,328,331,443,444,864,974*) are in general agreement on this conclusion. It should be noted, however, that most of the experimental evidence comes from $Ag^+$ TLC separations where the various saturated acids are not distinguished; and more detailed compositional data are required for full verification of the 1,3-random–2-random hypothesis for seed fats. A few noted exceptions to 1,3-random–2-random distribution have been reported (*345,580,862,865,875*), but these are all with seed fats of highly unusual fatty acid composition. Hence it seems reasonable to restrict application of the 1,3-random–2-random hypothesis to seed fats containing only common fatty acids.

The applicability of the 1-random–2-random–3-random hypothesis to seed fats cannot be evaluated at the present time since too few data are available on the enantiomeric forms of the triglycerides present. Sampugna and Jensen (*782*) and Schlenk (*803*) found that the monounsaturated triglycerides of cocoa butter are mostly racemic as assumed by the 1,3-random–2-random pattern; but Morris (*662*) found asymmetric SUU in both

TABLE 12-3

COMPARISON OF EXPERIMENTAL AND 1,3-RANDOM–2-RANDOM TRIGLYCERIDE COMPOSITIONS FOR THREE SEED FATS (MOLE PERCENT)

| | Sunflower seed oil | | Candlenut oil | | | Cottonseed oil | | |
|---|---|---|---|---|---|---|---|---|
| Triglyceride | Found[a,b] | 1,3-Random–2-Random | Triglyceride | Found[b,c] | 1,3-Random–2-Random | Triglyceride | Found[b,d] | 1,3-Random–2-Random |
| β-SOS | 0.3 | 0.5 | SSO | 1 | 1 | SSS | 0.5 | 0.6 |
| β-SSO | 0.2 | tr | SOO | 2 | 2 | β-SSO | 0.8 | 0.4 |
| β-SOO | 2.3 | 1.6 | OOO | 1 | 1 | β-SOS | 4.5 | 6.3 |
| β-OSO | 0.1 | tr | SSL | 1 | 1 | β-SOO | 4.8 | 3.5 |
| β-SSL | 0.3 | 0.2 | SOL | 5 | 5 | β-OSO | 0.3 | 0.1 |
| β-SLS | 2.2 | 1.7 | OOL | 5 | 5 | β-SSL | 0.6 | 1.2 |
| OOO | 1.3 | 1.2 | SLL | 6 | 4 | β-SLS | 12.4 | 12.5 |
| β-SOL | 4.4 | 4.2 | OLL | 8 | 9 | OOO | 0.8 | 0.5 |
| β-SLO | 4.9 | 5.3 | LLL | 6 | 5 | β-SOL | 9.4 | 12.5 |
| β-OSL | 0.5 | 0.2 | SOLn | 4 | 4 | β-SLO | 8.4 | 7.0 |
| β-OOL | 8.1 | 6.5 | OOLn | 5 | 5 | β-OSL | 0.6 | 0.3 |
| β-OLO | 3.1 | 4.2 | SLLn | 8 | 7 | β-OOL | 4.1 | 3.5 |
| β-SLL | 13.2 | 14.0 | OLLn | 14 | 16 | β-OLO | 1.6 | 1.0 |
| β-LSL | 1.3 | 0.3 | LLLn | 13 | 13 | β-SLL | 22.5 | 24.7 |
| β-OLL | 20.4 | 21.9 | SLnLn | 3 | 2 | β-LSL | 1.1 | 0.6 |
| β-LOL | 8.4 | 8.7 | OLnLn | 6 | 6 | β-OLL | 6.4 | 6.9 |
| LLL | 28.1 | 28.9 | LLnLn | 10 | 11 | β-LOL | 6.5 | 6.2 |
| Others | 0.9 | 0.6 | LnLnLn | 2 | 3 | LLL | 13.0 | 12.2 |

[a] From Juriens and Schouten (444).
[b] Triglycerides were first fractionated on the basis of unsaturation by thin-layer chromatography on AgNO$_3$-impregnated silicic acid, and then major fractions were analyzed by hydrolysis with pancreatic lipase.
[c] From Gunstone and Padley (328).
[d] From Juriens and Kroesen (443).

cocoa butter and malabar tallow. Aruga and Morrison (*33a*) report that the distribution of fatty acids in wheat flour triglycerides follows a 1-random–2-random pattern. Obviously, many more species of seed fats must be analyzed before any generalization can be reached.

Distribution hypotheses have proven less effective in predicting the triglyceride composition of animal fats, probably because of their non-homogeneous origin (Section II,D,1). This can be seen in the data of Privett *et al.* (*736*) who determined the triglyceride compositions of various body fats from rats maintained for six months on a fat-free diet (Table 12-4). In the epididymal fat pads and the kidney, triglyceride composition (enantiomorphs not distinguished) was very close to that predicted by a 1,3-random–2-random distribution. In the liver and plasma, however, there was considerable discrepancy between the found and predicted compositions. It seems logical to relate these findings to the homogeneous synthesis of the epididymal fat pad triglycerides in that tissue, while the liver and plasma triglycerides were probably of mixed origin (Section II,D,1).

One may logically ask how a 1,3-random–2-random calculation could possibly predict the composition of rat epididymal triglycerides, since the *sn*-1- and *sn*-3-positions in rat adipose tissue triglycerides are known to have distinctly different fatty acid compositions (*121*). The answer to this question, as pointed out by Vander Wal (*911*), is that the predicted triglyceride compositions from the 1,3-random–2-random and the 1-random–2-random–3-random hypotheses are usually very similar, *as long as enantiomorphic triglycerides are not distinguished* (Table 12-5).

The 1-random–2-random–3-random hypothesis has also proven of variable effectiveness in predicting the triglyceride composition of animal fats. Christie and Moore (*172*) have performed detailed analyses of the isomeric triglycerides from five different pig tissues and compared their results with the 1-random–2-random–3-random hypothesis (Table 12-6). Agreement was excellent for inner and outer back fats, acceptable with liver and milk fats, and less satisfactory for blood triglycerides. Chicken egg yolk triglycerides were similarly analyzed (*173*) and found to follow approximately a 1-random–2-random–3-random distribution. Analyses of rat liver triglycerides by various workers have shown moderate agreement in two cases (*12,843*) and poor agreement in another (*960*). Much of this experimental evidence comes from $Ag^+$ TLC separations, however, where the chain lengths of the various acids are not distinguished; and more data on the individual molecular species of triglycerides are definitely needed. When both sample heterogeneity considerations (Section II,D,1) and the limited experimental evidence are considered, it does not seem likely that the 1-random–2-random–3-random distribution hypothesis will find wide application in predicting the composition of animal triglyceride mixtures.

TABLE 12-4

COMPARISON OF EXPERIMENTAL AND 1,3-RANDOM–2-RANDOM TRIGLYCERIDE COMPOSITIONS OF VARIOUS BODY FATS FROM RATS RAISED ON A FAT-FREE DIET[a] (WEIGHT PERCENT)

| Triglyceride | Epididymal fat pads | | Liver | | Plasma | | Kidney | |
|---|---|---|---|---|---|---|---|---|
| | Found[b] | 1,3-Random–2-Random | Found | 1,3-Random–2-Random | Found | 1,3-Random–2-Random | Found | 1,3-Random–2-Random |
| 000     | 2.5  | 1.4  | 2.7  | 2.1  | 1.7  | 2.4  | 3.7  | 5.4  |
| β-010   | 12.9 | 12.0 | 15.0 | 20.9 | 11.6 | 18.9 | 29.0 | 30.4 |
| β-001   | 4.2  | 4.9  | 4.2  | 4.5  | 6.3  | 5.5  |      |      |
| β-011   | 40.1 | 41.5 | 57.1 | 45.5 | 50.6 | 44.1 | 47.3 | 44.6 |
| β-101   | 3.5  | 4.2  | 2.7  | 2.4  | 3.3  | 3.2  |      |      |
| 111     | 36.8 | 36.0 | 18.3 | 24.6 | 23.1 | 25.9 | 20.0 | 19.6 |
| 002     | —    | —    | —    | —    | 2.1  | —    | —    | —    |
| Others  | —    | —    | —    | —    | 1.3  | —    | —    | —    |

[a] From Privett et al. (736).
[b] Triglycerides were first fractionated on the basis of unsaturation by thin-layer chromatography on AgNO$_3$-impregnated silicic acid, and then major fractions were analyzed by hydrolysis with pancreatic lipase.

TABLE 12-5
COMPARISON OF THE TRIGLYCERIDE COMPOSITIONS PREDICTED BY THE 1-RANDOM–2-RANDOM–3-RANDOM AND 1,3-RANDOM–2-RANDOM FATTY ACID DISTRIBUTION HYPOTHESES

| Distribution hypothesis | Fatty acid | Fatty acid compositions at individual positions[a] (mole percent) | | | Predicted triglyceride composition (mole percent) | | | | | | |
|---|---|---|---|---|---|---|---|---|---|---|---|
| | | $sn$-1 | $sn$-2 | $sn$-3 | SSS | β-SSU | β-SUS | β-SUU | β-USU | UUU |
| 1-Random–2-Random–3-Random | S | 56 | 24 | 37 | 5 | 12 | 16 | 39 | 7 | 21 |
| | U | 44 | 76 | 63 | | | | | | |
| 1,3-Random–2-Random | S | 46[b] | 24 | 46 | 5 | 12 | 16 | 38 | 7 | 22 |
| | U | 54 | 76 | 54 | | | | | | |

[a] Stereospecific analysis data on human subcutaneous fat taken from Brockerhoff (110).
[b] For the calculation of a 1,3-random–2-random distribution, the fatty acid compositions of the $sn$-1- and the $sn$-3-positions are assumed to be equivalent.

TABLE 12-6

COMPARISON OF OBSERVED TRIGLYCERIDE COMPOSITIONS OF VARIOUS PIG TISSUES WITH COMPOSITIONS PREDICTED BY 1-RANDOM–2-RANDOM–3-RANDOM FATTY ACID DISTRIBUTION HYPOTHESIS[a]

| | Triglycerides | | | | | | | | | | | | | | | |
|---|---|---|---|---|---|---|---|---|---|---|---|---|---|---|---|---|
| Tissue | 000 | sn-001 | sn-010 | sn-100 | sn-011 | sn-101 | sn-110 | 111 | sn-002 | sn-020 | sn-200 | sn-012 + sn-021 | sn-102 + sn-120 | sn-201 + sn-210 | 112 | 022 | Others |
| Inner back fat | | | | | | | | | | | | | | | | | |
| Observed[b] | 6.8 | 24.1 | 1.1 | 7.8 | 5.1 | 23.0 | 0.3 | 5.5 | 6.2 | 0.2 | 1.0 | 2.3 | 6.1 | 3.2 | 2.7 | 1.3 | 3.3 |
| Predicted | 6.7 | 25.9 | 1.5 | 6.0 | 5.8 | 23.4 | 1.3 | 5.1 | 5.8 | 0.2 | 0.9 | 2.3 | 5.3 | 4.4 | 2.2 | 1.0 | 2.2 |
| Outer back fat | | | | | | | | | | | | | | | | | |
| Observed | 2.7 | 23.4 | 1.0 | 2.7 | 5.5 | 28.5 | 0.3 | 7.5 | 4.6 | 0.2 | 1.5 | 2.6 | 7.0 | 2.7 | 4.3 | 1.6 | 3.9 |
| Predicted | 2.4 | 22.8 | 0.6 | 3.1 | 5.3 | 30.6 | 0.7 | 7.2 | 5.6 | 0.2 | 0.4 | 2.3 | 7.6 | 3.7 | 3.9 | 1.2 | 2.4 |
| Liver | | | | | | | | | | | | | | | | | |
| Observed | 2.7 | 5.6 | 6.5 | 1.2 | 12.9 | 3.9 | 4.6 | 4.7 | 2.4 | 3.8 | 0.6 | 11.0 | 4.6 | 1.4 | 5.7 | 4.6 | 23.8 |
| Predicted | 3.4 | 5.3 | 7.6 | 1.3 | 12.1 | 2.1 | 3.0 | 4.7 | 2.4 | 4.2 | 0.4 | 12.1 | 2.5 | 1.5 | 6.1 | 3.8 | 27.5 |
| Blood | | | | | | | | | | | | | | | | | |
| Observed | 4.6 | 13.7 | 6.8 | 3.2 | 13.6 | 15.6 | 3.3 | 5.6 | 2.7 | 1.8 | 0.3 | 6.4 | 5.0 | 1.5 | 4.5 | 2.8 | 8.6 |
| Predicted | 5.8 | 13.5 | 5.6 | 4.0 | 12.9 | 9.2 | 3.8 | 8.9 | 3.5 | 1.9 | 0.7 | 7.7 | 3.7 | 2.1 | 6.8 | 1.7 | 8.2 |
| Milk | | | | | | | | | | | | | | | | | |
| Observed | 4.7 | 11.3 | 2.2 | 9.1 | 5.1 | 21.8 | 2.9 | 9.4 | 2.8 | 0.5 | 2.1 | 1.9 | 5.8 | 4.2 | 5.7 | 1.4 | 9.1 |
| Predicted | 5.0 | 12.6 | 1.9 | 9.1 | 4.8 | 22.9 | 3.5 | 8.8 | 2.3 | 0.6 | 1.8 | 2.5 | 5.4 | 5.3 | 6.3 | 1.4 | 5.8 |

[a] From Christie and Moore (172).
[b] Triglycerides were first fractionated on the basis of unsaturation by thin-layer chromatography on AgNO$_3$-impregnated silicic acid, and then major fractions were subjected to stereospecific analysis.

It may possibly prove useful, however, for describing the triglyceride composition of adipose tissue fats from animals maintained on a constant diet for 4–6 months before sacrifice.

Present evidence on the validity of fatty acid distribution hypotheses for predicting the triglyceride composition of natural fats can therefore be summarized as follows:

(a) The 1,3-random–2-random hypothesis has proven fairly accurate for predicting the triglyceride composition of seed fats containing only common fatty acids, provided enantiomorphic triglycerides are not considered.

(b) The 1-random–2-random–3-random hypothesis may possibly describe the triglyceride composition of adipose tissue fats from higher animals maintained on a constant diet for 4–6 months before sacrifice.

(c) With other natural fat triglyceride mixtures, the 1,3-random–2-random and 1-random–2-random–3-random hypotheses give only a rough guess of the triglycerides present. Such calculations provide "best guess" answers which can be useful in planning or interpreting experiments, but they do not approach the accuracy of direct analysis procedures.

Further study may well modify these conclusions, however.

## III. BIOSYNTHESIS OF TRIGLYCERIDES

Another approach to understanding the complex fatty acid distribution patterns in natural fats is to define the enzymatic processes for triglyceride biosynthesis in living tissues. Knowledge of the specificity of these reactions should help explain the preferential formation of certain species of triglycerides.

Three routes for triglyceride biosynthesis have been defined: the glycerol phosphate pathway (Fig. 12-11), the dihydroxyacetone phosphate pathway (Fig. 12-11), and the monoglyceride pathway (Fig. 12-12). Glycerol phosphate and dihydroxyacetone phosphate are the only acceptors for the initial acylation step; hence only these two pathways can account for *de novo* synthesis of triglycerides. The monoglyceride pathway is chiefly a mechanism for rebuilding glycerides which have been partially hydrolyzed by various lipases.

At least two and possibly all three pathways operate simultaneously in many tissues. During the intestinal absorption of triglycerides in mammals,

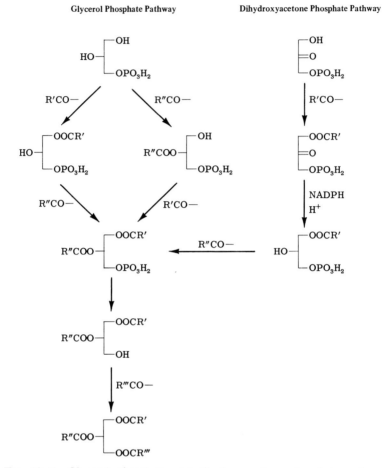

FIG. 12-11. Glycerol phosphate and dihydroxyacetone phosphate pathways for the *de novo* biosynthesis of triglycerides *(10,504,554,741,752,755,904,946)*.

pancreatic lipase hydrolyzes dietary fats to form 2-monoglycerides for passage through the intestinal wall. The monoglyceride pathway then rebuilds these 2-monoglycerides into triglycerides which pass into the lymph *(431,624)*. At the same time, triglycerides are synthesized *de novo* via the glycerol phosphate and dihydroxyacetone phosphate pathways in the intestinal wall *(624,751)*. Mattson and Volpenhein *(624)* have estimated that 78% of the triglycerides in rat lymph originate via the monoglyceride pathway while 22% come from *de novo* synthesis. The liver can synthesize triglyceride by all three pathways *(10,398,741,947)*. Adipose tissue triglycerides originate both by *de novo* synthesis via the glycerol phosphate

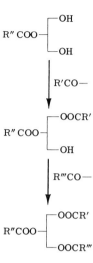

FIG. 12-12. Monoglyceride pathway for the biosynthesis of triglycerides (*431,432,621,867*).

pathway (*773,856*) and by the monoglyceride pathway (*824*). The mammary gland can also synthesize triglyceride by both the glycerol phosphate and monoglyceride pathways (*632*). Fat-producing plant tissues show evidence for triglyceride biosynthesis via the glycerol phosphate pathway (*57,373,377,629,630*), but the simultaneous appearance of significant amounts of monoglyceride may indicate that a monoglyceride-type pathway is also present.

If two or three pathways for triglyceride biosynthesis are operating simultaneously in the same tissue, then how can a single fatty acid distribution pattern such as the 1-random–2-random–3-random hypothesis be the final result? The answer to this puzzle must lie in the substrate specificities of acyl transferases which esterify the fatty acids to the glycerol. A 1-random–2-random–3-random pattern could only result if the acyl transferases adding the second and third acyl groups possess *no glyceride substrate specificity* (as distinguished from the acyl group specificity of these enzymes). Obviously the first acyl group esterified is a random addition, since it cannot be correlated with any other acyl chains in the molecule. The acyl transferase adding the second fatty acid, however, might do so by random selection of the monoacyl glyceride substrates. Similarly the acyl transferase adding the third fatty acid could make a random selection of the available *sn*-1,2-diglyceride substrates present.

The comparative reaction rates of different monoacyl glyceride substrates in the second acylation step has not yet been defined. Hence it remains uncertain whether the second fatty acid is esterified in a random or nonrandom manner.

On the other hand, the random nature of the third acylation step has been partially confirmed by *in vitro* enzymatic experiments. Åkesson *et al.* (*11,15*), Hill *et al.* (*370*), and de Kruyff *et al.* (*211*) report that the rate of synthesis of triglyceride from *sn*-1,2-diglycerides by rat liver is independent of the fatty acid composition of the diglycerides. If this nonspecificity proves generally true, then there is a biochemical basis for the 3-random pattern assumed by the 1-random–2-random–3-random distribution hypothesis, provided there is a free mixing of all diglyceride substrates from all three pathways. Evidence exists, however, that the diglyceride acylation steps in the glycerol phosphate and monoglyceride pathways are completely independent of one another in the intestinal mucosa (*433*).

For the present, therefore, the enzymatic specificities demonstrated in biochemical investigations correlate only partially with the 1-random–2-random–3-random and 1,3-random–2-random patterns derived from compositional studies. It must be kept in mind, however, that our current understanding of triglyceride biosynthesis is very incomplete, and many questions about the process remain to be answered. For example, what are the relative amounts of triglyceride coming from the three pathways in various tissues *in vivo?* Are most of the triglycerides synthesized by the glycerol phosphate and dihydroxyacetone phosphate pathways subsequently rearranged to different species by lipolysis and resynthesis through the monoglyceride pathway? Do all three biosynthetic pathways produce the same component triglycerides from the same pool of fatty acids? Are fatty acids ever desaturated or modified *after* they are esterified to the glycerol? These and many other problems must be investigated before the validity of fatty acid distribution hypotheses can be fully evaluated.

# 13

# COMBINING METHODS FOR DETAILED ANALYSIS OF COMPLEX TRIGLYCERIDE MIXTURES

The many experimental techniques for the separation and positional analysis of triglycerides have been discussed in Chapters 4 through 11. However, the selection of a single method or combination of methods to solve a specific analytical problem requires considerable forethought. Natural fat triglyceride mixtures contain so many molecular species (50 to 1000 or more) that rarely can a single analytical technique give the required results. Therefore one must resort to a consecutive series of separation techniques to accomplish fractionation, subfractionation, etc., followed by positional analysis of each final fraction. Complete analysis of all the triglyceride species in a natural fat is now possible in many cases. However, a prodigious amount of work is necessary to achieve this goal, and no complete analysis of any natural fat has yet been reported.

The purpose of this chapter is to compare the major separation and positional analysis techniques currently available and to show how they can be combined for a detailed analysis of the molecular species found in complex natural fat triglyceride mixtures. An understanding of how such detailed analyses are accomplished will aid the reader in selecting the best combination of techniques to solve the particular analytical problem he faces.

## I. COMBINING TRIGLYCERIDE ANALYSIS TECHNIQUES

For maximum compositional information, triglyceride mixtures should be fractionated prior to positional analysis, since positional analysis necessitates hydrolysis of the ester linkages.

### A. Separation Techniques

The four most useful separation techniques for triglyceride mixtures are listed in Table 13-1. Silver ion adsorption chromatography separates molecules on the basis of unsaturation, while gas–liquid chromatography fractionates molecules according to carbon number. Liquid–liquid partition chromatography resolves triglycerides on the basis of integral partition number, which is a function of both unsaturation and carbon number. Mass spectrometry, a technique still in its infancy, separates molecules according to their carbon number and the number of double bonds they contain. It is apparent, therefore, that there are only two fundamental bases for separating triglyceride mixtures: by degree of unsaturation and by carbon number.

The usefulness of the four separation techniques listed in Table 13-1 can be compared by constructing a three-way grid showing the separation of a simple triglyceride mixture by unsaturation, partition number, and carbon number. Figure 13-1 presents such a grid for the 20 different triglycerides containing 16:0, 18:0, 18:1, and 18:2 (positional isomers ignored). These 20 species can be resolved into 10 fractions by silver ion adsorption chromatography (including useful subfractionation in the two-, three-, and four-double-bond triglycerides), 7 fractions by liquid–liquid

TABLE 13-1
Major Separation Techniques for Triglycerides

| Method | Basis of separation | Sample recovery |
|---|---|---|
| Silver ion adsorption chromatography | Unsaturation | Nondestructive |
| Liquid–liquid partition chromatography | Partition number | Nondestructive |
| Gas–liquid chromatography | Carbon number | Destructive[a] |
| Mass spectrometry | Carbon number and unsaturation | Destructive |

[a] Preparative GLC of carbon numbers below $C_{40}$ is possible (Chapter 6, Section II,F).

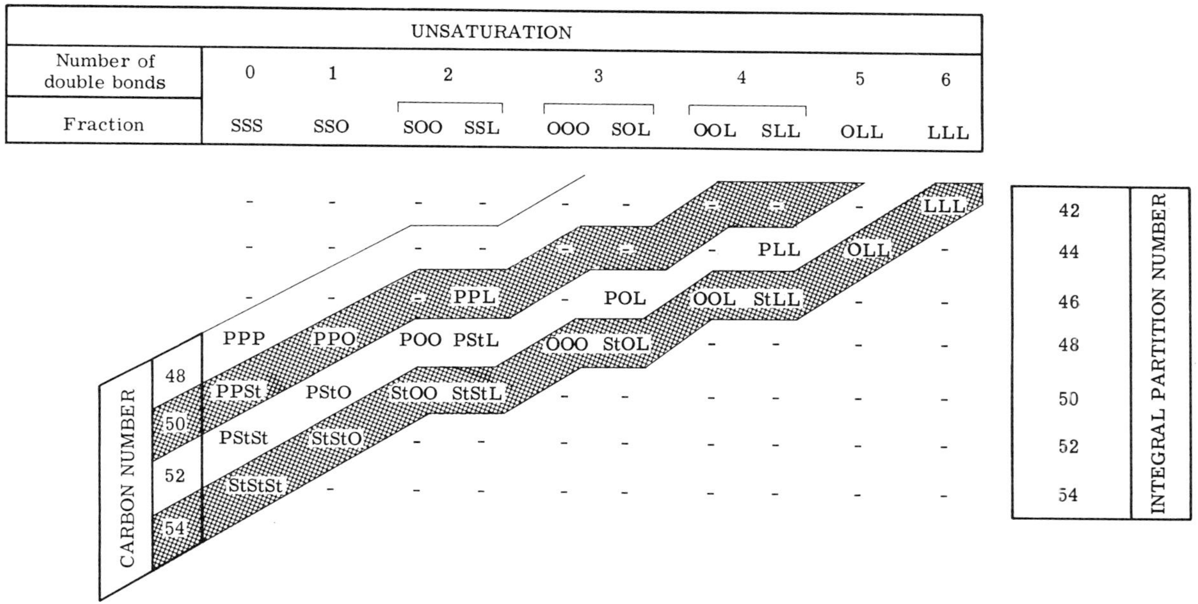

FIG. 13-1. Three-way grid defining the possible separations of a simple triglyceride mixture by unsaturation (Ag⁺ adsorption chromatography), integral partition number (liquid–liquid partition chromatography), and carbon number (gas–liquid chromatography). Mass spectrometry distinguishes both carbon number and the number of double bonds per molecule. The mixture considered here represents all possible triglycerides containing 16:0, 18:0, 18:1, and 18:2 (positional isomers ignored).

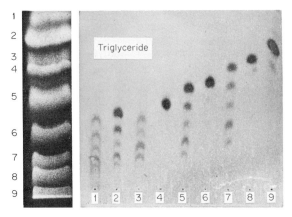

FIG. 13-2. Analysis of peanut oil triglycerides using Ag$^+$ adsorption thin-layer chromatography followed by thin-layer liquid–liquid partition chromatography of each fraction. Identity of fractions: 1, SSO; 2, SOO; 3, SSL; 4, OOO; 5, SOL; 6, OOL; 7, SLL; 8, OLL; 9, LLL. Peanut oil contains 16:0, 18:0, 20:0, 22:0, and 24:0; hence all Ag$^+$ TLC fractions containing saturated acyl groups can be further subfractionated by liquid–liquid partition chromatography. *Operating conditions for Ag$^+$ TLC:* 100 × 400 mm TLC plate coated with 0.6 mm layer of AgNO$_3$/Kieselgel G 10/90; sample size, 12 mg; single ascending development with benzene/diethyl ether 80/20; bands visualized under ultraviolet light after spraying with 2′,7′-dichlorofluorescein solution. *Operating conditions for liquid–liquid partition TLC:* 200 × 200 mm TLC plate coated with Kieselgel G and impregnated with paraffin oil; sample size, ~15 µg; double development with acetone/acetonitrile 80/20; spots located with iodine vapor/α-cyclodextrin. From Wessels and Rajagopal (*951*).

partition chromatography, 4 fractions by gas–liquid chromatography, and 16 fractions by mass spectrometry. To resolve all 20 species, it is necessary to use two consecutive techniques, one of which must be silver ion adsorption chromatography. There is no need to apply three separation techniques to the same mixture, since maximum resolution can be accomplished with only two.

The methods employed for any specific triglyceride analysis problem must, of course, depend on the results required, the work involved, and the equipment available. Many individual and combination techniques are now possible. For most complex natural triglyceride mixtures, however, consecutive silver ion adsorption and liquid–liquid partition chromatography is definitely the method of choice. Although fairly time-consuming, this combination generally yields the maximum possible resolution, gives the highest quantitative accuracy, and produces subfractions that can be used for further positional analyses (see below). For *rapid* analysis of complex triglyceride mixtures, however, either mass spectrometry or gas–liquid chromatography is the method of choice; but in both techniques the sample is destroyed during analysis. Mass spectrometry quantitates the greatest number of species of any single analytical technique; but gas–liquid chromatography is more widely available in most laboratories.

The separation of peanut oil triglycerides by consecutive silver ion adsorption and liquid–liquid partition chromatography is illustrated in Fig. 13-2. This combination of techniques resolved peanut oil into 31 distinct groups of triglycerides. Figure 13-3 shows the fractionation of ucuhuba seed triglycerides by consecutive silver ion adsorption and gas–liquid chromatography to resolve 24 groups of triglycerides. Consecutive liquid–liquid partition and gas–liquid chromatography has been used by Litchfield (567) to determine the very complex triglyceride composition of *Ephedra nevadensis* seed fat.

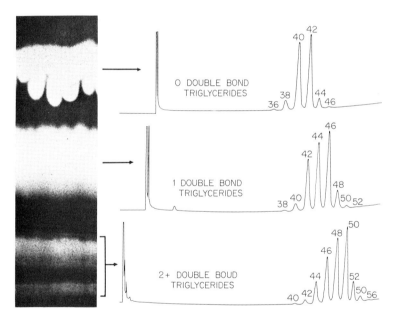

FIG. 13-3. Analysis of ucuhuba (*Virola surinamensis*) seed triglycerides using Ag$^+$ adsorption thin-layer chromatography followed by gas–liquid chromatography of the intact triglycerides in each fraction. *Operating conditions for TLC:* 200 × 200 mm TLC plate coated with 1.0 mm layer of AgNO$_3$/Silica Gel G 6/94; sample size, ~100 mg; single ascending development with CHCl$_3$/C$_2$H$_5$OH 99.9/0.1; bands visualized under ultraviolet light after spraying with 2′,7′-dichlorofluorescein solution. *Operating conditions for GLC:* 610 × 3.0 mm i.d. glass column packed with 3.0% JXR on Gas Chrom Q; 100 ml/minute N carrier gas; column temperature programmed 170° → 305° at 4°/minute; peaks labeled according to carbon number. From Culp *et al.* (203).

## B. Positional Analysis Techniques

After a complex triglyceride mixture has been separated as much as possible by chromatographic techniques, four types of fractions are usually obtained: (i) monoacid triglycerides, (ii) diacid triglycerides, (iii) triacid triglycerides, and various mixtures of (i), (ii), and (iii). The next step in a detailed analysis of component triglycerides is to determine the isomeric molecular species present using one of the positional analysis techniques listed in Table 13-2.

Monoacid triglycerides have no isomeric forms and are easily identified by a simple fatty acid analysis.

The isomer composition of diacid triglycerides follows directly from a stereospecific analysis. The amounts of *sn*-OLL, *sn*-LOL, and *sn*-LLO in the OLL fraction from safflower oil, for example, would correspond to the amounts of oleic acid at the *sn*-1-, 2-, and 3-positions, respectively. If it is not necessary to distinguish optical isomers, then the content of β-OLL and β-LOL is readily determined by pancreatic lipolysis.

The six positional isomers of a triacid triglyceride such as POSt can be characterized as β-POSt, β-OPSt, and β-PStO by hydrolysis with pancreatic lipase. However, it is not possible to quantitate all six positional isomers by direct stereospecific analysis (see Chapter 10, Section II,B). For complete analysis of all six molecular species, one must resort to fractionation of the derived diglycerides (Figs. 10-4 and 10-5).

The isomeric content of mixtures of mono-, di-, and triacid triglycerides can sometimes be determined through the use of simultaneous equations derived from positional analysis results (Chapter 10, Section II,B). If the mixture is too complex, however, one must also resort to the fractionation of derived diglycerides prior to positional analysis as described below.

TABLE 13-2
MAJOR POSITIONAL ANALYSIS TECHNIQUES FOR TRIGLYCERIDES

| Method | Positions distinguished |
|---|---|
| Pancreatic lipase | $sn$-1 + $sn$-3 |
|  | $sn$-2 |
| Grignard reagent | $sn$-1 + $sn$-3 |
|  | $sn$-2 |
| Stereospecific analysis | $sn$-1 |
|  | $sn$-2 |
|  | $sn$-3 |

Positional analyses with pancreatic lipase or by stereospecific analysis are most frequently employed with fractions from silver ion adsorption chromatography where 10–20 mg size fractions are readily obtained. Separation by both Ag⁺ adsorption and liquid–liquid partition chromatography would obviously yield more detailed results, but this requires a large amount of additional work to obtain sufficient material for the positional analyses. Table 12-3 illustrates the consecutive use of Ag⁺ TLC and pancreatic lipase to determine the triglyceride compositions of sunflower and cottonseed oils. A similar combination of silver ion adsorption chromatography with stereospecific analysis yields even more detailed data on isomer content, as shown with various pig fats in Table 12-6.

## II. USE OF DERIVED DIGLYCERIDES FOR ANALYSIS OF UNRESOLVABLE TRIGLYCERIDE MIXTURES

Natural fat triglyceride mixtures are often so complex that the application of all available separation and positional analysis techniques still fails to distinguish all the individual molecular species. For example, this problem occurs with the saturated $C_{38}$ triglycerides of coconut oil (see Fig. 13-6) and with the **011** subfraction from the partition number 46 triglycerides of horse fat (Fig. 13-7). It has already been shown (Chapter 10, Section II,B) how the six positional isomers of a triacid triglyceride can be distinguished by chromatographic fractionation of the derived diglycerides prior to stereospecific analysis. This same approach is also generally applicable to most unresolvable mixtures of mono-, di-, and triacid triglycerides.

Hammond (340) has shown how the fractionation of derived diglycerides could be used to obtain a complete analysis of all the saturated $C_{38}$ triglycerides obtained from coconut oil by consecutive silver ion adsorption chromatography (to isolate SSS) and liquid–liquid partition chromatography (to isolate $C_{38}$). This extremely complex fraction contains three diacid and three triacid triglycerides (DDSt + DMM + LaLaM + OcLaSt + OcMP + DLaP = 27 possible molecular species) and cannot be further resolved by chromatographic techniques or by mass spectrometry.

Hammond's solution to this analytical problem is illustrated in Fig. 13-6. Random deacylation with a Grignard reagent yields representative $sn$-1,3-diglycerides, which are first isolated as a group and then further resolved into $C_{20}$, $C_{22}$, $C_{24}$, $C_{26}$, $C_{28}$, and $C_{30}$ fractions by liquid–liquid partition chromatography. All the diglycerides that have lost 18:0 will appear in the $C_{20}$ fraction, all that have lost 16:0 will appear in the $C_{22}$ fraction, etc., so that there will be a different diglyceride fraction for each of the fatty acids lost from the 2-position. Stereospecific analysis of each $sn$-1,3-diglyceride fraction readily quantitates all the diglyceride isomers present, since each different acyl group occurs only once at the $sn$-1- and only once at the $sn$-3-position in any one fraction. Once the diglyceride composition of each fraction is determined, the amount of each corresponding triglyceride isomer in the original sample is easily calculated using the known identity of the acid removed from the 2-position.

The same results can also be obtained by working with the $sn$-1,2(2,3)-diglycerides from Grignard or lipase deacylation (Fig. 13-6), even though the $sn$-1,2(2,3)-diglycerides are twice as numerous as the $sn$-1,3-isomers. The $sn$-1,2(2,3)-diglycerides are first isolated as a group and then divided into $C_{20}$, $C_{22}$, $C_{24}$, $C_{26}$, $C_{28}$, and $C_{30}$ fractions by liquid–liquid partition chromatography. Stereospecific analysis of each fraction identifies all the diglyceride isomers present in the same manner as with the $sn$-1,3-isomers; and these data can then be used to quantitate the corresponding triglyceride isomers. Note, however, that the amount of each triglyceride species is the sum of both the $sn$-1,2- and $sn$-2,3-diglycerides that originate from it (i.e., $sn$-MMD = $sn$-MM- + $sn$--MD).

Similar analytical procedures for the complete analysis of complex mixtures of di- and triacid triglycerides are also included in Figs. 13-5 and 13-7. In some cases, however, it may be necessary to fractionate the derived diglycerides by two consecutive chromatographic techniques before a complete analysis of the original triglycerides can be obtained (Fig. 13-7).

When analyzing such mixtures of derived diglycerides, it is useful to convert the free hydroxyl groups to acetate esters by reacting the diglycerides with acetic anhydride (Chapter 3, Section III,A) as soon as they are isolated. This prevents undesirable acyl migration and also facilitates subsequent fractions.

## III. MAXIMUM ANALYSIS OF COMPLEX TRIGLYCERIDE MIXTURES

A general procedure for the maximum analysis of all molecule species in complex triglyceride mixtures has been proposed by Hammond (*340*). His approach to this problem is outlined in Fig. 13-4 and consists of four major steps:

(a) Maximum chromatographic fractionation of triglycerides
(b) Deacylation of each triglyceride fraction to representative diglycerides

## III. MAXIMUM ANALYSIS

(c) Maximum chromatographic fractionation of diglycerides
(d) Stereospecific analysis of each diglyceride fraction

By fractionating the derived diglycerides with the same separation techniques that were used on the triglycerides, complete analysis of individual molecular species is often possible. The specific chromatographic separa-

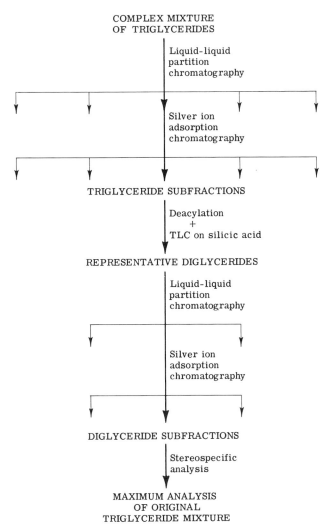

FIG. 13-4. Outline of Hammond's general procedure for maximum analysis of molecular species in complex mixtures of triglycerides (*340*). See Figs. 13-5, 13-6, and 13-7 for applications to specific examples.

tion techniques selected depend on the sample being analyzed, and it is not always necessary to utilize every step shown in Fig. 13-4 on each fraction in order to obtain a complete analysis.

Figure 13-5 illustrates the use of Hammond's procedure for the complete analysis of all 64 possible molecular species in a hypothetical triglyceride mixture containing only palmitic, stearic, oleic, and linoleic acids. An initial fractionation by silver ion adsorption chromatography yields 10 fractions, 4 of which are further resolved by liquid–liquid partition chromatography. Deacylation of each resultant fraction (except PPP, StStSt, OOO, and LLL) to representative sn-1,3-diglycerides is followed by stereospecific analysis. When a triacid triglyceride is being analyzed, however, the sn-1,3-diglyceride mixture is resolved by silver ion adsorption chromatography prior to stereospecific analysis in order to obtain complete quantitation of all isomers. A total of 5 silver ion separations, 4 liquid–liquid partition separations, 14 deacylation reactions, and 18 stereospecific analyses are necessary for the procedures shown. However, the number of deacylations can be reduced to 8 and the number of stereospecific analyses can be reduced to 10 by combining easily differentiated fractions before these procedures are initiated (see caption for Fig. 13-5).

The analysis illustrated in Fig. 13-5 is relatively simple since only 4 acids (2 saturated, 1 monoene, 1 diene) are present. Figure 13-6 shows a schematic diagram for the complete analysis of all the molecular species in a much more complex mixture, coconut oil, using the Hammond approach. Coconut oil contains 8 acids (6 saturated, 1 monoene, 1 diene) and is considered one of the most difficult fats for triglyceride analysis (*190*). The mixture is first separated by silver ion adsorption chromatography, followed by further subfractionation with liquid–liquid partition chromatography. After random deacylation of each subfraction, representative diglycerides are isolated by TLC on silicic acid. Either the sn-1,3- or the sn-1,2(2,3)-diglycerides may be used for further analysis; both yield the same results. The diglyceride mixture from each triglyceride subfraction is then resolved by liquid–liquid partition chromatography, followed by stereospecific analysis of each diglyceride fraction separated (Chapter 10, Section II,C). A typical scheme for complete analysis of all the molecular species in coconut oil triglycerides in this manner requires 35 silver ion separations, 11 liquid–liquid partition separations, 49 deacylation reactions, and 111 stereospecific analyses. The entire analysis is too lengthy to fully outline in Fig. 13-6; but the procedure for complete analysis of the saturated $C_{38}$ triglycerides, one of the most complicated triglyceride subfractions, is given in detail.

The resolution of another type of complex triglyceride mixture, horse adipose tissue fat containing substantial 18:3, is shown in Fig. 13-7. As

## III. MAXIMUM ANALYSIS

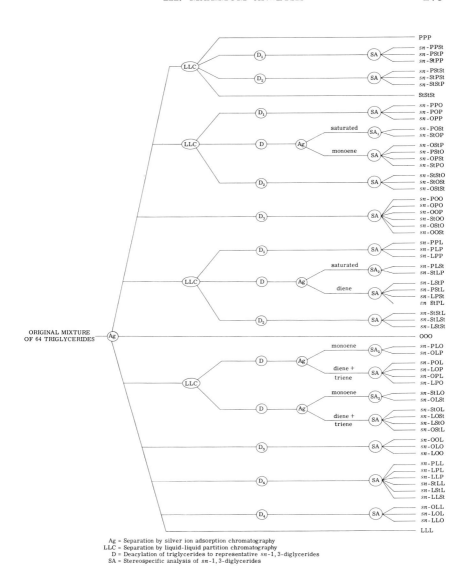

Ag = Separation by silver ion adsorption chromatography
LLC = Separation by liquid-liquid partition chromatography
D = Deacylation of triglycerides to representative $sn$-1,3-diglycerides
SA = Stereospecific analysis of $sn$-1,3-diglycerides

FIG. 13-5. Schematic diagram of one possible procedure for analysis of all molecular species of triglycerides containing palmitic, stearic, oleic, and linoleic acids. A total of 5 Ag, 4 LLC, 14 D, and 18 SA analyses are necessary for the procedure shown. However, the number of D can be reduced to 8 and the number of SA can be reduced to 10 by combining easily differentiated fractions before these procedures are initiated. For example, the three $D_1$ analyses could be combined, and only one SA would be necessary to distinguish all nine molecular species. Similarly, the $D_2$, $D_3$, $D_4$, $SA_1$, and $SA_2$ groups of analyses could be combined to save work.

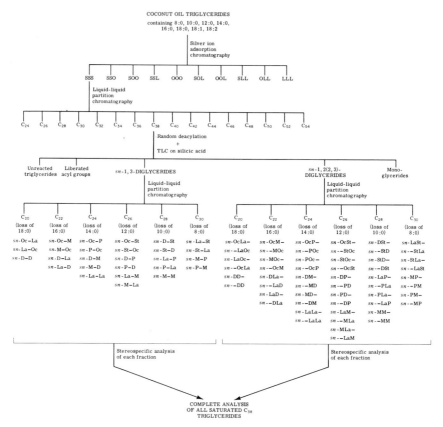

FIG. 13-6. Schematic diagram showing analytical procedures for complete analysis of all saturated $C_{38}$ triglycerides from coconut oil using the derived sn-1,3- or sn-1,2(2,3)-diglycerides. Complete analysis of all the molecular species of triglycerides in coconut oil would utilize similar analytical procedures for all the triglyceride subfractions. Adapted from Hammond (340).

already noted (Chapter 4, Section II,A,2), the presence of linolenic acid in natural fat triglycerides makes resolution by silver ion adsorption chromatography more difficult. Triglyceride groups containing five to nine double bonds per molecule often overlap and are not completely resolved. For this reason, it is better to fractionate horse fat triglycerides first by liquid–liquid partition chromatography, which yields good resolution of highly unsaturated triglyceride mixtures. The resultant fractions are much simpler, and thus they are presumably more easily resolved by silver ion chromatography. Using this approach, horse fat triglycerides are separated into 10 fractions by liquid–liquid partition chromatography, and each of these is then further subfractionated by silver ion adsorption chromatography. Random deacylation of each subfraction to representative $sn$-1,3- or $sn$-1,2(2,3)-diglycerides is followed by suitable diglyceride fractionation and stereospecific analysis procedures as before. In the case of the [MOO + PPoO + PoPoSt] subfraction (Fig. 13-7), the $sn$-1,3-diglycerides need only be resolved once by liquid–liquid partition chromatography, while the $sn$-1,2(2,3)-diglycerides must be separated by consecutive liquid–liquid partition and silver ion adsorption chromatography before

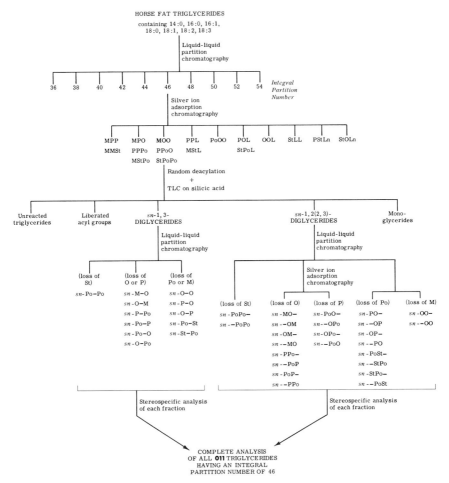

FIG. 13-7. Schematic diagram showing analytical procedures for complete analysis of all **011** triglycerides having an integral partition number of 46 from horse fat using the derived $sn$-1,3- or $sn$-1,2(2,3)-diglycerides. Complete analysis of all the molecular species of triglycerides in horse fat would require similar analytical procedures for all triglyceride subfractions.

stereospecific analysis will determine all positional isomers. Assuming efficient silver ion fractionation of the linolenic triglycerides, complete analysis of all the molecular species of horse fat triglycerides can be achieved using 29 silver ion separations, 7 liquid–liquid partition separations, 43 deacylation reactions, and 76 stereospecific analyses. Only the analytical scheme for the [MOO + PPoO + PoPoSt] subfraction is outlined in Fig. 13-7, but other triglyceride subfractions are treated similarly.

If a triglyceride mixture contains acyl groups that differ only in the position of their double bonds (such as petroselinic, oleic, and vaccinic acids), then complete analysis of all molecular species is not possible using only the techniques outlined above. Complete analysis should be possible, however, if one resorts to the fractionation of $KMnO_4/KIO_4$ oxidized triglycerides and if each isomer yields a unique dicarboxylic acid remaining attached to the glycerol (but note limitations of the oxidation reaction mentioned in Chapter 3, Section I,B). Such a procedure for the determination of all 12 isomers in a POO + POV + PVV mixture is outlined in Fig. 13-8.

There remain a number of highly complex natural fats for which a complete analysis of all molecular species of triglycerides is not possible with present techniques. Samples from aquatic animal and seed fats containing tetraene, pentaene, and hexaene acids fall into this extremely difficult category. The behavior of triglycerides containing four-, five-, and six-double-bond acids in silver ion adsorption and liquid–liquid partition chromatography has not been thoroughly investigated, but preliminary evidence (Chapter 4, Section II,A,2 and Chapter 5, Sections II,A,1 and II,A,2) indicates that systematic resolution according to the number of double bonds per molecule or according to integral partition number cannot be fully achieved. Some fractionation is possible, of course, but this is probably insufficient for complete analysis of all triglyceride species.

The diagrams in Figs. 13-4, 13-5, 13-6, and 13-7 clearly demonstrate that complete analysis of all the molecular species in a single natural fat triglyceride mixture using present techniques constitutes a staggering amount of tedious work. At the time this chapter is written, no complete analysis of any natural fat triglyceride mixture has yet been reported in the literature, even though such an analysis is theoretically possible. However, in many cases partial analysis is satisfactory to solve the problem at hand. The relative abilities of various individual and combinations of analytical methods to determine the component triglycerides in typical natural fats are compared in Table 13-3. The increased information obtained when two or more analytical techniques are combined in tandem is clearly evident, but each researcher must decide whether the usefulness of these additional data justifies the increased effort of obtaining them.

### III. MAXIMUM ANALYSIS

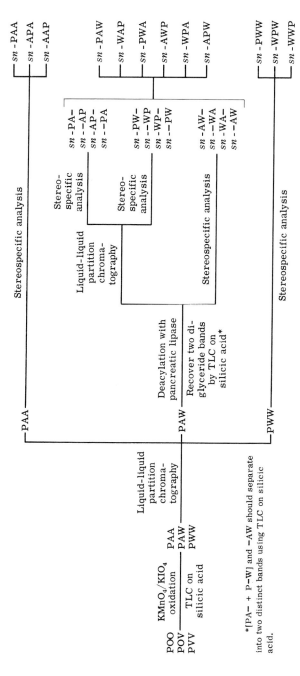

FIG. 13-8. Possible analytical procedure for the separation of oxidized triglycerides and the derived oxidized diglycerides to achieve complete analysis of all triglyceride species containing two isomeric fatty acids such as oleic and vaccinic, V, vaccinic acid; A, azelaic acid; W, undecanedioic acid.

TABLE 13-3
Relative Ability of Various Combinations of Analytical Techniques to Determine the Composition of Complex Triglyceride Mixtures

| Analytical techniques[a] | Number of triglyceride groups distinguished | | |
|---|---|---|---|
| | All species containing 16:0, 18:0, 18:1, 18:2 (Fig. 13-5) | Coconut oil triglycerides (Fig. 13-6) | Horse fat triglycerides (Fig. 13-7) |
| *Total possible molecular species* | *64* | *512* | *343* |
| Silver ion adsorption chromatography (Ag) | 13 | 25 | 36 |
| Liquid–liquid partition chromatography (LLC) | 11 | 35 | 20 |
| Gas–liquid chromatography (GLC) | 4 | 16 | 7 |
| Mass spectrometry | 16 | 52 | 37 |
| Deacylation[b] (D) | 0 | 0 | 0 |
| D + stereospecific analysis (SA) | 0 | 0 | 0 |
| LLC + Ag | 20 | 102 | 84 |
| LLC + Ag + D | 40 | 222 | 190 |
| LLC + Ag + D + SA | 52 | 258 | 234 |
| LLC + Ag + D + LLC + SA | 64 | 512 | 343 |
| LLC + Ag + D + LLC + Ag + SA | 64 | 512 | 343 |

[a] Including fatty acid analyses after each separation where possible. This frequently permits the quantitation of separate groups of triglycerides within each fraction. For example, the SOO fraction from coconut oil can be subdivided into OcOO + DOO + LaOO + MOO + POO + StOO by analysis of its fatty acid composition.

[b] Random deacylation of triglycerides followed by isolation of representative $sn$-1,3-diglycerides.

## IV. MAJOR UNSOLVED PROBLEMS OF TRIGLYCERIDE ANALYSIS

This chapter summarizes how modern chromatographic and enzymatic techniques now make it possible to determine the complete triglyceride composition of many natural fats. However, such a review also reveals the various unsolved problems in this field, and it would be well to mention them here as goals for future research.

Coleman (*190*) has pointed out that the complete analysis of aquatic animal triglycerides with their wide range of four-, five-, and six-double-bond acids will clearly require radically different methods from those now

## IV. MAJOR UNSOLVED PROBLEMS

in prospect. Although such samples can be partially resolved by silver ion adsorption and liquid–liquid partition chromatography, available evidence (Chapter 4, Section II,A,2 and Chapter 5, Sections II,A,1 and II,A,2) indicates that systematic resolution according to the number of double bonds per molecule or according to partition number cannot be fully accomplished. Hammond (*340*) has suggested a procedure similar to that shown in Fig. 13-7 for the complete analysis of all molecular species in a hypothetical marine oil, but this assumes that silver ion chromatography can resolve molecules containing 0 through 18 double bonds into 19 discrete fractions, which is not possible at present. Clearly new approaches are needed to handle triglyceride mixtures containing four-, five-, and six-double-bond acids.

Another major unsolved problem of triglyceride analysis is the automation of chromatographic separation techniques. If complex analyses such as outlined in Figs. 13-5, 13-6, and 13-7 are to be undertaken on a routine basis, the same level of automation currently available for gas–liquid chromatography analyses must be developed for silver ion adsorption and liquid–liquid partition chromatography procedures. This would undoubtedly involve automated operation of column chromatography including automatic monitoring and fraction collecting devices. These objectives are certainly possible with currently available techniques, but a fully automated liquid chromatography system for triglycerides remains to be developed.

A rapid technique for stereospecific analysis of triglycerides is another urgent need; present procedures require 3–5 days to perform. A method which can be completed in less than 8 hours is an absolute necessity if the dozens of stereospecific analyses required for a complete analysis of molecular species are to be performed on a routine basis. One possible approach might be a random deacylation yielding representative *sn*-1- and *sn*-3-monoglycerides, which would then be resolved as disastereoisomer derivatives on an optically active GLC stationary phase; each fatty acid would produce two peaks corresponding to the compositions at the *sn*-1- and *sn*-3-positions. Another delightful, though remote, possibility would be the discovery of a stereospecific lipase. Clearly such novel approaches to stereospecific analysis are called for if rapid analyses are to be achieved.

# REFERENCES

1. Aasen, A. J., Lauer, W. M., and Holman, R. T., *Lipids* **5,** 869 (1970).
2. Abel, E. W., Pollard, F. H., Uden, P. C., and Nickless, G., *J. Chromatogr.* **22,** 23 (1966).
3. Achaya, K. T., Craig, B. M., and Youngs, C. G., *J. Amer. Oil Chem. Soc.* **41,** 783 (1964).
4. Achaya, K. T., and Hilditch, T. P., *Proc. Roy. Soc., Ser. B* **137,** 187 (1950).
5. Ackman, R. G., *in* "Methods in Enzymology" (J. M. Lowenstein, ed.), Vol. 14, pp. 329–381. Academic Press, New York, 1969.
6. Ackman, R. G., *Progr. Chem. Fats Other Lipids* **12,** 165 (1972).
7. Ackman, R. G., and Burgher, R. D., *J. Lipid Res.* **5,** 130 (1964).
8. Ackman, R. G., and Sipos, J. C., *J. Amer. Oil Chem. Soc.* **41,** 377 (1964).
9. Addison, R. F., and Ackman, R. G., *Anal. Biochem.* **28,** 515 (1969).
10. Agranoff, B. W., and Hajra, A. K., *Proc. Nat. Acad. Sci. U.S.* **68,** 411 (1971).
11. Åkesson, B., *Eur. J. Biochem.* **9,** 406 (1969).
12. Åkesson, B., *Eur. J. Biochem.* **9,** 463 (1969).
13. Åkesson, B., *Biochim. Biophys. Acta* **218,** 57 (1970).
14. Åkesson, B., Elovson, J., and Arvidson, G., *Biochim. Biophys. Acta* **210,** 15 (1970).
15. Åkesson, B., Elovson, J., and Arvidson, G., *Biochim. Biophys. Acta* **218,** 44 (1970).
16. Albrink, M. J., *J. Lipid Res.* **1,** 53 (1959).
17. Alford, J. A., Pierce, D. A., and Suggs, F. G., *J. Lipid Res.* **5,** 390 (1964).
18. Alford, J. A., and Smith, J. L., *J. Amer. Oil Chem. Soc.* **42,** 1038 (1965).
19. Amat, F., Marquinez, E., Utrilla, R. M., and Martin, D., *Grasas Aceites* **17,** 47 (1966).
20. Amberger, C., *Z. Unters. Nahr.- Genussm. Gebranchsgegenstaende* **40,** 192 (1920).
21. Amberger, C., and Bromig, K., *Z. Unters. Nahr.- Genussm. Gebranchsgegentstaende* **42,** 193 (1921).

22. Amenta, J. S., *J. Lipid Res.* **5,** 270 (1964).
23. American Oil Chemists' Society, "Official and Tentative Methods of the American Oil Chemists' Society," Method Cc-1-25. Amer. Oil Chem. Soc., Chicago, Illinois, 1962.
24. American Oil Chemists' Society, "Official and Tentative Methods of The American Oil Chemists' Society," Method Aa-4-38. Amer. Oil Chem. Soc., Chicago, Illinois, 1962.
25. Anderson, R. E., Bottino, N. R., and Reiser, R., *Lipids* **2,** 440 (1967).
26. Anderson, R. E., Bottino, N. R., and Reiser, R., *Lipids* **5,** 161 (1970).
27. Anker, L., and Sonanini, D., *Pharm. Acta Helv.* **37,** 360 (1962).
28. Anonymous, *Gas-Chrom Newslett.* (Applied Science Laboratories) **2,** No. 4, 1 (1961).
29. Anonymous, *Gas-Chrom Newslett.* (Applied Science Laboratories) **3,** No. 4, 1 (1962).
30. Anonymous, *Gas-Chrom Newslett.* (Applied Science Laboratories) **6,** No. 4, 2 (1965).
30a. Antonis, A., Platt, D. S., and Thorp, J. M., *J. Lipid Res.* **6,** 301 (1965).
31. Aparicio, M., *Proc. Int. Dairy Congr., 16th, 1962* Sect. B, pp. 5–10 (1963).
32. Appelqvist, L.-A., and Dowdell, R. J., *Ark. Kemi* **28,** 539 (1968).
33. Archibald, F. M., and Skipski, V. P., *J. Lipid Res.* **7,** 442 (1966).
33a. Arunga, R. O., and Morrison, W. R., *Lipids* **6,** 768 (1971).
34. Aue, W. A., and Hastings, C. R., *J. Chromatogr.* **42,** 319 (1969).
35. Augustin, M. P., *Oleagineux* **22,** 99 (1967).
36. Baer, E., and Fischer, H. O. L., *J. Biol. Chem.* **128,** 475 (1939).
37. Baer, E., and Kates, M., *J. Amer. Chem. Soc.* **72,** 942 (1950).
38. Baer, E., and Mahadevan, V., *J. Amer. Chem. Soc.* **81,** 2494 (1959).
39. Bagby, M. O., and Smith, C. R., Jr., *Biochim. Biophys. Acta* **137,** 475 (1967).
40. Bailey, A. E., "Melting and Solidification of Fats," pp. 123–138. Wiley (Interscience), New York, 1950.
41. Bailey, A. E., "Melting and Solidification of Fats," pp. 153–166. Wiley (Interscience), New York, 1950.
42. Balatre, P., Bertin, P., and Traisnel, M., *Bull. Soc. Pharm. Lille* No. 2, p. 89 (1967).
43. Balls, A. K., and Matlack, M. B., *J. Biol. Chem.* **123,** 679 (1938).
44. Balls, A. K., Matlack, M. B., and Tucker, I. W., *J. Biol. Chem.* **122,** 125 (1937).
45. Bandi, Z. L., and Mangold, H. K., *Separ. Sci.* **4,** 83 (1969).
46. Bandyopadhyay, C., *J. Chromatogr.* **37,** 123 (1968).
47. Barber, M., Chapman, J. R., and Wolstenholme, W. A., *Int. J. Mass Spectrom. Ion Phys.* **1,** 98 (1968).
48. Barber, M., Merren, T. O., and Kelley, W., *Tetrahedron Lett.* No. 18, p. 1063 (1964).
49. Barford, R. A., Herb, S. F., Luddy, F. E., Magidman, P., and Riemenschneider, R. W., *J. Amer. Oil Chem. Soc.* **40,** 136 (1963).
50. Barford, R. A., Luddy, F. E., Herb, S. F., Magidman, P., and Riemenschneider, R. W., *J. Amer. Oil Chem. Soc.* **42,** 446 (1965).
51. Barford, R. A., Luddy, F. E., and Magidman, P., *Lipids* **1,** 287 (1966).
51a. Barford, R. A., Rothbart, H. L., and Bertsch, R. J., *Separ. Sci.* **6,** 175 (1971).
52. Barker, C., and Hilditch, T. P., *J. Oil Colour Chem. Ass.* **33,** 6 (1950).
53. Barr, J. K., and Sawyer, D. T., *Anal. Chem.* **36,** 1753 (1964).

54. Barrall, E. M., and Guffy, J. C., *Advan. Chem.* **63**, 1 (1967).
55. Barrett, C. B., Dallas, M. S. J., and Padley, F. B., *Chem. Ind.* (*London*) p. 1050 (1962).
56. Barrett, C. B., Dallas, M. S. J., and Padley, F. B., *J. Amer. Oil Chem. Soc.* **40**, 580 (1963).
57. Barron, E. J., and Stumpf, P. K., *Biochim. Biophys. Acta* **60**, 329 (1962).
58. Baskys, B., Klein, E., and Lever, W. F., *Arch. Biochem. Biophys.* **102**, 201 (1963).
59. Baumann, W. J., Schmid, H. H. O., Ulshöfer, H. W., and Mangold, H. K., *Biochim. Biophys. Acta* **144**, 355 (1967).
60. Beeson, J. H., and Pecsar, R. E., *Anal. Chem.* **41**, 1678 (1969).
61. Begemann, P. H., Keppler, J. G., and Boekenoogen, H. A., *Rec. Trav. Chim. Pays. Bas* **69**, 439 (1950).
62. Belfrage, P., Wiebe, T., and Lundquist, A., *Scand. J. Clin. Lab. Invest.* **26**, 53 (1970).
63. Bell, J., "Parliamentary Papers," No. 293, June 15, 1876.
64. Bell, J., "The Chemistry of Foods," Vol. 2, pp. 44–46. Chapman & Hall, London, 1883.
64a. Bell, J. L., Atkinson, S. M., and Baron, D. N., *J. Clin. Pathol.* **23**, 509 (1970).
65. Bentz, A. P., and Breidenbach, B. G., *J. Amer. Oil Chem. Soc.* **46**, 60 (1969).
66. Benzonana, G., *Biochim. Biophys. Acta* **151**, 137 (1968).
67. Benzonana, G., Entressangles, B., Marchis-Mouren, G., Pasero, L., Sarda, L., and Desnuelle, P., *in* "Metabolism and Physiological Significance of Lipids" (R. M. C. Dawson and D. N. Rhodes, eds.), pp. 141–154. Wiley, New York, 1964.
68. Bergelson, L. D., Vaver, V. A., Prokazova, N. V., Ushakov, A. N., and Popkova, G. A., *Biochim. Biophys. Acta* **116**, 511 (1966).
69. Berger, K. G., and Akehurst, E. E., *J. Food Technol.* **1**, 237 (1966).
70. Bergmann, L., *Z. Instrumentenkunde* **65**, 2 (1957).
71. Bergström, S., Borgström, B., Tryding, N., and Westöö, G., *Biochem. J.* **58**, 604 (1954).
72. Berner, D. L., and Hammond, E. G., *Lipids* **5**, 558 (1970).
73. Bernstein, I. M., *J. Polym. Sci.* **1**, 495 (1946).
74. Beroza, M., and Bowman, M. C., *Anal. Chem.* **43**, 808 (1971).
75. Berthelot, M., *Ann. Chim. Phys.* [3] **41**, 216 (1854).
76. Bezard, J., Bugaut, M., and Clement, G., *J. Amer. Oil Chem. Soc.* **48**, 134 (1971).
77. Bhattacharya, R., and Hilditch, T. P., *Proc. Roy. Soc., Ser. A* **129**, 468 (1930).
78. Biernoth, G., *Fette, Seifen, Anstrichm.* **70**, 402 (1968).
79. Bird, P. R., de Haas, G. H., Heemskerk, C. H. T., and van Deenen, L. L. M., *Biochim. Biophys. Acta* **98**, 566 (1965).
80. Black, B. C., and Hammond, E. G., *J. Amer. Oil Chem. Soc.* **40**, 575 (1963).
81. Black, H. C., and Overley, C. A., *J. Amer. Chem. Soc.* **61**, 3051 (1939).
82. Blank, M. L., and Privett, O. S., *J. Dairy Sci.* **47**, 481 (1964).
83. Blank, M. L., and Privett, O. S., *Lipids* **1**, 27 (1966).
84. Blank, M. L., Schmit, J. A., and Privett, O. S., *J. Amer. Oil Chem. Soc.* **41**, 371 (1964).
85. Blank, M. L., Verdino, B., and Privett, O. S., *J. Amer. Oil Chem. Soc.* **42**, 87 (1965).

## REFERENCES

86. Bligh, E. G., and Dyer, W. J., *Can. J. Biochem. Physiol.* **37**, 911 (1959).
86a. Block, W. D., and Jarrett, K. J., Jr., *Amer. J. Med. Technol.* **35**, 93 (1969).
87. Blyth, A. W., and Robertson, G. H., *Proc. Chem. Soc., London* **5**, 5 (1889).
88. Bobbitt, J. M., "Thin-Layer Chromatography," Van Nostrand-Reinhold, Princeton, New Jersey, 1963.
89. Bombaugh, K. J., Dark, W. A., and King, R. N., *Res. Develop.* No. 9, 28 (1968).
90. Bombaugh, K. J., Dark, W. A., and Levangie, R. F., *Z. Anal. Chem.* **236**, 443 (1968).
91. Bombaugh, K. J., Dark, W. A., and Levangie, R. F., *J. Chromatogr. Sci.* **7**, 42 (1969).
92. Bömer, A., *Z. Unters. Nahr.- Genussm. Gebranchsgegenstaende* **14**, 90 (1907).
93. Bömer, A., *Z. Unters. Nahr.- Genussm. Gebranchsgegenstaende* **17**, 353 (1909).
94. Bonar, A. R., *Chem. Ind. (London)* p. 221 (1965).
95. Bonsen, P. P. M., and de Haas, G. H., *Chem. Phys. Lipids* **1**, 100 (1967).
96. Borgström, B., *Acta Chem. Scand.* **7**, 557 (1953).
97. Borgström, B., *Biochim. Biophys. Acta* **13**, 149 (1954).
98. Borgström, B., *Biochim. Biophys. Acta* **13**, 491 (1954).
99. Borgström, B., *J. Lipid Res.* **5**, 522 (1964).
100. Bottino, N. R., *J. Lipid Res.* **12**, 24 (1971).
101. Bottino, N. R., personal communication (1971).
102. Bottino, N. R., Vanderburg, G. A., and Reiser, R., *Lipids* **2**, 489 (1967).
103. Boucrot, P., and Clement, J., *Arch. Sci. Physiol.* **22**, 313 (1968).
104. Bougault, J., and Schuster, G., *C. R. Acad. Sci.* **192**, 953 (1931).
105. Bowie, J. H., and Cameron, D. W., *J. Chem. Soc., London* p. 5651 (1965).
106. Braae, B., *Anal. Chem.* **21**, 1461 (1949).
107. Braconnot, H., *Ann. Chim. (Paris)* [1] **93**, 225 (1815).
108. Breckenridge, W. C., and Kuksis, A., *J. Lipid Res.* **9**, 388 (1968).
109. Breckenridge, W. C., and Kuksis, A., *Lipids* **5**, 342 (1970).
110. Brockerhoff, H., *Arch. Biochem. Biophys.* **110**, 586 (1965).
111. Brockerhoff, H., *J. Lipid Res.* **6**, 10 (1965).
112. Brockerhoff, H., *Comp. Biochem. Physiol.* **19**, 1 (1966).
113. Brockerhoff, H., *Lipids* **1**, 162 (1966).
114. Brockerhoff, H., *J. Lipid Res.* **8**, 167 (1967).
115. Brockerhoff, H., personal communication (1968).
116. Brockerhoff, H., *Biochim. Biophys. Acta* **212**, 92 (1970).
117. Brockerhoff, H., *Lipids* **6**, 942 (1971).
118. Brockerhoff, H., and Hoyle, R. J., *Biochim. Biophys. Acta* **98**, 435 (1965).
119. Brockerhoff, H., Hoyle, R. J., and Hwang, P. C., *Can. J. Biochem.* **44**, 1519 (1966).
120. Brockerhoff, H., Hoyle, R. J., Hwang, P. C., and Litchfield, C., *Lipids* **3**, 24 (1968).
121. Brockerhoff, H., Hoyle, R. J., and Wolmark, N., *Biochim. Biophys. Acta* **116**, 67 (1966).
122. Brockerhoff, H., and Yurkowski, M., *J. Lipid Res.* **7**, 62 (1966).
123. Brown, C. A., Sethi, S. C., and Brown, H. C., *Anal. Chem.* **39**, 823 (1967).
124. Brown, J. L., and Johnston, J. M., *Biochim. Biophys. Acta* **84**, 448 (1964).
125. Bugaut, M., and Bezard, J., *J. Chromatogr. Sci.* **8**, 380 (1970).
126. Burchfield, H. P., and Storrs, E. E., "Biochemical Applications of Gas Chromatography." Academic Press, New York, 1962.

127. Burns, D. T., Stretton, R. J., Shepherd, G. F., and Dallas, M. S. J., *J. Chromatogr.* **44,** 399 (1969).
128. Buteau, G. H., Jr., and Fairbairn, D., *Exp. Parasitol.* **25,** 265 (1969).
129. Butterfield, R. O., and Dutton, H. J., *Anal. Chem.* **36,** 903 (1964).
130. Buziassy, C., and Nawar, W. W., *J. Food Sci.* **33,** 305 (1968).
131. Callery, I. M., *J. Chromatogr. Sci.* **8,** 408 (1970).
132. Capella, P., Fedeli, E., Cirimele, M., and Jacini, G., *Riv. Ital. Sostanze Grasse* **41,** 635 (1964).
133. Carlson, L. A., *J. Atheroscler. Res.* **3,** 334 (1963).
134. Carracedo, C. F., and Prieto, A., *Grasas Aceites* **20,** 289 (1969).
135. Carreau, J.-P., and Raulin, J., *Rev. Fr. Corps Gras* **12,** 87 (1965).
136. Carroll, K. K., *J. Lipid Res.* **2,** 135 (1961).
137. Carroll, K. K., and Serdarevich, B., in "Lipid Chromatographic Analysis" (G. V. Marinetti, ed.), Vol. 1, pp. 205–237. Dekker, New York, 1967.
138. Cavina, G., Moretti, G., Mollica, A., Moretta, L., and Siniscalchi, P., *J. Chromatogr.* **44,** 493 (1969).
139. Chacko, G. K., and Perkins, E. G., *J. Amer. Oil Chem. Soc.* **42,** 1121 (1965).
140. Chakrabarty, M. M., Bandyopadhyay, C., Bhattacharyya, D., and Gayen, A. K., *J. Chromatogr.* **36,** 84, 1968.
141. Chakrabarty, M. M., and Bhattacharyya, D., *J. Oil Technol. Ass. India* **18,** 317 (1963).
142. Chakrabarty, M. M., and Bhattacharyya, D., *J. Chromatogr.* **31,** 556 (1967).
143. Chakrabarty, M. M., Bhattacharyya, D., and Gupta, A., *J. Chromatogr.* **22,** 84 (1966).
144. Chakrabarty, M. M., Bhattacharyya, D., and Mondal, B., *Indian J. Technol.* **1,** 473 (1963).
145. Chandan, R. C., and Shahani, K. M., *J. Dairy Sci.* **46,** 275 (1963).
146. Chapman, D., *J. Amer. Oil Chem. Soc.* **37,** 73 (1960).
147. Chapman, D., *Chem. Rev.* **62,** 433 (1962).
148. Chapman, D., *J. Chem. Soc., London* p. 131 (1963).
149. Chapman, D., *J. Amer. Oil Chem. Soc.* **42,** 353 (1965).
150. Chapman, D., "The Structure of Lipids," pp. 52–132. Methuen, London, 1965.
151. Chapman, D., "The Structure of Lipids," pp. 133–159. Methuen, London, 1965.
152. Chapman, D., "The Structure of Lipids," pp. 160–207. Methuen, London, 1965.
153. Chapman, D., "The Structure of Lipids," pp. 221–315. Methuen, London, 1965.
154. Chapman, D., Crossley, A., and Davies, A. C., *J. Chem. Soc., London* p. 1502 (1957).
155. Chapman, D., Richards, R. E., and Yorke, R. W., *J. Chem. Soc., London* p. 436 (1960).
156. Chen, P. C., and deMan, J. M., *J. Dairy Sci.* **49,** 612 (1966).
157. Cherayil, G. D., and Scaria, K. S., *J. Lipid Res.* **11,** 378 (1970).
158. Chernick, S. S., in "Methods in Enzymology" (J. M. Lowenstein, ed.), Vol. 14, pp. 627–630. Academic Press, New York, 1969.
159. Chevreul, M. E., *Ann. Chim. (Paris)* [1] **94,** 113 (1815).
160. Chevreul, M. E., "Recherches chimiques sur les corps gras d'origine animale." Levrault, Paris, 1823.
161. Chin, H. P., El-Meguid, S. S. A., and Blankenhorn, D. H., *Clin. Chim. Acta* **31,** 381 (1971).
162. Chino, H., and Gilbert, L. I., *Anal. Biochem.* **10,** 395 (1965).

163. Chobanov, D., *C. R. Acad. Bulg. Sci.* **14,** 27 (1961).
164. Chobanov, D., *C. R. Acad. Bulg. Sci.* **14,** 587 (1961).
165. Chobanov, D., and Popov, A., *C. R. Acad. Bulg. Sci.* **14,** 171 (1961).
166. Christian, B. C., and Hilditch, T. P., *Analyst* **55,** 75 (1930).
167. Christie, W. W., *J. Chromatogr.* **34,** 405 (1968).
168. Christie, W. W., *Biochim. Biophys. Acta* **187,** 1 (1969).
169. Christie, W. W., *Topics Lipid Chem.* **3,** 171 (1972).
170. Christie, W. W., and Moore, J. H., *Biochim. Biophys. Acta* **176,** 445 (1969).
171. Christie, W. W., and Moore, J. H., *Lipids* **4,** 345 (1969).
172. Christie, W. W., and Moore, J. H., *Biochim. Biophys. Acta* **210,** 46 (1970).
173. Christie, W. W., and Moore, J. H., *Biochim. Biophys. Acta* **218,** 83 (1970).
174. Christie, W. W., and Moore, J. H., *Lipids* **5,** 921 (1970).
175. Christie, W. W., and Moore, J. H., *J. Sci. Food Agr.* **22,** 120 (1971).
176. Christie, W. W., Noble, R. D., and Moore, J. H., *Analyst* **95,** 940 (1970).
177. Christopherson, S. W., and Glass, R. L., *J. Dairy Sci.* **52,** 1289 (1969).
178. Claesson, S., *Ark. Kemi, Mineral. Geol.* **15A,** No. 9 (1942).
179. Clement, G., Belleville, J., Loriette, C., and Raulin, J., *Bull. Soc. Chim. Biol.* **45,** 1433 (1963).
180. Clement, J., Boucrot, P., Loriette, C., and Raulin, J., *Bull. Soc. Chim. Biol.* **45,** 1031 (1963).
181. Clement, G., and Clement, J., *J. Physiol., Paris* **50,** 244 (1958).
182. Clement, G., Clement, J., and Bezard, J., *Arch. Sci. Physiol.* **16,** 213 (1962).
183. Clement, J., Lavoue, G., and Clement, G., *J. Amer. Oil Chem. Soc.* **42,** 1035 (1965).
184. Clement, J., and Rigollot, B., *Nutr. Dieta* **6,** 61 (1964).
185. Coenen, J. W. E., Boerma, H., Linsen, B. G., and de Vries, B., *Proc. Int. Congr. Catal., 3rd, 1964* pp. 1387–1399 (1965).
186. Coleman, M. H., *Advan. Lipid Res.* **1,** 1 (1963).
187. Coleman, M. H., *J. Amer. Oil Chem. Soc.* **40,** 568 (1963).
188. Coleman, M. H., *J. Amer. Oil Chem. Soc.* **41,** 247 (1964).
189. Coleman, M. H., *J. Amer. Oil Chem. Soc.* **42,** 751 (1965).
190. Coleman, M. H., *J. Amer. Oil Chem. Soc.* **42,** 1040 (1965).
191. Coleman, M. H., and Fulton, W. C., in "Enzymes of Lipid Metabolism" (P. Desnuelle, ed.), pp. 127–137. Pergamon, Oxford, 1961.
192. Collin, G., and Hilditch, T. P., *Biochem. J.* **23,** 1273 (1929).
193. Collin, G., Hilditch, T. P., and Lea, C. H., *J. Soc. Chem. Ind.* **48,** 46T (1929).
194. Conacher, H. B. S., Gunstone, F. D., Hornby, G. M., and Padley, F. B., *Lipids* **5,** 434 (1970).
195. Cotgreave, T., *Chem. Ind. (London)* p. 689 (1966).
196. Cotgreave, T., and Lynes, A., *J. Chromatogr.* **30,** 117 (1967).
197. Craig, B. M., Tulloch, A. P., and Murty, N. L., *J. Amer. Oil Chem. Soc.* **40,** 61 (1963).
198. Cramp, D. G., and Robertson, G., *Anal. Biochem.* **25,** 246 (1968).
199. Crawford, R. V., and Hilditch, T. P., *J. Sci. Food Agr.* **1,** 230 (1950).
200. Crider, Q. E., Alaupovic, P., Hillsberry, J., Yen, C., and Bradford, R. H., *J. Lipid Res.* **5,** 479 (1964).
201. Crossley, A., Freeman, I. P., Hudson, B. J. F., and Pierce, J. H., *J. Chem. Soc., London* p. 760 (1959).

202. Cubero, J. M., and Mangold, H. K., *Microchem. J.* **9,** 227 (1965).
203. Culp, T. W., Harlow, R. D., Litchfield, C., and Reiser, R., *J. Amer. Oil Chem. Soc.* **42,** 974 (1965).
204. Dalgliesh, C. E., Horning, E. C., Horning, M. G., Knox, K. L., and Yarger, K., *Biochem. J.* **101,** 792 (1966).
205. Dallas, M. S. J., *J. Chromatogr.* **33,** 58 (1968).
205a. Dalton, C., and Mallon, J. P., *Advan. Automat. Anal., Technicon Int. Congr., 1969,* Vol. 2, 183 (1970).
206. Dasso, I., and Cattaneo, P., *An. Asoc. Quim. Argent.* **59,** 35 (1971).
207. Daubert, B. F., *J. Amer. Oil Chem. Soc.* **26,** 556 (1949).
208. Day, A. J., and Fidge, N. H., *J. Lipid Res.* **5,** 163 (1964).
209. de Haas, G. H., and van Deenen, L. L. M., *Biochim. Biophys. Acta* **84,** 467 (1964).
210. de Haas, G. H., and van Deenen, L. L. M., *Biochim. Biophys. Acta* **106,** 315 (1965).
211. de Kruyff, B., van Golde, L. M. G., and van Deenen, L. L. M., *Biochim. Biophys. Acta* **210,** 425 (1970).
212. de la Roche, I. A., Weber, E. J., and Alexander, D. E., *Lipids* **6,** 531 (1971).
213. den Boer, F. C., *Z. Anal. Chem.* **205,** 308 (1964).
214. Desnuelle, P., *Advan. Enzymol.* **23,** 129 (1961).
215. Desnuelle, P., and Naudet, M., *Bull. Soc. Chim. Fr.* [5] p. 90 (1946).
216. Desnuelle, P., Naudet, M., and Constantin, M. J., *Biochim. Biophys. Acta* **5,** 561 (1950).
217. Desnuelle, P., and Savary, P., *J. Lipid Res.* **4,** 369 (1963).
218. de Vries, B., *Chem. Ind. (London)* p. 1049 (1962).
219. de Vries, B., *J. Amer. Oil Chem. Soc.* **40,** 184 (1963).
220. de Vries, B., *J. Amer. Oil Chem. Soc.* **41,** 403 (1964).
221. de Vries, B., and Jurriens, G., *Fette, Seifen, Anstrichm.* **65,** 725 (1963).
222. de Vries, B., and Jurriens, G., *J. Chromatogr.* **14,** 525 (1964).
223. Distler, E., and Baur, F. J., *J. Ass. Offic. Agr. Chem.* **48,** 444 (1965).
224. Dittmer, J. C., and Wells, M. A., *in* "Methods in Enzymology" (J. M. Lowenstein, ed.), Vol. 14, pp. 513–514. Academic Press, New York, 1969.
225. Dixon, C. W., and Schmit, J. A., *Anal. Advan.* (Hewlett-Packard Co.) **1,** No. 1, 17 (1968).
226. Doerschuk, A. P., and Daubert, B. F., *J. Amer. Oil Chem. Soc.* **25,** 425 (1948).
227. Dolendo, A. L., Means, J. C., Tobias, J., and Perkins, E. G., *J. Dairy Sci.* **52,** 21 (1969).
228. Dolev, A., and Olcott, H. S., *J. Amer. Oil Chem. Soc.* **42,** 624 (1965).
229. Dolev, A., and Olcott, H. S., *J. Amer. Oil Chem. Soc.* **42,** 1046 (1965).
230. Downey, R. K., and Craig, B. M., *J. Amer. Oil Chem. Soc.* **41,** 475 (1964).
231. Downey, W. K., and Andrews, P., *Biochem. J.* **112,** 559 (1969).
232. Downey, W. K., Murphy, R. F., and Keogh, M. K., *J. Chromatogr.* **46,** 120 (1970).
233. Downing, D. T., *J. Chromatogr.* **38,** 91 (1968).
234. Duffy, P., *J. Chem. Soc., London* **5,** 197 (1853).
235. Duncan, W. R. H., and Garton, G. A., *J. Sci. Food Agr.* **18,** 99 (1967).
236. Duthie, A. H., and Atherton, H. V., *J. Chromatogr.* **51,** 319 (1970).
237. Dutton, H. J., *J. Amer. Oil Chem. Soc.* **32,** 652 (1955).
238. Dutton, H. J., and Cannon, J. A., *J. Amer. Oil Chem. Soc.* **33,** 46 (1956).

239. Dutton, H. J., Lancaster, C. R., and Brekke, O. L., *J. Amer. Oil Chem. Soc.* **27**, 25 (1950).
240. Dutton, H. J., and Scholfield, C. R., *Progr. Chem. Fats Other Lipids* **6**, 314 (1963).
241. Dutton, H. J., Scholfield, C. R., and Mounts, T. L., *J. Amer. Oil Chem. Soc.* **38**, 96 (1961).
242. Ebing, W., *J. Gas Chromatogr.* **5**, No. 9, 20A (1967).
243. Eckey, E. W., "Vegetable Fats and Oils," pp. 278–286. Van Nostrand-Reinhold, Princeton, New Jersey, 1954.
244. Eggstein, M., and Kreutz, F. H., *Klin. Wochenschr.* **44**, 262 (1966).
245. Eibner, A., and Schmidinger, K., *Chem. Umsch. Geb. Fette, Oele, wachse Harze* **30**, 293 (1923).
246. Eibner, A., Widenmayer, L., and Schild, E., *Chem. Umsch. Geb. Fette, Oele, wachse Harze* **34**, 312 (1927).
247. Ellingboe, J., Nyström, E., and Sjövall, J., *J. Lipid Res.* **11**, 266 (1970).
248. Entenman, C., *in* "Methods in Enzymology" (S. P. Colowick and N. O. Kaplan, eds.), Vol. 3, pp. 299–317. Academic Press, New York, 1957.
249. Entenman, C., *J. Amer. Oil Chem. Soc.* **38**, 534 (1961).
250. Entressangles, B., Sari, H., and Desnuelle, P., *Biochim. Biophys. Acta* **125**, 597 (1966).
251. Entressangles, B., Savary, P., Constantin, M. J., and Desnuelle, P., *Biochim. Biophys. Acta* **84**, 140 (1964).
252. Erlanson, C., and Borgström, B., *Scand. J. Gastroenterol.* **5**, 395 (1970).
253. Eshelman, L. R., and Hammond, E. G., *J. Amer. Oil Chem. Soc.* **35**, 230 (1958).
254. Eshelman, L. R., Manzo, E. Y., Marcus, S. J., Decoteau, A. E., and Hammond, E. G., *Anal. Chem.* **32**, 844 (1960).
255. Ettre, L. S., and Zlatkis, A., eds., "The Practice of Gas Chromatography." Wiley (Interscience), New York, 1967.
256. Evans, C. D., McConnell, D. G., Hoffmann, R. L., and Peters, H., *J. Amer. Oil Chem. Soc.* **44**, 281 (1967).
257. Evans, C. D., McConnell, D. G., List, G. R., and Scholfield, C. R., *J. Amer. Oil Chem. Soc.* **46**, 421 (1969).
258. Evans, C. D., McConnell, D. G., Scholfield, C. R., and Dutton, H. J., *J. Amer. Oil Chem. Soc.* **43**, 345 (1966).
258a. Fales, H. M., and Milne, G. W. A., *J. Amer. Oil Chem. Soc.* **48**, 333A (1971).
259. Farquhar, J. W., Insull, W., Jr., Rosen, P., Stoffel, W., and Ahrens, E. H., Jr., *Nutr. Rev.* **17**, August Suppl., pp. 1–30 (1959).
260. Fedeli, E., *Riv. Ital. Sostanze Grasse* **44**, 220 (1967).
261. Fedeli, E., and Camurati, F., *Riv. Ital. Sostanze Grasse* **46**, 97 (1969).
262. Fedeli, E., Tarenghi, A., and Jacini, G., *Riv. Ital. Sostanze Grasse* **44**, 391 (1967).
263. Feuge, R. O., and Lovegren, N. V., *J. Amer. Oil Chem. Soc.* **33**, 367 (1956).
264. Filer, L. J., Jr., Mattson, F. H., and Fomon, S. J., *J. Nutr.* **99**, 293 (1969).
265. Fillerup, D. L., and Mead, J. F., *Proc. Soc. Exp. Biol. Med.* **83**, 574 (1953).
265a. Finch, R. W., *Analabs Res. Notes* **10**, No. 1, 1 (1970).
266. Findley, T. W., Swern, D., and Scanlan, J. T., *J. Amer. Chem. Soc.* **67**, 412 (1945).

267. Fioriti, J. A., Buide, N., and Sims, R. J., *J. Amer. Oil Chem. Soc.* **46,** 108 (1969).
268. Fioriti, J. A., Buide, N., and Sims, R. J., *Lipids* **4,** 142 (1969).
269. Fioriti, J. A., Kanuk, M. J., and Sims, R. J., *J. Chromatogr. Sci.* **7,** 448 (1969).
270. Fischer, G. A., and Kabara, J. J., *Anal. Biochem.* **25,** 432 (1968).
271. Fischer, R., *Fette, Seifen, Anstrichm.* **67,** 748 (1965).
272. Fischer, R., and Horner, J., *Mikrochim. Acta* No. 4, 386 (1953).
273. Fletcher, M. J., *Clin. Chim. Acta* **22,** 393 (1968).
274. Fodor, P. J., *Arch. Biochem.* **25,** 223 (1950).
275. Fodor, P. J., *Arch. Biochem.* **26,** 307 (1950).
276. Folch, J., Lees, M., and Stanley, G. H. S., *J. Biol. Chem.* **226,** 497 (1957).
276a. Fontell, K., Holman, R. T., and Lambertsen, G., *J. Lipid Res.* **1,** 391 (1960).
277. Fosslien, E., and Musil, F., *J. Lipid Res.* **11,** 605 (1970).
278. Fox, P. F., and Tarassuk, N. P., *J. Dairy Sci.* **51,** 826 (1968).
279. Frankel, E. N., and Tarassuk, N. P., *J. Dairy Sci.* **39,** 1517 (1956).
280. Frankel, E. N., and Tarassuk, N. P., *J. Dairy Sci.* **39,** 1523 (1956).
281. Frankel, E. N., and Tarassuk, N. P., *J. Dairy Sci.* **42,** 409 (1959).
282. Franzke, C., Heims, K.-O., and Vollgraf, I., *Nahrung* **11,** 515 (1967).
283. Franzke, C., Kretzschmann, F., Rüstow, B., and Rugenstein, H., *Pharmazie* **22,** 487 (1967).
284. Frazer, A. C., and Walsh, V. G., *J. Physiol. (London)* **78,** 467 (1933).
285. Freeman, N. K., *J. Lipid Res.* **5,** 236 (1964).
286. Freeman, N. K., Lindgren, F. T., Ng, Y. C., and Nichols, A. V., *J. Biol. Chem.* **227,** 449 (1957).
287. Friedrich, J. P., *Anal. Chem.* **33,** 974 (1961).
288. Fritz, J. S., and Wood, G. E., *Anal. Chem.* **40,** 134 (1968).
289. Fritz, P. J., and Melius, P., *Can. J. Biochem. Physiol.* **41,** 719 (1963).
290. Fryer, F. H., Ormand, W. L., and Crump, G. B., *J. Amer. Oil Chem. Soc.* **37,** 589 (1960).
291. Fuller, G., Diamond, M. J., and Applewhite, T. H., *J. Amer. Oil Chem. Soc.* **44,** 264 (1967).
292. Gaffney, P. J., Jr., Harper, W. J., and Gould, I. A., *J. Dairy Sci.* **49,** 921 (1966).
293. Gaffney, P. J., Jr., Harper, W. J., and Gould, I. A., *J. Dairy Sci.* **51,** 1161 (1968).
294. Galanos, D. S., Aïvazis, G. A. M., and Kapoulas, V. M., *J. Lipid Res.* **5,** 242 (1964).
295. Galanos, D. S., Kapoulas, V. M., and Voudouris, E. C., *J. Amer. Oil Chem. Soc.* **45,** 825 (1968).
296. Galletti, F., *Clin. Chim. Acta* **15,** 184 (1967).
297. Galoppini, C., and Lotti, G., *Chim. Ind. (Milan)* **45,** 812 (1963).
298. Gander, G. W., and Jensen, R. G., *J. Dairy Sci.* **43,** 1762 (1960).
299. Gander, G. W., Jensen, R. G., and Sampugna, J., *J. Dairy Sci.* **44,** 1980 (1961).
300. Garner, C. W., and Smith, L. C., *Arch. Biochem. Biophys.* **140,** 503 (1970).
301. Garner, C. W., and Smith, L. C., *Biochem. Biophys. Res. Commun.* **39,** 672 (1970).
302. Gessmann, G. W., "Die Geheimsymbole der Alchymie, Arzneikunde und Astrologie des Mittelalters," Table VI. Arkana-Verlag, Ulm/Donau, 1964.

303. Gidez, L. I., *J. Lipid. Res.* **9,** 794 (1968).
304. Glass, R. L., Jenness, R., and Lohse, L. W., *Comp. Biochem. Physiol.* **28,** 783 (1969).
305. Gödicke, W., and Gerike, U., *Clin. Chim. Acta* **30,** 727 (1970).
306. Golborn, P., *J. Amer. Oil Chem. Soc.* **46,** 385 (1969).
307. Gold, M., *Lipids* **3,** 539 (1968).
308. Gold, M., *Lipids* **4,** 288 (1969).
309. Goldman, M. L., Burton, T. H., and Rayman, M. M., *Food Res.* **19,** 503 (1954).
310. Golikova, V. S., Mitrofanova, T. K., Shvets, V. I., Zubov, P. I., and Preobrazhenskii, N. A., *Zh. Org. Khim.* **1,** 433 (1965) (available in English translation).
311. Goodman, L. P., and Dugan, L. R., Jr. *Lipids* **5,** 362 (1970).
312. Gorbach, G., *Fette Seifen* **47,** 499 (1940).
313. Gordis, E., *J. Clin. Invest.* **44,** 1451 (1965).
314. Gordis, E., *J. Clin. Invest.* **44,** 1978 (1965).
315. Gouw, T. H., and Vlugter, J. C., *Fette, Seifen, Anstrichm.* **68,** 544 (1966).
316. Gouw, T. H., and Vlugter, J. C., *Fette, Seifen, Anstrichm.* **69,** 159 (1967).
317. Gouw, T. H., and Vlugter, J. C., *Fette, Seifen, Anstrichm.* **69,** 223 (1967).
318. Gruger, E. H., Jr., Malins, D. C., and Gauglitz, E. J., Jr., *J. Amer. Oil Chem. Soc.* **37,** 214 (1960).
319. Grynberg, H., Ceglowska, K., and Szczepanska, H., *Rev. Fr. Corps Gras* **13,** 595 (1966).
320. Grynberg, H., and Szczepanska, H., *J. Amer. Oil Chem. Soc.* **43,** 151 (1966).
321. Grynberg, H., Szczepanska, H., and Beldowicz, M., *Oleagineux* **17,** 875 (1962).
322. Gunde, B. G., and Hilditch, P. T., *J. Soc. Chem. Ind.* **59,** 47 (1940).
323. Gunstone, F. D., *Chem. Ind. (London)* p. 1214 (1962).
324. Gunstone, F. D., Hamilton, R. J., Padley, F. B., and Qureshi, M. I., *J. Amer. Oil Chem. Soc.* **42,** 965 (1965).
325. Gunstone, F. D., Hamilton, R. J., and Qureshi, M. I., *J. Chem. Soc., London* p. 319 (1965).
326. Gunstone, F. D., Hilditch, T. P., and Riley, J. P., *J. Soc. Chem. Ind.* **66,** 293 (1947).
327. Gunstone, F. D., Ismail, I. A., and Lie Ken Jie, M., *Chem. Phys. Lipids* **1,** 376 (1967).
328. Gunstone, F. D., and Padley, F. B., *J. Amer. Oil Chem. Soc.* **42,** 957 (1965).
329. Gunstone, F. D., and Padley, F. B., *Chem. Phys. Lipids* **1,** 110 (1967).
330. Gunstone, F. D., Padley, F. B., and Qureshi, M. I., *Chem. Ind. (London)* p. 483 (1964).
331. Gunstone, F. D., and Qureshi, M. I., *J. Amer. Oil Chem. Soc.* **42,** 961 (1965).
332. Gunstone, F. D., and Qureshi, M. I., *J. Sci. Food Agr.* **19,** 386 (1968).
333. Gupta, S. S., and Hilditch, T. P., *Biochem. J.* **48,** 137 (1951).
334. Haab, W., Smith, L. M., and Jack, E. L., *J. Dairy Sci.* **42,** 454 (1959).
335. Haahti, E., Vihko, R., Jaakonmäki, I., and Evans, R. S., *J. Chromatogr. Sci.* **8,** 370 (1970).
336. Hagemann, J. W., personal communication (1968).
337. Hagony, P. L., *Olaj, Szappan, Kozmet.* **16,** 9 (1967).
338. Haighton, A. J., van Beers, G. J., and Hannewijk, J., *Riv. Ital. Sostanze Grasse* **39,** 130 (1962).
339. Hamilton, J. G., and Holman, R. T., *J. Amer. Chem. Soc.* **76,** 4107 (1954).

340. Hammond, E. G., *Lipids* **4,** 246 (1969).
341. Hammond, E. G., and Jones, G. V., *J. Amer. Oil Chem. Soc.* **37,** 376 (1960).
342. Hammonds, T. W., and Shone, G., *J. Chromatogr.* **15,** 200 (1964).
343. Hanahan, D. J., Brockerhoff, H., and Barron, E. J., *J. Biol. Chem.* **235,** 1917 (1960).
343a. Harding, U., and Heinzel, G., *Z. Klin. Chem. Klin. Biochem.* **7,** 356 (1969).
344. Harlow, R. D., personal communication (1968).
345. Harlow, R. D., Litchfield, C., Fu, H.-C., and Reiser, R., *J. Amer. Oil Chem. Soc.* **42,** 747 (1965).
346. Harlow, R. D., Litchfield, C., and Reiser, R., *Lipids* **1,** 216 (1966).
347. Harris, W. E., and Habgood, H. W., "Programmed Temperature Gas Chromatography." Wiley, New York, 1966.
348. Hashi, K., *Nippon Kagaku Zasshi* **31,** 117 (1928).
349. Hastings, C. R., Aue, W. A., and Augl, J. M., *J. Chromatogr.* **53,** 487 (1970).
349a. Hastings, C. R., Aue, W. A., and Larsen, F. N., *J. Chromatogr.* **60,** 329 (1971).
350. Haux, P., and Natelson, S., *Microchem. J.* **16,** 68 (1971).
351. Heise, R., *Arb. Kaiserl. Gesundheitsamte* **12,** 540 (1896).
352. Heise, R., *Arb. Kaiserl. Gesundheitsamte* **13,** 302 (1897).
353. Herb, S. F., and Martin, V. G., *J. Amer. Oil Chem. Soc.* **47,** 415 (1970).
354. Hilditch, T. P., and Jones, E. C., *J. Soc. Chem. Ind.* **53,** 13T (1934).
355. Hilditch, T. P., and Lea, C. H., *J. Chem. Soc., London* p. 3106 (1927).
356. Hilditch, T. P., and Maddison, L., *J. Soc. Chem. Ind.* **59,** 162 (1940).
357. Hilditch, T. P., and Meara, M. L., *J. Soc. Chem. Ind.* **61,** 117 (1942).
358. Hilditch, T. P., and Saletore, S. A., *J. Soc. Chem. Ind.* **50,** 468T (1931).
359. Hilditch, T. P., and Seavell, A. J., *J. Oil Colour Chem. Ass.* **33,** 24 (1950).
360. Hilditch, T. P., and Shrivastava, R. K., *J. Amer. Oil Chem. Soc.* **26,** 1 (1949).
361. Hilditch, T. P., and Stainsby, W. J., *Biochem. J.* **29,** 90 (1935).
362. Hilditch, T. P., and Stainsby, W. J., *Biochem. J.* **29,** 599 (1935).
363. Hilditch, T. P., and Williams, P. N., "The Chemical Constitution of Natural Fats," 4th ed., p. 19. Chapman & Hall, London, 1964.
364. Hilditch, T. P., and Williams, P. N., "The Chemical Constitution of Natural Fats," 4th ed., pp. 358–423. Chapman & Hall, London, 1964.
365. Hilditch, T. P., and Williams, P. N., "The Chemical Constitution of Natural Fats," 4th ed., pp. 358–527. Chapman & Hall, London, 1964.
366. Hilditch, T. P., and Williams, P. N., "The Chemical Constitution of Natural Fats," 4th ed., pp. 700–701. Chapman & Hall, London, 1964.
367. Hilditch, T. P., and Williams, P. N., "The Chemical Constitution of Natural Fats," 4th ed., pp. 701–702. Chapman & Hall, London, 1964.
368. Hilditch, T. P., and Zaky, Y. A. H., *Biochem. J.* **35,** 940 (1941).
369. Hill, E. E., Husbands, D. R., and Lands, W. E. M., *J. Biol. Chem.* **243,** 4440 (1968).
370. Hill, E. E., Lands, W. E. M., and Slakey, P. M., *Lipids* **3,** 411 (1968).
371. Hirayama, O., *Nippon Nogei Kagaku Kaishi* **35,** 437 (1961).
372. Hirayama, O., *Agr. Biol. Chem.* **28,** 193 (1964).
373. Hirayama, O., and Hujii, K., *Agr. Biol. Chem.* **29,** 1 (1965).
374. Hirayama, O., and Inouye, Y., *J. Agr. Chem. Soc. Jap.* **35,** 367 (1961).
375. Hirayama, O., and Inouye, Y., *J. Agr. Chem. Soc. Jap.* **35,** 372 (1961).
376. Hirayama, O., and Nakae, T., *Agr. Biol. Chem.* **28,** 201 (1964).
377. Hirayama, O., and Ohama, S., *Agr. Biol. Chem.* **29,** 111 (1965).

378. Hirsch, J., in "Digestion, Absorption Intestinale, et Transport des Glycerides chez les Animaux Superieurs" (P. Desnuelle, ed.), pp. 11–33. CNRS, Paris, 1961.
379. Hirsch, J., J. Lipid Res. **4,** 1 (1963).
380. Hirsch, J., and Ahrens, E. H., Jr., J. Biol. Chem. **233,** 311 (1958).
381. Hirschmann, H., J. Biol. Chem. **235,** 2762 (1960).
382. Hites, R. A., Anal. Chem. **42,** 1736 (1970).
383. Ho, R. J., Anal. Biochem. **36,** 105 (1970).
384. Ho, R. J., and Meng, H. C., Anal. Biochem. **31,** 426 (1969).
385. Hoefnagel, M. A., van Veen, A., and Verkade, P. E., Rec. Trav. Chim. Pays-Bas **81,** 461 (1962).
386. Hofmann, A. F., Biochim. Biophys. Acta **70,** 306 (1963).
387. Holla, K. S., Horrocks, L. A., and Cornwell, D. G., J. Lipid Res. **5,** 263 (1964).
388. Hollenberg, C. H., J. Lipid Res. **6,** 84 (1965).
389. Hollingsworth, C. A., Taber, J. J., and Daubert, B. F., Anal. Chem. **28,** 1901 (1956).
389a. Holub, B. J., Breckenridge, W. C., and Kuksis, A., Lipids **6,** 307 (1971).
390. Holub, B. J., and Kuksis, A., Lipids **4,** 466 (1969).
391. Hopkins, C. Y., and Bernstein, H. J., Can. J. Chem. **37,** 775 (1959).
392. Horning, E. C., Ahrens, E. H., Jr., Lipsky, S. R., Mattson, F. H., Mead, J. F., Turner, D. A., and Goldwater, W. H., J. Lipid Res. **5,** 20 (1964).
393. Horning E. C., Moscatelli, E. A., and Sweeley, C. C. Chem. Ind. (London) p. 751 (1959).
394. Horning, M. G., Casparrini, G., and Horning, E. C., J. Chromatogr. Sci. **7,** 267 (1969); Am. J. Clin. Nutr. **24,** 1086 (1971); Anal. Letters **1,** 481 (1968).
395. Horning, M. G., Williams, E. A., and Horning, E. C., J. Lipid Res. **1,** 482 (1960).
396. Hornstein, I., Crowe, P. F., and Ruck, J. B., Anal. Chem. **39,** 352 (1967).
397. Horvath, W. L., and Pieringer, R. A., Lipids **5,** 994 (1970).
398. Hubscher, G., Biochim. Biophys. Acta **52,** 582 (1961).
399. Huebner, V. R., J. Amer. Oil Chem. Soc. **36,** 262 (1959).
400. Huebner, V. R., Pap. Los Angeles Meet. Amer. Oil Chem. Soc., Sept. 1959 Pap. No. 4 (1959).
401. Huebner, V. R., quoted in Fontell et al. (276a).
402. Huebner, V. R., J. Amer. Oil Chem. Soc. **38,** 628 (1961).
403. Husbands, D. R., Biochem. J. **120,** 365 (1970).
404. Hustad, G. O., Richardson, T., Winder, W. C., and Dean, M. P., J. Dairy Sci. **53,** 1525 (1970).
405. Inkpen, J. A., and Quackenbush, F. W., Toronto Meet. Amer. Oil Chem. Soc., Oct. 1962 Pap. No. 23 (1962).
406. IUPAC-IUB Commission on Biochemical Nomenclature, J. Lipid Res. **8,** 523 (1967); Biochim. Biophys. Acta **152,** 1 (1968).
407. Jack, E. L., Freeman, C. P., Smith, L. M., and Mickle, J. B., J. Dairy Sci. **46,** 284 (1963).
408. Jack, R. C. M., Contrib. Boyce Thompson Inst. **22,** 335 (1964).
409. Jackson, J. E., and Lundberg, W. O., J. Amer. Oil Chem. Soc. **40,** 276 (1963).
410. Jagannathan, S. N., Can. J. Biochem. **42,** 566 (1964).
411. James, A. T., and Martin, A. J. P., Biochem. J. **50,** 679 (1952).

412. Jamieson, G. R., *Topics Lipid Chem.* **1**, 107 (1970).
412a. Jeejeebhoy, K. N., Ahmad, S., and Kozak, G., *Clin. Biochem.* **3**, 157 (1970).
413. Jellum, E., and Björnstad, P., *J. Lipid Res.* **5**, 314 (1964).
414. Jensen, R. G., *Progr. Chem. Fats Other Lipids* **11**, 347 (1971).
415. Jensen, R. G., Duthie, A. H., Gander, G. W., and Morgan, M. E., *J. Dairy Sci.* **43**, 96 (1960).
416. Jensen, R. G., Gander, G. W., Sampugna, J., and Forster, T. L., *J. Dairy Sci.* **44**, 943 (1961).
417. Jensen, R. G., Marks, T. A., Sampugna, J., Quinn, J. G., and Carpenter, D. L., *Lipids* **1**, 451 (1966).
418. Jensen, R. G., Pitas, R. E., Quinn, J. G., and Sampugna, J., *Lipids* **5**, 580 (1970).
419. Jensen, R. G., Quinn, J. G., Carpenter, D. L., and Sampugna, J., *J. Dairy Sci.* **50**, 119 (1967).
420. Jensen, R. G., Sampugna, J., Parry, R. M., Jr., and Forster, T. L., *J. Dairy Sci.* **45**, 842 (1962).
421. Jensen, R. G., Sampugna, J., Parry, R. M., Jr., and Shahani, K. M., *J. Dairy Sci.* **46**, 907 (1963).
422. Jensen, R. G., Sampugna, J., Parry, R. M., Jr., Shahani, K. M., and Chandan, R. C., *J. Dairy Sci.* **45**, 1527 (1962).
423. Jensen, R. G., Sampugna, J., and Pereira, R. L., *Biochim. Biophys. Acta* **84**, 481 (1964).
424. Jensen, R. G., Sampugna, J., and Pereira, R. L., *J. Dairy Sci.* **47**, 727 (1964).
425. Jensen, R. G., Sampugna, J., Pereira, R. L., Chandan, R. C., and Shahani, K. M., *J. Dairy Sci.* **47**, 1012 (1964).
426. Jensen, R. G., Sampugna, J., and Quinn, J. G., *Lipids* **1**, 294 (1966).
427. Jensen, R. G., Sampugna, J., Quinn, J. G., Carpenter, D. L., Marks, T. A., and Alford, J. A., *J. Amer. Oil Chem. Soc.* **42**, 1029 (1965).
428. Jezyk, P. F., *Can. J. Biochem.* **46**, 1167 (1968).
429. Johnson, A. R., Murray, K. E., Fogerty, A. C., Kennett, B. H., Pearson, J. A., and Shenstone, F. S., *Lipids* **2**, 316 (1967).
430. Johnson, C. B., and Holman, R. T., *Lipids* **1**, 371 (1966).
431. Johnston, J. M., *in* "Handbook of Physiology" (Amer. Physiol. Soc., J. Field, ed.), Sect. 6, Vol. III, pp. 1353–1375. Williams & Wilkins, Baltimore, Maryland, 1968.
432. Johnston, J. M., Paultauf, F., Schiller, C. M., and Schultz, L. D., *Biochim. Biophys. Acta* **218**, 124 (1970).
433. Johnston, J. M., Rao, G. A., and Lowe, P. A., *Biochim. Biophys. Acta* **137**, 578 (1967).
434. Johnson, L. F., and Shoolery, J. N., *Anal. Chem.* **34**, 1136 (1962).
435. Jones, G. V., and Hammond, E. G., *J. Amer. Oil Chem. Soc.* **38**, 69 (1961).
436. Joustra, M., Söderqvist, B., and Fischer, L., *J. Chromatogr.* **28**, 21 (1967).
437. Jover, A., *J. Lipid Res.* **4**, 228 (1963).
438. Jurriens, G., *Chem. Weekbl.* **61**, 257 (1965).
439. Jurriens, G., *Anal. Character. Oils, Fats, Fat Prod.* **2**, 237 (1968).
440. Jurriens, G., *Anal. Character. Oils, Fats, Fat Prod.* **2**, 273 (1968).
441. Jurriens, G., de Vries, B., and Schouten, L., *J. Lipid Res.* **5**, 267 (1964).
442. Jurriens, G., de Vries, B., and Schouten, L., *J. Lipid Res.* **5**, 366 (1964).
443. Jurriens, G., and Kroesen, A. C. J., *J. Amer. Oil Chem. Soc.* **42**, 9 (1965).
444. Jurriens, G., and Schouten, L., *Rev. Fr. Corps Gras* **12**, 505 (1965).

445. Kaimal, T. N. B., and Lakshminarayana, G., *J. Amer. Oil Chem. Soc.* **47**, 193 (1970).
446. Kaimal, T. N. B., and Lakshminarayana, G., *J. Amer. Oil Chem. Soc.* **47**, 316A (1970).
447. Karlsson, K.-A., Nilsson, K., and Pascher, I., *Lipids* **3**, 389 (1968).
448. Karlsson, K.-A., Norrby, A., and Samuelsson, B., *Biochim. Biophys. Acta* **144**, 162 (1967).
449. Karmen, A., *Separ. Sci.* **2**, 387 (1967).
450. Karnovsky, M. L., and Wolff, D., *in* "Biochemistry of Lipids" (G. Popjak, ed.), pp. 53–59. Pergamon, Oxford, 1960.
451. Kartha, A. R. S., *J. Amer. Oil Chem. Soc.* **30**, 280 (1953).
452. Kartha, A. R. S., *J. Amer. Oil Chem. Soc.* **30**, 326 (1953).
453. Kartha, A. R. S., *J. Amer. Oil Chem. Soc.* **31**, 85 (1954).
454. Kartha, A. R. S., *J. Sci. Ind. Res., Sect. A* **13**, 471 (1954).
455. Kartha, A. R. S., *J. Amer. Oil Chem. Soc.* **39**, 478 (1962).
456. Kartha, A. R. S., *J. Sci. Ind. Res., Sect. A* **21**, 577 (1962).
457. Kartha, A. R. S., *J. Sci. Food Agr.* **14**, 515 (1963).
458. Kartha, A. R. S., *Indian J. Chem.* **2**, 199 (1964).
459. Kartha, A. R. S., *J. Amer. Oil Chem. Soc.* **41**, 456 (1964).
460. Kartha, A. R. S., *J. Amer. Oil Chem. Soc.* **46**, 56 (1969).
461. Kartha, A. R. S., *J. Amer. Oil Chem. Soc.* **46**, 632 (1969).
462. Kartha, A. R. S., *J. Amer. Oil Chem. Soc.* **47**, 366 (1970).
463. Kartha, A. R. S., and Narayanan, R., *J. Sci. Food Agr.* **13**, 411 (1962).
464. Kartha, A. R. S., and Narayanan, R., *J. Sci. Ind. Res., Sect. B* **21**, 494 (1962).
465. Kartha, A. R. S., and Narayanan, R., *Indian J. Chem.* **4**, 544 (1966).
466. Kartha, A. R. S., and Narayanan, R., *J. Amer. Oil Chem. Soc.* **44**, 350 (1967).
467. Kartha, A. R. S., and Narayanan, R., *J. Amer. Oil Chem. Soc.* **44**, 733 (1967).
468. Kartha, A. R. S., and Selvaraj, Y., *Indian J. Agr. Sci.* **39**, 633 (1969).
469. Kartha, A. R. S., and Selvaraj, Y., *J. Amer. Oil Chem. Soc.* **46**, 685 (1969).
470. Kartha, A. R. S., and Upadhyay, G. S., *J. Amer. Oil Chem. Soc.* **45**, 750 (1968).
470a. Kashket, S., *Anal. Biochem.* **41**, 166 (1971).
471. Kates, M., *in* "Lipide Metabolism" (K. Bloch, ed.), pp. 165–237. Wiley, New York, 1960.
472. Kates, M., *J. Lipid Res.* **5**, 132 (1964).
473. Katz, I., and Keeney, M., *Anal. Chem.* **36**, 231 (1964).
474. Kaufmann, H. P., "Studien auf dem Fettgebiet," pp. 23–24. Verlag Chemie, Weinheim, 1935.
475. Kaufmann, H. P., *Riv. Ital. Sostanze Grasse* **41**, 188 (1964).
476. Kaufmann, H. P., and Aparicio, M., *Fette, Seifen, Anstrichm.* **61**, 768 (1959).
477. Kaufmann, H. P., Budwig, J., and Schmidt, C. W., *Fette, Seifen, Anstrichm.* **55**, 85 (1953).
478. Kaufmann, H. P., and Das, B., *Fette, Seifen, Anstrichm.* **64**, 214 (1962).
479. Kaufmann, H. P., and Hennig, H. J., *Mikrochim. Acta* No. 2, 333 (1961).
480. Kaufmann, H. P., and Khoe, T. H., *Fette, Seifen, Anstrichm.* **64**, 81 (1962).
481. Kaufmann, H. P., and Khoe, T. H., *Fette, Seifen, Anstrichm.* **66**, 590 (1964).
482. Kaufmann, H. P., and Makus, Z., *Fette, Seifen, Anstrichm.* **61**, 631 (1959).

483. Kaufmann, H. P., and Makus, Z., *Fette, Seifen, Anstrichm.* **62,** 1014 (1960).
484. Kaufmann, H. P., and Makus, Z., *Fette, Seifen, Anstrichm.* **63,** 125 (1961).
485. Kaufmann, H. P., Makus, Z., and Das, B., *Fette, Seifen, Anstrichm.* **63,** 807 (1961).
486. Kaufmann, H. P., Makus, Z., and Khoe, T. H., *Fette, Seifen, Anstrichm.* **63,** 689 (1961).
487. Kaufmann, H. P. Makus, Z., and Khoe, T. H., *Fette, Seifen, Anstrichm.* **64,** 1 (1962).
488. Kaufmann, H. P., and Mukherjee, K. D., *Fette, Seifen, Anstrichm.* **67,** 183 (1965).
489. Kaufman, H. P., and Mukherjee, K. D., *Fette, Seifen, Anstrichm.* **71,** 11 (1969).
490. Kaufmann, H. P., and Schnurbusch, H., *Fette, Seifen, Anstrichm.* **61,** 523 (1959).
491. Kaufmann, H. P., Seher, A., and Mankel, G., *Fette, Seifen, Anstrichm.* **64,** 501 (1962).
492. Kaufmann, H. P., and Viswanathan, C. V., *Fette, Seifen, Anstrichm.* **65,** 538 (1963).
493. Kaufmann, H. P., and Viswanathan, C. V., *Fette, Seifen, Anstrichm.* **65,** 607 (1963).
494. Kaufmann, H. P., and Wessels, H., *Fette, Seifen, Anstrichm.* **66,** 13 (1964).
495. Kaufmann, H. P., and Wessels, H., *Fette, Seifen, Anstrichm.* **66,** 81 (1964).
496. Kaufmann, H. P., and Wessels, H., *Fette, Seifen, Anstrichm.* **68,** 249 (1966).
497. Kaufmann, H. P., and Wessels, H., *Fette, Seifen, Anstrichm.* **69,** 338 (1967).
498. Kaufmann, H. P., Wessels, H., and Das, B., *Fette, Seifen, Anstrichm.* **64,** 723 (1962).
499. Kaufmann, H. P., Wessels, H., and Viswanathan, C. V., *Fette, Seifen, Anstrichm.* **64,** 509 (1962).
500. Kaufmann, H. P., and Wolf, W., *Fette Seifen* **50,** 519 (1943).
501. Kay, H. D., *Nature, London* **157,** 511 (1946).
502. Keeney, P. G., *J. Amer. Oil Chem. Soc.* **39,** 304 (1962).
503. Kelley, T. F., *J. Lipid Res.* **9,** 799 (1968).
504. Kennedy, E. P., *Fed. Proc., Fed. Amer. Soc. Exp. Biol.* **20,** 934 (1961).
505. Kerkhoven, E., and deMan, J. M., *J. Chromatogr.* **24,** 56 (1966).
506. Kircher, H. W., *J. Amer. Oil Chem. Soc.* **42,** 899 (1965).
507. Kirkland, J. J., *J. Chromatogr. Sci.* **9,** 206 (1971).
508. Kirkland, J. J., and DeStefano, J. J., *J. Chromatogr. Sci.* **8,** 309 (1970).
509. Kleiman, R., Earle, F. R., Tallent, W. H., and Wolff, I. A., *Lipids* **5,** 515 (1970).
510. Kleiman, R., Earle, F. R., and Wolff, I. A., *Lipids* **4,** 317 (1969).
511. Kleiman, R., Miller, R. W., Earle, F. R., and Wolff, I. A., *Lipids* **1,** 286 (1966).
512. Kleiman, R., Miller, R. W., Earle, F. R., and Wolff, I. A., *Lipids* **2,** 473 (1967).
513. Kleiman, R., Smith, C. R., Jr., Yates, S. G., and Jones, Q., *J. Amer. Oil Chem. Soc.* **42,** 169 (1965).
514. Klein, E., Lyman, R. B., Jr., Peterson, L., and Berger, R. I., *Life Sci.* **6,** 1305 (1967).
515. Klein, R. A., *J. Lipid Res.* **12,** 123 (1971).

516. Klimont, J., *Monatsh. Chem.* **24,** 408 (1903).
517. Klimont, J., *Monatsh. Chem.* **25,** 929 (1904).
518. Knittle, J. L., and Hirsch, J., *J. Lipid Res.* **6,** 565 (1965).
518a. Ko, H., and Royer, M. E., *Anal. Biochem.* **26,** 18 (1968).
519. Koch, R., "The Book of Signs," p. 68. Dover, New York, 1955.
520. Kolloff, R. H., *Anal. Chem.* **34,** 1840 (1962).
521. Komarek, R. J., Jensen, R. G., and Pickett, B. W., *J. Lipid Res.* **5,** 268 (1964).
521a. Kraml, M., and Cosyns, L., *Clin. Biochem.* **2,** 373 (1969).
522. Krehl, W. A., Lopez-S., A., and Good, E. I., *Amer. J. Clin. Nutr.* **20,** 968 (1967).
523. Krell, K., and Hashim, S. A., *J. Lipid Res.* **4,** 407 (1963).
524. Kresze, G., Bederke, K., and Schäuffelhut, F., *Z. Anal. Chem.* **209,** 329 (1965).
525. Krewson, C. F., Ard, J. S., and Riemenschneider, R. W., *J. Amer. Oil Chem. Soc.* **39,** 334 (1962).
526. Krukovsky, V. N., and Sharp, P. F., *J. Dairy Sci.* **23,** 1119 (1940).
527. Kruppa, R. F., Henly, R. S., and Smead, D. L., *Anal. Chem.* **39,** 851 (1967).
528. Kuemmel, D. F., *J. Amer. Oil Chem. Soc.* **48,** 184 (1971).
529. Kuksis, A., *Can. J. Biochem.* **42,** 419 (1964).
530. Kuksis, A., *J. Amer. Oil Chem. Soc.* **42,** 269 (1965).
531. Kuksis, A., *in* "Lipid Chromatographic Analysis" (G. V. Marinetti ed.), Vol. 1, pp. 239–337. Dekker, New York, 1967.
531a. Kuksis, A., *Can. J. Biochem.* **49,** 1245 (1971).
532. Kuksis, A., *Fette, Seifen, Anstrichm.* **73,** 130 (1971).
533. Kuksis, A., *Fette, Seifen, Anstrichm.* **73,** 332 (1971).
533a. Kuksis, A., *J. Chromatogr. Sci.* **10,** 53 (1972).
534. Kuksis, A., and Breckenridge, W. C., *J. Amer. Oil Chem. Soc.* **42,** 978 (1965).
535. Kuksis, A., and Breckenridge, W. C., *J. Lipid Res.* **7,** 576 (1966).
536. Kuksis, A., and Breckenridge, W. C., *in* "Dairy Lipids and Lipid Metabolism" (M. F. Brink and D. Kritchevsky, eds.), pp. 28–98. Avi, Westport, Connecticut, 1968.
537. Kuksis, A., Breckenridge, W. C., Marai, L., and Stachnyk, O., *J. Amer. Oil Chem. Soc.* **45,** 537 (1968).
538. Kuksis, A., Breckenridge, W. C., Marai, L., and Stachnyk, O., *J. Lipid Res.* **10,** 25 (1969).
539. Kuksis, A., and Ludwig, J., *Lipids* **1,** 202 (1966).
540. Kuksis, A., and McCarthy, M. J., *Can. J. Biochem. Physiol.* **40,** 679 (1962).
541. Kuksis, A., McCarthy, M. J., and Beveridge, J. M. R., *J. Amer. Oil Chem. Soc.* **40,** 530 (1963).
542. Kuksis, A., McCarthy, M. J., and Beveridge, J. M. R., *J. Amer. Oil Chem. Soc.* **41,** 201 (1964).
543. Kuksis, A., and Marai, L., *Lipids* **2,** 217 (1967).
544. Kuksis, A., Marai, L., Breckenridge, W. C., Gornall, D. A., and Stachnyk, O., *Can. J. Physiol. Pharmacol.* **46,** 511 (1968).
545. Kuksis, A., Marai, L., and Gornall, D. A., *J. Lipid Res.* **8,** 352 (1967).
546. Kuksis, A., Stachnyk, O., and Holub, B. J., *J. Lipid Res.* **10,** 660 (1969).
547. Kwapniewski, Z., and Sliwiok, J., *Mikrochim. Acta* No. 5, 616 (1964).
548. Laboureur, P., and Labrousse, M., *Bull. Soc. Chim. Biol.* **48,** 747 (1966).
549. Laboureur, P., and Labrousse, M., *Bull. Soc. Chim. Biol.* **50,** 2179 (1968).
550. Lakshminarayana, G., *J. Sci. Ind. Res.* **23,** 506 (1964).

551. Lakshminarayana, G., and Rebello, D., *J. Sci. Ind. Res., Sect. B* **14**, 189 (1955).
552. Lakshminarayana, G., and Rebello, D., *J. Amer. Oil Chem. Soc.* **37**, 274 (1960).
553. Lakshminarayana, G., and Rebello, D., *J. Amer. Oil Chem. Soc.* **40**, 300 (1963).
554. Lamb, R. G., and Fallon, H. J., *J. Biol. Chem.* **245**, 3075 (1970).
555. Lands, W. E. M., and Hart, P., *J. Lipid Res.* **5**, 81 (1964).
556. Lands, W. E. M., Pieringer, R. A., Slakey, P. M., and Zschocke, A., *Lipids* **1**, 444 (1966).
557. Lands, W. E. M., and Slakey, P. M., *Lipids* **1**, 295 (1966).
557a. Laudat, P., and Wolf, L.-M., *Biochim. Biophys. Acta* **176**, 425 (1969).
558. Lauer, W. M., Aasen, A. J., Graff, G., and Holman, R. T., *Lipids* **5**, 861 (1970).
559. Laurell, S., *Biochim. Biophys. Acta* **152**, 75 (1968).
559a. Laurell, S., and Tibbling, G., *Clin. Chim. Acta* **16**, 57 (1967).
560. Lavery, H., *J. Amer. Oil Chem. Soc.* **35**, 418 (1958).
561. Lawson, D. D., and Getz, H. R., *Chem. Ind. (London)* p. 1404 (1961).
562. Lea, C. H., *J. Soc. Chem. Ind.* **48**, 41T (1929).
563. Lederkremer, J. M., and Johnson, R. M., *J. Lipid Res.* **6**, 572 (1965).
564. Leeder, L. G., and Clark, D. A., *Microchem. J.* **12**, 396 (1967).
564a. Leegwater, D. C., and van Gend, H. W., *Fette, Seifen, Anstrichm.* **67**, 1 (1967).
565. Lefort, D., Perron, R., Pourchez, A., Madelmont, C., and Petit, J., *J. Chromatogr.* **22**, 266 (1966).
566. Levitskii, A. P., *Biokhimiya* **30**, 45 (1965) (available in English translation).
567. Litchfield, C., *Lipids* **3**, 170 (1968).
568. Litchfield, C., *Lipids* **3**, 417 (1968).
569. Litchfield, C., unpublished observation (1969).
570. Litchfield, C., *Chem. Phys. Lipids* **4**, 96 (1970).
571. Litchfield, C., *Lipids* **5**, 144 (1970).
572. Litchfield, C., *Chem. Phys. Lipids* **6**, 200 (1971).
573. Litchfield, C., *J. Amer. Oil Chem. Soc.* **48**, 467 (1971).
574. Litchfield, C., *Fette, Seifen, Anstrichm.* in press.
575. Litchfield, C., Ackman, R. G., Sipos, J. C., and Eaton, C. A., *Lipids* **6**, 674 (1971).
576. Litchfield, C., Farquhar, M., and Reiser, R., *J. Amer. Oil Chem. Soc.* **41**, 588 (1964).
577. Litchfield, C., and Harlow, R. D., unpublished data (1968).
578. Litchfield, C., Harlow, R. D., and Reiser, R., *J. Amer. Oil Chem. Soc.* **42**, 849 (1965).
579. Litchfield, C., Harlow, R. D., and Reiser, R., *Lipids* **2**, 363 (1967).
580. Litchfield, C., Miller, E., Harlow, R. D., and Reiser, R., *Lipids* **2**, 345 (1967).
581. Litchfield, C., and Reiser, R., *J. Amer. Oil Chem. Soc.* **42**, 757 (1965).
582. Littlewood, A. B., "Gas Chromatography: Principles, Techniques, and Applications," 2nd ed. Academic Press, New York, 1970.
583. Lloyd, M. R., and Goldrick, R. B., *Med. J. Aust.* **2**, 493 (1968).
583a. Lofland, H. B., Jr., *Anal. Biochem.* **9**, 393 (1964).
584. Longenecker, H. E., *Biol. Symp.* **5**, 99 (1941).
585. Lovern, J. A., *Compr. Biochem.* **6**, 63–80 (1965).

586. Luddy, F. E., Barford, R. A., Herb, S. F., and Magidman, P., *J. Amer. Oil Chem. Soc.* **45,** 549 (1968).
587. Luddy, F. E., Barford, R. A., Herb, S. F., Magidman, P., and Riemenschneider, R. W., *J. Amer. Oil Chem. Soc.* **41,** 693 (1964).
588. Luddy, F. E., Menna, A. J., and Calhoun, R. R., Jr., *J. Amer. Oil Chem. Soc.* **46,** 505 (1969).
589. Luhtala, A., and Antila, M., *Fette, Seifen, Anstrichm.* **70,** 280 (1968).
590. Lutton, E. S., *J. Amer. Chem. Soc.* **68,** 676 (1946).
591. Lutton, E. S., *J. Amer. Oil Chem. Soc.* **34,** 521 (1957).
592. Lutton, E. S., *J. Amer. Oil Chem. Soc.* **43,** 509 (1966).
593. Lutton, E. S., *J. Amer. Oil Chem. Soc.* **44,** 303 (1967).
594. Lutton, E. S., and Fehl, A. J., *Lipids* **5,** 90 (1970).
595. Lutton, E. S., and Jackson, F. L., *J. Amer. Chem. Soc.* **72,** 3254 (1950).
596. Maerker, G., Haeberer, E. T., and Herb, S. F., *J. Amer. Oil Chem. Soc.* **43,** 505 (1966).
597. Magnusson, J. R., and Hammond, E. G., *J. Amer. Oil Chem. Soc.* **36,** 339 (1959).
598. Maier, R., and Holman, R. T., *Biochemistry* **3,** 270 (1964).
599. Malangeau, P., and Pays, M., *Ann. Biol. Clin. (Paris)* **25,** 845 (1967).
600. Malins, D. C., and Mangold, H. K., *J. Amer. Oil Chem. Soc.* **37,** 576 (1960).
601. Malkin, T., *Progr. Chem. Fats Other Lipids* **2,** 1 (1954).
602. Mallon, J. P., and Dalton, C., *Anal. Biochem.* **40,** 174 (1971).
603. Mangold, H. K., *Fette, Seifen, Anstrichm.* **61,** 877 (1959).
604. Mangold, H. K., Lamp, B. G., and Schlenk, H., *J. Amer. Chem. Soc.* **77,** 6070 (1955).
605. Mani, V. V. S., and Lakshminarayana, G., *Indian J. Technol.* **3,** 339 (1965).
606. Marinetti, G. V., *J. Lipid Res.* **7,** 786 (1966).
607. Marks, P. A., Gellhorn, A., and Kidson, C., *J. Biol. Chem.* **235,** 2579 (1960).
608. Marks, T. A., Quinn, J. G., Sampugna, J., and Jensen, R. G., *Lipids* **3,** 143 (1968).
609. Marsh, J. B., and Weinstein, D. B., *J. Lipid Res.* **7,** 574 (1966).
610. Martin, A. J., Bennett, C. E., and Martinez, F. W., Jr., paper presented at the 3rd Symposium on Gas Chromatography, June 8–10, 1960, Edinburgh, Scotland.
611. Martinek, R. G., *J. Amer. Med. Technol.* **30,** 274 (1968).
612. Marzo, A., Ghirardi, P., Sardini, D., and Meroni, G., *Clin. Chem.* **17,** 145 (1971).
613. Masoro, E. J., "Physiological Chemistry of Lipids in Mammals," pp. 179–210. Saunders, Philadelphia, Pennsylvania, 1968.
614. Mattil, K. F., and Norris, F. A., *Science* **105,** 257 (1947).
615. Mattson, F. H., and Beck, L. W., *J. Biol. Chem.* **214,** 115 (1955).
616. Mattson, F. H., and Beck, L. W., *J. Biol. Chem.* **219,** 735 (1956).
617. Mattson, F. H., Benedict, J. H., Martin, J. B., and Beck, L. W., *J. Nutr.* **48,** 335 (1952).
618. Mattson, F. H., and Lutton, E. S., *J. Biol. Chem.* **233,** 868 (1958).
619. Mattson, F. H., and Volpenhein, R. A., *J. Biol. Chem.* **236,** 1891 (1961).
620. Mattson, F. H., and Volpenhein, R. A., *J. Lipid Res.* **2,** 58 (1961).
621. Mattson, F. H., and Volpenhein, R. A., *J. Biol. Chem.* **237,** 53 (1962).
622. Mattson, F. H., and Volpenhein, R. A., *J. Lipid Res.* **3,** 281 (1962).
623. Mattson, F. H., and Volpenhein, R. A., *J. Lipid Res.* **4,** 392 (1963).

624. Mattson, F. H., and Volpenhein, R. A., *J. Biol. Chem.* **239**, 2772 (1964).
625. Mattson, F. H., and Volpenhein, R. A., *J. Amer. Oil Chem. Soc.* **43**, 286 (1966).
626. Mattson, F. H., and Volpenhein, R. A., *J. Lipid Res.* **7**, 536 (1966).
627. Mattson, F. H., and Volpenhein, R. A., *J. Lipid Res.* **9**, 79 (1968).
628. Mattson, F. H., Volpenhein, R. A., and Lutton, E. S., *J. Lipid Res.* **5**, 363 (1964).
629. Mazliak, P., *Phytochemistry* **6**, 941 (1967).
630. Mazliak, P., *Phytochemistry* **6**, 957 (1967).
631. Mazuelos-Vela, F., *Grasas Aceites* **19**, 13 (1968).
632. McBride, O. W., and Korn, E. D., *J. Lipid Res.* **5**, 448 (1964).
633. McCarthy, M. J., and Kuksis, A., *J. Amer. Oil Chem. Soc.* **41**, 527 (1964).
634. McCarthy, M. J., Kuksis, A., and Beveridge, J. M. R., *Can. J. Biochem. Physiol.* **40**, 1693 (1962).
634a. McLellan, G. H., *Clin. Chem.* **17**, 535 (1971).
635. McNair, H. M., and Bonelli, E. J., "Basic Gas Chromatography." Varian Aerograph, Walnut Creek, California, 1967.
636. Meara, M. L., *J. Chem. Soc., London* p. 22 (1945).
637. Meara, M. L., *J. Chem. Soc., London* p. 23 (1945).
638. Meara, M. L., *J. Chem. Soc., London* p. 773 (1947).
639. Meara, M. L., *J. Chem. Soc., London* p. 2154 (1949).
640. Meara, M. L., *J. Chem. Soc., London* p. 1337 (1950).
641. Meara, M. L., and Steiner, E. H., "Studies on the Fatty Acid and Glyceride Composition of Cottonseed Oil and the Crystallising Behavior of Some of the Major Components," Final Rep., Pub. Law 480 Proj. No. UR-E29-(40)-26, Brit. Food Mfg. Ind. Res. Ass., Leatherhead, England, 1966.
642. Mehlenbacher, V. C., "The Analysis of Fats and Oils," p. 325. Garrard, Champaign, Illinois, 1960.
643. Mendelsohn, D., and Antonis, A., *J. Lipid Res.* **2**, 45 (1961).
644. Metcalfe, L. D., and Schmitz, A. A., *Anal. Chem.* **33**, 363 (1961).
645. Mhaskar, V. V., Phalnikar, N. L., and Bhide, B. V., *J. Univ. Bombay* **18A**, 28 (1950).
646. Michalec, C., Sulc, M., and Mestan, J., *Nature (London)* **193**, 63 (1962).
647. Michalski, E., and Stapor, M., *Lodz. Tow. Nauk., Pr. Wydz. 3* **11**, 25 (1966).
648. Michel, G., *C. R. Acad. Sci.* **244**, 2529 (1957).
649. Mikolajczak, K. L., and Smith, C. R., Jr., *Lipids* **2**, 261 (1967).
650. Mikolajczak, K. L., and Smith, C. R., Jr., *Biochim. Biophys. Acta* **152**, 244 (1968).
651. Mikolajczak, K. L., Smith, C. R., Jr., and Tjarks, L. W., *Lipids* **5**, 812 (1970).
652. Mikolajczak, K. L., Smith, C. R., Jr., and Wolff, I. A., *Lipids* **3**, 215 (1968).
653. Miller, R. W., Earle, F. R., Wolff, I. A., and Jones, Q., *J. Amer. Oil Chem. Soc.* **42**, 817 (1965).
654. Miwa, T. K., Kwolek, W. F., and Wolff, I. A., *Lipids* **1**, 152 (1966).
655. Miwa, T. K., Mikolajczak, K. L., Earle, F. R., and Wolff, I. A., *Anal. Chem.* **32**, 1739 (1960).
656. Montgomery, M. W., and Forster, T. L., *J. Dairy Sci.* **44**, 721 (1961).
657. Morgan, R. G. H., Barrowman, J., Filipek-Wender, H., and Borgström, B., *Biochim. Biophys. Acta* **146**, 314 (1967).
658. Morris, L. J., *Chem. Ind. (London)* p. 1238 (1962).
659. Morris, L. J., *in* "Metabolism and Physiological Significance of Lipids" (R.

M. C. Dawson and D. N. Rhodes, ed.), pp. 641–650. Wiley, New York, 1964.
660. Morris, L. J., *in* "New Biochemical Separations" (A. T. James and L. J. Morris, eds.), pp. 305–306. Van Nostrand-Reinhold, Princeton, New Jersey, 1964.
661. Morris, L. J., *Biochem. Biophys. Res. Commun.* **18,** 495 (1965). The steric configurations shown in Fig. 1 of this paper are incorrect; the starting compound for the synthesis should be *sn*-1,2-isopropylidene glycerol rather than the *sn*-2,3-isomer shown [L. J. Morris, personal communication (1967)].
662. Morris, L. J., *Biochem. Biophys. Res. Commun.* **20,** 340 (1965).
663. Morris, L. J., *J. Lipid Res.* **7,** 717 (1966).
664. Morris, L. J., and Hall, S. W., *Lipids* **1,** 188 (1966).
665. Morris, L. J., Holman, R. T., and Fontell, K., *J. Lipid Res.* **2,** 68 (1961).
666. Morris, L. J., Wharry, D. M., and Hammond, E. W., *J. Chromatogr.* **31,** 69 (1967).
667. Morrison, A., Barratt, M. D., and Aneja, R., *Chem. Phys. Lipids* **4,** 47 (1970).
668. Morrison, W. R., and Smith, L. M., *J. Lipid Res.* **5,** 600 (1964).
669. Muldrey, J. E., *J. Amer. Oil Chem. Soc.* **43,** 138A (1966).
670. Nakajima, H., *J. Phys. Soc. Jap.* **16,** 1778 (1961).
671. Nelson, J. H., Glass, R. L., and Geddes, W. F., *Cereal Chem.* **40,** 343 (1963).
672. Neudoerffer, T. S., and Lea, C. H., *J. Chromatogr.* **21,** 138 (1966).
673. Nickell, E. C., and Privett, O. S., *Separ. Sci.* **2,** 307 (1967).
674. Nickell, E. C., and Privett, O. S., *Washington, D.C. Meet. Amer. Oil Chem. Soc., April 1968* Pap. No. 121 (1968).
674a. Nicolosi, R. J., Smith, S. C., and Santerre, R. F., *J. Chromatogr.* **60,** 111 (1971).
675. Nikkari, T., and Haahti, E., *Biochim. Biophys. Acta* **164,** 294 (1968).
676. Noble, A. C., Buziassy, C., and Nawar, W. W., *Lipids* **2,** 435 (1967).
677. Noble, R. P., and Campbell, F. M., *Clin. Chem.* **16,** 166 (1970).
678. Noda, M., *Sci. Rep. Kyoto Prefect. Univ. Agr.* **11,** 169 (1959).
679. Noda, M., and Hirayama, O., *Yukagaku* **10,** 24 (1961).
680. Noma, A., and Borgström, B., *Biochim. Biophys. Acta* **227,** 106 (1971).
680a. Noma, A., and Borgström, B., *Scand. J. Gastroenterol.* **6,** 217 (1971).
681. Norris, F. A., and Mattil, K. F., *J. Amer. Oil Chem. Soc.* **24,** 274 (1947).
682. Notarnicola, L., *Riv. Ital. Sostanze Grasse* **44,** 72 (1967).
683. Novitskaya, G. V., Kaverina, A. V., and Vereshchagin, A. G., *Biokhimiya* **30,** 1260 (1965) (available in English translation).
684. Novitskaya, G. V., and Mal'tseva, V. I., *Biokhimiya* **31,** 953 (1966) (available in English translation).
685. Nutter, L. J., and Privett, O. S., *Lipids* **1,** 258 (1966).
686. Nutter, L. J., and Privett, O. S., *J. Dairy Sci.* **50,** 298 (1967).
687. Nutter, L. J., and Privett, O. S., *J. Dairy Sci.* **50,** 1194 (1967).
688. Nutter, L. J., and Privett, O. S., *J. Chromatogr.* **35,** 519 (1968).
689. Nystrom, E., and Sjövall, J., *Anal. Biochem.* **12,** 235 (1965).
690. O'Brien, J. F., and Klopfenstein, W. E., *Chem. Phys. Lipids* **6,** 1 (1971).
691. O'Connor, R. T., *in* "Fatty Acids" (K. S. Markley, ed.), 2nd ed., pp. 323–351. Wiley (Interscience), New York, 1960.
692. O'Connor, R. T., DuPre, E. F., and Feuge, R. O., *J. Amer. Oil Chem. Soc.* **32,** 88 (1955).
693. Oette, K., and Ahrens, E. H., Jr., *Anal. Chem.* **33,** 1847 (1961).

694. Olney, C. E., Jensen, R. G., Sampugna, J., and Quinn, J. G., *Lipids* **3,** 498 (1968).
695. Ord, W. O., and Bamford, P. C., *Chem. Ind. (London)* p. 1681 (1966).
696. Ord, W. O., and Bamford, P. C., *Chem. Ind. (London)* p. 277 (1967).
697. Ory, R. L., Bickford, W. G., and Dieckert, J. W., *Anal. Chem.* **31,** 1447 (1959).
698. Pack, F. C., Planck, R. W., and Dollear, F. G., *J. Amer. Oil Chem. Soc.* **29,** 227 (1952).
699. Padley, F. B., Ph.D. Dissertation, University of St. Andrews (1965).
700. Padley, F. B., *Chromatogr. Rev.* **8,** 208 (1966).
701. Padley, F. B., *J. Chromatogr.* **39,** 37 (1969).
702. Pande, S. V., Khan, R. P., and Venkitasubramanian, T. A., *Anal. Biochem.* **6,** 415 (1963).
703. Parijs, J., Barbier, F., and Vermeire, P., *Z. Klin. Chem. Klin. Biochem.* **6,** 331 (1968).
704. Pays, M., Malangeau, P., and Bourdon, R., *Ann. Pharm. Fr.* **25,** 29 (1967).
705. Peisker, K. V., *J. Amer. Oil Chem. Soc.* **41,** 87 (1964).
706. Pelick, N., Supina, W. R., and Rose, A., *J. Amer. Oil Chem. Soc.* **38,** 506 (1961).
707. Pelouze, J., and Boudet, F., *Ann. Chim. Phys.* [2] **69,** 43 (1838).
708. Perkins, E. G., and Hanson, A. V., *J. Amer. Oil Chem. Soc.* **42,** 1032 (1965).
709. Perkins, E. G., and Johnston, P. V., *Lipids* **4,** 301 (1969).
710. Perron, R., and Auffret, M., *Oleagineux* **20,** 379 (1965).
711. Perron, R., Gardant, D., and Perichon, J., *Rev. Fr. Corps Gras* **14,** 5 (1967).
712. Perron, R., Mathieu, A., and Paquot, C., *Bull. Soc. Chim. Fr.* [5] p. 2085 (1962).
713. Perron, R., Mathieu, A., and Paquot, C., *Fette, Seifen, Anstrichm.* **68,** 530 (1966).
714. Persmark, U., and Töregård, B., *J. Chromatogr.* **37,** 121 (1968).
714a. Pfeffer, P. E., and Rothbart, H. L., *Tetrahedron Letters* p. 2533 (1972).
715. Philip, K. J., Venkatarao, P., and Achaya, K. T., *Indian J. Technol.* **1,** 427 (1963).
716. Phillips, B. E., and Smith, C. R., Jr., *Biochim. Biophys. Acta* **218,** 71 (1970).
717. Phillips, B. E., Smith, C. R., Jr., and Hagemann, J. W., *Lipids* **4,** 473 (1969).
718. Phillips, B. E., Smith, C. R., Jr., and Tallent, W. H., *Lipids* **6,** 93 (1971).
719. Pieringer, R. A., and Kunnes, R. S., *J. Biol. Chem.* **240,** 2833 (1965).
720. Piguelevsky, G. V., and Saprokhina, A. E., *Zh. Prikl. Khim. (Leningrad)* **30,** 1104 (1957) (available in English translation).
721. Piguelevsky, G. V., and Starostina, T. A., *Dokl. Akad. Nauk SSSR* **79,** 261 (1951).
722. Pinter, J. K., Hayashi, J. A., and Watson, J. A., *Arch. Biochem. Biophys.* **121,** 404 (1967).
723. Pinter, K. G., Hamilton, J. G., and Miller, O. N., *Anal. Biochem.* **8,** 158 (1964).
724. Pitas, R. E., Sampugna, J., and Jensen, R. G., *J. Dairy Sci.* **50,** 1332 (1967).
725. Pokorny, J., Hladik, J., and Zeman, I., *Pharmazie* **23,** 332 (1968).
726. Pokorny, J., and Prochazkova, O., *Sb. Vys. Sk. Chem.-Technol. v. Praze, Potravinarska Technol.* **8,** 93 (1964).
727. Poukka, R., Vasenius, L., and Turpeinen, O., *J. Lipid Res.* **3,** 128 (1962).
728. Powell, R. G., Kleiman, R., and Smith, C. R., Jr., *Lipids* **4,** 450 (1969).

729. Prada, D., Carracedo, C. F., Montenegro, L., and Prieto, A., *Grasas Aceites* **21,** 261 (1970).
730. Priori, O., *Olii Miner., Grassi Saponi, Colori Vernici* **33,** 23 (1956).
731. Privett, O. S., *Progr. Chem. Fats Other Lipids* **9,** 91 (1966).
732. Privett, O. S., and Blank, M. L., *J. Lipid Res.* **2,** 37 (1961).
733. Privett, O. S., and Blank, M. L., *J. Amer. Oil Chem. Soc.* **40,** 70 (1963).
734. Privett, O. S., Blank, M. L., and Romanus, O., *J. Lipid Res.* **4,** 260 (1963).
735. Privett, O. S., Blank, M. L., and Schmit, J. A., *J. Food Sci.* **27,** 463 (1962).
736. Privett, O. S., Blank, M. L., and Verdino, B., *J. Nutr.* **84,** 187 (1965).
737. Privett, O. S., and Nickell, E. C., *J. Amer. Oil Chem. Soc.* **40,** 189 (1963).
738. Privett, O. S., and Nickell, E. C., *J. Amer. Oil Chem. Soc.* **41,** 72 (1964).
739. Privett, O. S., and Nickell, E. C., *J. Amer. Oil Chem. Soc.* **43,** 393 (1966).
740. Privett, O. S., and Nutter, L. J., *Lipids* **2,** 149 (1967).
741. Puleo, L. E., Rao, G. A., and Reiser, R., *Lipids* **5,** 770 (1970).
742. Purdy, S. J., and Truter, E. V., *Analyst* **87,** 802 (1962).
743. Quinlin, P., and Weiser, H. J., *J. Amer. Oil Chem. Soc.* **35,** 325 (1958).
744. Radin, N. S., in "Methods in Enzymology" (J. M. Lowenstein, ed.), Vol. 14, pp. 245–254. Academic Press, New York, 1969.
745. Radin, N. S., in "Methods in Enzymology" (J. M. Lowenstein, ed.), Vol. 14, pp. 268–272. Academic Press, New York, 1969.
746. Rajiah, A., Subbaram, M. R., and Achaya, K. T., *J. Chromatogr.* **38,** 35 (1968).
747. Raju, P. K., and Reiser, R., *Lipids* **1,** 10 (1966).
748. Ramachandran, S., Rao, P. V., and Cornwell, D. G., *J. Lipid Res.* **9,** 137 (1968).
749. Ramachandran, S., Yip, Y. K., and Wagle, S. R., *Eur. J. Biochem.* **12,** 201 (1970).
750. Randrup, A., *Scand. J. Clin. Lab. Invest.* **12,** 1 (1960).
751. Rao, G. A., Sorrels, M. F., and Reiser, R., *Lipids* **5,** 762 (1970).
752. Rao, G. A., Sorrels, M. F., and Reiser, R., *Biochem. Biophys. Res. Commun.* **44,** 1279 (1971).
752a. Regouw, B. J. M., Cornelissen, P. J. H. C., Helder, R. A. P., Spijkers, J. B. F., and Weeber, Y. M. M., *Clin. Chim. Acta* **31,** 187 (1971).
753. Reinbold, C. L., and Dutton, H. J., *J. Amer. Oil Chem. Soc.* **25,** 117 (1948).
754. Reiser, R., *Southwest Retort* **17,** No. 5, 7 (1965).
755. Reiser, R., Bryson, M. J., Carr, M. J., and Kuiken, K. A., *J. Biol. Chem.* **194,** 131 (1952).
756. Reiser, R., and Fu, H.-C., *Biochim. Biophys. Acta* **116** 563 (1966).
757. Reiser, R., Williams, M. C., and Sorrels, M. F., *J. Lipid Res.* **1,** 241 (1960).
758. Renkonen, O., *J. Lipid Res.* **3,** 181 (1962).
759. Renkonen, O., *Ann. Med. Exp. Biol. Fenn.* **43,** 194 (1965).
760. Renkonen, O., *J. Amer. Oil Chem. Soc.* **42,** 298 (1965).
761. Renkonen, O., *Ann. Med. Exp. Biol. Fenn.* **44,** 356 (1966).
762. Renkonen, O., *Biochim. Biophys. Acta* **125,** 288 (1966).
763. Renkonen, O., *Advan. Lipid Res.* **5,** 329 (1967).
764. Renkonen, O., *Biochim. Biophys. Acta* **137,** 575 (1967).
765. Renkonen, O., *Biochim. Biophys. Acta* **152,** 114 (1968).
766. Renkonen, O., *Lipids* **3,** 191 (1968).
767. Renkonen, O., Renkonen, O.-V., and Hirvisalo, E. L., *Acta Chem. Scand.* **17,** 1465 (1963).

768. Renkonen, O., and Rikkinen, L., *Acta Chem. Scand.* **21,** 2282 (1967).
769. Rheineck, A. E., Koley, S. N., and Parsons, J. L., *Paint Varn. Prod.* **57,** 34 (1967).
770. Riiner, Ü., *J. Amer. Oil Chem. Soc.* **47,** 129 (1970).
771. Roberts, R. N., *in* "Lipid Chromatographic Analysis" (G. V. Marinetti, ed.), Vol. 1, pp. 447–463. Dekker, New York, 1967.
772. Roehm, J. N., and Privett, O. S., *Lipids* **5,** 353 (1970).
772a. Rohwedder, W. K., *Lipids* **6,** 906 (1971).
773. Roncari, D. A. K., and Hollenberg, C. H., *Biochim. Biophys. Acta* **137,** 446 (1967).
774. Rowe, C. E., *J. Neurochem.* **16,** 205 (1969).
775. Royer, M. E., and Ko, H., *Anal. Biochem.* **29,** 405 (1969).
776. Ryhage, R., and Stenhagen, E., *J. Lipid Res.* **1,** 361 (1960).
777. Sahasrabudhe, M. R., and Chapman, D. G., *J. Amer. Oil Chem. Soc.* **38,** 88 (1961).
778. Sahasrabudhe, M. R., and Legari, J. J., *J. Amer. Oil Chem. Soc.* **44,** 379 (1967).
779. Sahasrabudhe, M. R., Legari, J. J., and McKinley, W. P., *J. Ass. Offic. Anal. Chem.* **49,** 337 (1966).
780. Sampugna, J., and Jensen, R. G., *J. Dairy Sci.* **50,** 386 (1967).
781. Sampugna, J., and Jensen, R. G., *Lipids* **3,** 519 (1968).
782. Sampugna, J., and Jensen, R. G., *Lipids* **4,** 444 (1969).
783. Sampugna, J., Jensen, R. G., Parry, R. M., Jr., and Krewson, C. F., *J. Amer. Oil Chem. Soc.* **41,** 132 (1964).
784. Sampugna, J., Pitas, R. E., and Jensen, R. G., *J. Dairy Sci.* **49,** 1462 (1966).
785. Sampugna, J., Quinn, J. G., Pitas, R. E., Carpenter, D. L., and Jensen, R. G., *Lipids* **2,** 397 (1967).
786. Sand, J. R., and Huber, C. O., *Talanta* **14,** 1309 (1967).
787. Saran, B., and Singh, B. K., *Proc. Nat. Acad. Sci., India* **12,** 219 (1942).
788. Sarda, L., and Desnuelle, P., *Biochim. Biophys. Acta* **30,** 513 (1958).
789. Sarda, L., Marchis-Mouren, G., and Desnuelle, P., *Biochim Biophys. Acta* **24,** 425 (1957).
790. Sarda, L., Marchis-Mouren, G., and Desnuelle, P., *in* "The Enzymes of Lipid Metabolism" (P. Desnuelle, ed.), pp. 20–21. Pergamon, Oxford, 1961.
791. Sardesai, V. M., and Manning, J. A., *Clin. Chem.* **14,** 156 (1968).
792. Sarycheva, I. K., Vargaftik, M. N., Utkina, O. V., and Preobrazhenskii, N. A., *Zh. Obshch. Khim.* **30,** 1048 (1960) (available in English translation).
793. Sato, K., Matsui, M., and Ikekawa, N., *Bunseki Kagaku* **15,** 954 (1966) (English translation available from National Translations Center, John Crerar Library, 35 W. 33rd St., Chicago, Ill. 60616).
794. Sato, K., Matsui, M., and Ikekawa, N., *Bunseki Kagaku* **16,** 1160 (1967).
795. Savary, P., and Desnuelle, P., *C. R. Acad. Sci.* **240,** 2571 (1955).
796. Savary, P., and Desnuelle, P., *Biochim. Biophys. Acta* **21,** 349 (1956).
797. Savary, P., and Desnuelle, P., *Biochim. Biophys. Acta* **50,** 319 (1961).
798. Schlenk, H., *J. Amer. Oil Chem. Soc.* **38,** 728 (1961).
799. Schlenk, H., and Gellerman, J. L., *Anal. Chem.* **32,** 1412 (1960).
800. Schlenk, H., Gellerman, J. L., Tillotson, J. A., and Mangold, H. K., *J. Amer. Oil Chem. Soc.* **34,** 377 (1957).
801. Schlenk, W., Jr., *Festschr. Carl Wurster 60. Geburtstag* pp. 105–111 (1960); *Chem. Abstr.* **57,** 14930g (1962).

802. Schlenk, W., Jr., *Angew. Chem.* **76**, 161 (1964); *Angew. Chem., Int. Ed. Engl.* **4**, 139 (1965).
803. Schlenk, W., Jr., *J. Amer. Oil Chem. Soc.* **42**, 945 (1965).
804. Schlierf, G., and Wood, P., *J. Lipid Res.* **6**, 317 (1965).
805. Schmid, H. H. O., and Mangold, H. K., *Biochem. Z.* **346**, 13 (1966).
806. Schmid, H. H. O., Mangold, H. K., and Lundberg, W. O., *Michrochem. J.* **7**, 287 (1963).
807. Schmid, H. H. O., Mangold, H. K., and Lundberg, W. O., *Microchem. J.* **7**, 297 (1963).
808. Schmid, H. H. O., Mangold, H. K., and Lundberg, W. O., *J. Amer. Oil Chem. Soc.* **42**, 372 (1965).
809. Schmidt, F. H., and von Dahl, K., *Z. Klin. Chem. Klin. Biochem.* **6**, 156 (1968).
810. Schmit, J. A., and Wynne, R. B., *J. Gas Chromatogr.* **4**, 325 (1966).
811. Scholfield, C. R., *J. Amer. Oil Chem. Soc.* **38**, 562 (1961).
812. Scholfield, C. R., *in* "Fatty Acids" (K. S. Markley, ed.), 2nd ed., pp. 2283–2307. Wiley (Interscience), New York, 1964.
813. Scholfield, C. R., and Dutton, H. J., *J. Amer. Oil Chem. Soc.* **35**, 493 (1958).
814. Scholfield, C. R., and Dutton, H. J., *J. Amer. Oil Chem. Soc.* **36**, 325 (1959).
815. Scholfield, C. R., and Hicks, M. A., *J. Amer. Oil Chem. Soc.* **34**, 77 (1957).
816. Scholfield, C. R., Jones, E. P., Butterfield, R. O., and Dutton, H. J., *Anal. Chem.* **35**, 1588 (1963).
817. Scholfield, C. R., Nowakowska, J., and Dutton, H. J., *J. Amer. Oil Chem. Soc.* **38**, 175 (1961).
818. Schønheyder, F., and Volqvartz, K., *Enzymologia* **11**, 178 (1944).
819. Schønheyder, F., and Volqvartz, K., *Acta Physiol. Scand.* **10**, 62 (1945).
820. Schønheyder, F., and Volqvartz, K., *Biochim. Biophys. Acta* **8**, 407 (1952).
821. Schønheyder, F., and Volqvartz, K., *Biochim. Biophys. Acta* **15**, 288 (1954).
822. Schoor, W. P., and Melius, P., *Biochim. Biophys. Acta* **187**, 186 (1969).
823. Schoor, W. P., and Melius, P., *Biochim. Biophys. Acta* **212**, 173 (1970).
824. Schultz, F. M., and Johnston, J. M., *J. Lipid Res.* **12**, 132 (1971).
825. Schwartz, D. P., Gould, I. A., and Harper, W. J., *J. Dairy Sci.* **39**, 1364 (1956).
826. Schwartz, D. P., Gould, I. A., and Harper, W. J., *J. Dairy Sci.* **39**, 1375 (1956).
827. Scott, R. P. W., and Lawrence, J. G., *J. Chromatogr. Sci.* **8**, 65 (1970).
828. Sedlak, M., *Anal. Chem.* **38**, 1503 (1966).
829. Semeriva, M., Benzonana, G., and Desnuelle, P., *Biochim. Biophys. Acta* **144**, 703 (1967).
830. Semeriva, M., Benzonana, G., and Desnuelle, P., *Bull. Soc. Chim. Biol.* **49**, 71 (1967).
831. Semeriva, M., Benzonana, G., and Desnuelle, P., *Biochim. Biophys. Acta* **191**, 598 (1969).
832. Serck-Hanssen, K., *Acta Chem. Scand.* **21**, 301 (1967).
833. Sezille, G., Biserte, G., Jaillard, J., and Scherpereel, P., *Eur. J. Clin. Biol. Res.* **15**, 1122 (1970).
834. Sezille, G., Jaillard, J., Scherpereel, P., and Biserte, G., *Clin. Chim. Acta* **29**, 335 (1970).
835. Sgoutas, D., and Kummerow, F. A., *J. Amer. Oil Chem. Soc.* **40**, 138 (1963).
836. Sheath, J., *Aust. J. Exp. Biol. Med. Sci.* **43**, 563 (1965).

837. Shkuropatova, Z. I., Sokolova, A. E., and Rzhekhin, V. P., *Tr. Vses. Nauch.-Issled. Inst. Zhirov* **26**, 440 (1967).
838. Shrivastava, R. K., and Bhutey, P. G., *Indian Oil Soap J.* **31**, 264 (1966).
839. Sinclair, R. G., Hinnekamp, E. R., Boni, K. A., and Berry, D. A., *J. Chromatogr. Sci.* **9**, 126 (1971).
840. Sink, J. D., Watkins, J. L., Ziegler, J. H., and Miller, R. C., *J. Anim. Sci.* **23**, 121 (1964).
841. Skidmore, W. D., and Entenman, C., *J. Lipid Res.* **3**, 356 (1962).
842. Skipski, V. P., and Barclay, M., *in* "Methods in Enzymology" (J. M. Lowenstein, ed.), Vol. 14, pp. 530–598. Academic Press, New York, 1969.
843. Slakey, P. M., and Lands, W. E. M., *Lipids* **3**, 30 (1968).
844. Sliwiok, J., *Mikrochim. Acta* No. 2, 294 (1965).
845. Sliwiok, J., and Kwapniewski, Z., *Mikrochim. Acta* No. 1, 1 (1965).
846. Smith, E. D., and Sorrells, K. E., *J. Chromatogr. Sci.* **9**, 15 (1971).
847. Smith, L. M., Freeman, C. P., and Jack, E. L., *J. Dairy Sci.* **48**, 531 (1965).
848. Snyder, F., and Piantadosi, C., *Biochim. Biophys. Acta* **152**, 794 (1968).
848a. Soloni, F. G., *Clin. Chem.* **17**, 529 (1971).
849. Sowden, J. C., and Fischer, H. O. L., *J. Amer. Chem. Soc.* **63**, 3244 (1941).
850. Spinella, C. J., and Mager, M., *J. Lipid Res.* **7**, 167 (1966).
851. Sprecher, H. W., Maier, R., Barber, M., and Holman, R. T., *Biochemistry* **4**, 1856 (1965).
852. Stadhouders, J., and Mulder, H., *Neth. Milk Dairy J.* **13**, 122 (1959).
853. Stahl, E., "Thin-Layer Chromatography. A Laboratory Handbook," 2nd. ed. Allen & Unwin, London, 1969.
854. Stein, R. A., and Nicolaides, N., *J. Lipid Res.* **3**, 476 (1962).
855. Stein, R. A., Slawson, V., and Mead, J. F., *in* "Lipid Chromatographic Analysis" (G. V. Marinetti, ed.), Vol. 1, 361–400. Dekker, New York, 1967.
856. Steinberg, D., Vaughan, M., and Margolis, S., *J. Biol. Chem.* **236**, 1631 (1961).
857. Steiner, E. H., and Bonar, A. R., *J. Sci. Food Agr.* **12**, 247 (1961).
858. Steiner, E. H., and Bonar, A. R., *Rev. Int. Choc.* **20**, 248 (1965).
859. Steinhagen, E., *Acta Chem. Scand.* **5**, 805 (1951).
860. Stenhagen, E., *Anal. Character. Oils, Fats, Fat Prod.* **2**, 26 (1968).
861. Stinson, C. G., deMan, J. M., and Bowland, J. P., *J. Amer. Oil Chem. Soc.* **44**, 253 (1967).
862. Subbaram, M. R., Chakrabarty, M. M., Youngs, C. G., and Craig, B. M., *J. Amer. Oil Chem. Soc.* **41**, 691 (1964).
863. Subbaram, M. R., and Youngs, C. G., *J. Amer. Oil Chem. Soc.* **41**, 445 (1964).
864. Subbaram, M. R., and Youngs, C. G., *J. Amer. Oil Chem. Soc.* **41**, 595 (1964).
865. Subbaram, M. R., and Youngs, C. G., *J. Amer. Oil Chem. Soc.* **44**, 425 (1967).
866. Sun, K. K., and Holman, R. T., *J. Amer. Oil Chem. Soc.* **45**, 810 (1968).
867. Sundler, R., and Åkesson, B., *Biochim. Biophys. Acta* **218**, 89 (1970).
868. Supina, W. R., Henly, R. S., and Kruppa, R. F., *J. Amer. Oil Chem. Soc.* **43**, 202A (1966).
869. Suzuki, B., and Yokoyama, Y., *Proc. Imp. Acad. (Tokyo)* **3**, 526 (1927).
869a. Swell, L., *Anal. Biochem.* **16**, 70 (1966).
870. Swell, L., Dailey, R. E., Field, H., Jr., and Treadwell, C. R., *Arch. Biochem. Biophys.* **59**, 393 (1955).

871. Sylvester, N. D., *Chem. Ind. (London)* p. 994 (1965).
872. Sylvester, N. D., Ainsworth, A. N., and Hughes, E. B., *Analyst* **70,** 295 (1945).
873. Szakasits, J. J., Peurifoy, P. V., and Woods, L. A., *Anal. Chem.* **42,** 351 (1970).
874. Takahashi, Y., *Yukagaku* **17,** 492 (1968).
875. Tallent, W. H., Cope, D. G., Hagemann, J. W., Earle, F. R., and Wolff, I. A., *Lipids* **1,** 335 (1966).
876. Tallent, W. H., Harris, J., Spencer, G. F., and Wolff, I. A., *Lipids* **3,** 425 (1968).
877. Tallent, W. H., and Kleiman, R., *J. Lipid Res.* **9,** 146 (1968).
878. Tallent, W. H., Kleiman, R., and Cope, D. G., *J. Lipid Res.* **7,** 531 (1966).
879. Tamaki, Y., Loschiavo, S. R., and McGinnis, A. J., *J. Agr. Food Chem.* **19,** 285 (1971).
879a. Tamaki, Y., Loschiavo, S. R., and McGinnis, A. J., *J. Insect. Physiol.* **17,** 1239 (1971).
880. Tamsma, A., Kurtz, F. E., Rainey, N., and Pallansch, M. J., *J. Gas Chromatogr.* **5,** 271 (1967).
881. Tattrie, N. H., *J. Lipid Res.* **1,** 60 (1959).
882. Tattrie, N. H., Bailey, R. A., and Kates, M., *Arch. Biochem. Biophys.* **78,** 319 (1958).
883. Tels, M., Kruidenier, A. J., Boelhouwer, C., and Waterman, H. I., *J. Amer. Oil Chem. Soc.* **35,** 163 (1958).
884. Teupel, M., and Pollerberg, J., *Tenside* **5,** 275 (1968).
885. Therriault, D. G., *J. Amer. Oil Chem. Soc.* **40,** 395 (1963).
886. Thompson, M. P., Brunner, J. R., and Stine, C. M., *J. Dairy Sci.* **42,** 1651 (1959).
887. Timmen, H., Dimick, P. S., Patton, S., and Pohanka, D. S., *Milchwissenschaft* **25,** 217 (1970).
888. Timms, A. R., Kelly, L. A., Spirito, J. A., and Engstrom, R. G., *J. Lipid Res.* **9,** 675 (1968).
889. Tomarelli, R. M., Mayer, B. J., Weaber, J. R., and Bernhart, F. W., *J. Nutr.* **95,** 583 (1968).
889a. Tove, S. B., *J. Nutr.* **75,** 361 (1961).
890. Trappe, W., *Biochem. Z.* **306,** 316 (1940).
891. Trenchant, J., ed., "Practical Manual of Gas Chromatography." American Elsevier, New York, 1969.
892. Trevelyan, W. E., *J. Lipid Res.* **7,** 445 (1966).
893. Trippel, A. I., *Sb. Tr., Leningr. Inst. Sov. Torg.* No. 23, 151 (1964); *Chem. Abstr.* **63,** 18497f (1965).
894. Trowbridge, J. R., Herrick, A. B., and Bauman, R. A., *J. Amer. Oil Chem. Soc.* **41,** 306 (1964).
895. Tryding, N., *Acta Physiol. Scand.* **40,** 232 (1957).
896. Tsuda, S., *Yukagaku* **11,** 408 (1962) (English translation available from National Translations Center, John Crerar Library, 35 W. 33rd St., Chicago, Ill. 60616).
897. Tsuda, S., *Yukagaku* **17,** 26 (1968).
898. Tsuda, S., *Yukagaku* **19,** 572 (1970).
899. Tucknott, O. G., and Williams, A. A., *Anal. Chem.* **41,** 2086 (1969).
900. Tulloch, A. P., and Craig, B. M., *J. Amer. Oil Chem. Soc.* **41,** 322 (1964).

901. Tyutyunnikov, B. N., and Mastruk, S., *Maslob-Zhir. Prom.* **35**, No. 6, 17 (1969).
902. Vahouny, G. V., Weersing, S., and Treadwell, C. R., *Biochim. Biophys. Acta* **98**, 607 (1965).
903. van Deenen, L. L. M., and de Haas, G. H., *Biochim. Biophys. Acta* **70**, 538 (1963).
904. van den Bosch, H., and Vagelos, P. R., *Biochim. Biophys. Acta* **218**, 233 (1970).
905. Vandenheuvel, F. A., *Anal. Chem.* **24**, 847 (1952).
906. van den Tempel, M., de Bruyne, P., and Mank, A. P. J., *Rec. Trav. Chim. Pays-Bas* **81**, 1075 (1962).
907. van der Ven, B., *Rec. Trav. Chim. Pays-Bas* **83**, 976 (1964).
908. van der Ven, B., Begemann, P. H., and Schogt, J. C. M., *J. Lipid Res.* **4**, 91 (1963).
909. Vander Wal, R. J., *Progr. Chem. Fats Other Lipids* **3**, 327 (1955).
910. Vander Wal, R. J., *J. Amer. Oil Chem. Soc.* **37**, 18 (1960).
911. Vander Wal, R. J., *Advan. Lipid Res.* **2**, 1 (1964).
912. Vander Wal, R. J., *J. Amer. Oil Chem. Soc.* **42**, 754 (1965).
913. Vander Wal, R. J., *J. Amer. Oil Chem. Soc.* **42**, 1155 (1965).
914. Vander Wal, R. J., personal communication (1968).
915. van Golde, L. M. G., and van Deenen, L. L. M., *Biochim. Biophys. Acta* **125**, 496 (1966).
916. van Golde, L. M. G., and van Deenen, L. L. M., *Chem. Phys. Lipids* **1**, 157 (1967).
917. van Handel, E., *Clin. Chem.* **7**, 249 (1961).
918. Vaver, V. A., Ushakov, A. N., Sitnikova, M. L., Kolesova, N. P., and Bergelson, L. D., *Izv. Akad. Nauk SSSR., Ser. Khim.* p. 918 (1971) (available in English translation).
919. Venkatarao, C., Narasingarao, M., and Venkateswarlu, A., *J. Indian Chem. Soc.* **21**, 249 (1944).
920. Verdino, B., Blank, M. L., and Privett, O. S., *J. Lipid Res.* **6**, 356 (1965).
921. Vereshchagin, A. G., *Biokhimiya* **27**, 866 (1962) (available in English translation).
922. Vereshchagin, A. G., *Dokl. Akad. Nauk SSSR* **152**, 221 (1963) (available in English translation).
923. Vereshchagin, A. G., *J. Chromatogr.* **14**, 184 (1964).
924. Vereshchagin, A. G., *J. Chromatogr.* **17**, 382 (1965).
925. Vereshchagin, A. G., personal communication (1971).
926. Vereshchagin, A. G., and Novitskaya, G. V., *J. Amer. Oil Chem. Soc.* **42**, 970 (1965).
927. Vereshchagin, A. G., and Skvortsova, S. V., *Dokl. Akad. Nauk SSSR* **157**, 699 (1964) (available in English translation).
928. Vereshchagin, A. G., Skvortsova, S. V., and Iskhakov, N. I., *Biokhimiya* **28**, 868 (1963) (available in English translation).
929. Verger, R., de Haas, G. H., Sarda, L., and Desnuelle, P., *Biochim. Biophys. Acta* **188**, 272 (1969).
930. Vidyarthi, N. L., *J. Indian Chem. Soc.* **20**, 45 (1943).
931. Vidyarthi, N. L., and Mallya, M. V., *J. Indian Chem. Soc.* **17**, 87 (1940).
932. Vinkenborg, C., *J. Amer. Oil Chem. Soc.* **47**, 314A (1970).
933. Vioque, E., and Holman, R. T., *J. Amer. Oil Chem. Soc.* **39**, 63 (1962).

934. Vioque, E., and Maza, M. P., *Grasas Aceites* **22**, 25 (1971).
935. Vioque, E., Maza, M. P., and Calderon, M., *Grasas Aceites* **15**, 173 (1964).
936. Vogelberg, K. H., and Gries, F. A., *Klin. Wochenschr.* **48**, 227 (1970).
937. von Rudloff, E., *Can. J. Chem.* **34**, 1413 (1956).
938. Vorob'ev, N. V., *Sb. Rab. Maslich. Kul't.* No. 2, 28 (1967) (English translation available from National Translations Center, John Crerar Library, 35 W. 33rd St., Chicago, Illinois 60616).
939. Walker, F. T., *J. Oil Colour Chem. Ass.* **28**, 119 (1945).
940. Walker, F. T., and Mills, M. R., *J. Soc. Chem. Ind.* **61**, 125 (1942).
941. Walker, F. T., and Mills, M. R., *J. Soc. Chem. Ind.* **62**, 106 (1943).
942. Waters, W. A., *in* "Organic Chemistry, An Advanced Treatise" (H. Gilman, ed.), Vol. IV, p. 1121. Wiley, New York, 1953.
943. Watts, R., and Dils, R., *J. Lipid Res.* **9**, 40 (1968).
944. Watts, R., and Dils, R., *J. Lipid Res.* **10**, 33 (1969).
945. Weber, E. J. de la Roche, I. A., and Alexander, D. E., *Lipids* **6**, 525 (1971).
946. Weiss, S. R., and Kennedy, E. P., *J. Amer. Chem. Soc.* **78**, 3550 (1956).
947. Weiss, S. B., Kennedy, E. P., and Kiyasu, J. Y., *J. Biol. Chem.* **235**, 40 (1960).
948. Werthessen, N. T., Beall, J. R., and James, A. T., *J. Chromatogr.* **46**, 149 (1970).
949. Wessels, H., *Fette, Seifen, Anstrichm.* **72**, 937 (1970).
950. Wessels, H., private communication (1970).
951. Wessels, H., and Rajagopal, N. S., *Fette, Seifen, Anstrichm.* **71**, 543 (1969).
952. White, H. B., Jr., and Quackenbush, F. W., *J. Amer. Oil Chem. Soc.* **39**, 517 (1962).
953. White, H. B., Jr., Quackenbush, F. W., and Probst, A. H., St. Louis Meet. Amer. Oil Chem. Soc., 1961 Abstr. 24 (1961).
954. Whitner, V. S., Grier, O. T., Mann, A. N., and Witter, R. F., *J. Amer. Oil Chem. Soc.* **42**, 1154 (1965).
955. Whittle, K. J., Dunphy, P. J., and Pennock, J. F., *Chem. Ind. (London)* p. 1303 (1966).
956. Wills, E. D., *in* "Enzymes of Lipid Metabolism" (P. Desnuelle, ed.), pp. 13–19. Pergamon, Oxford, 1961.
957. Wills, E. D., *Advan. Lipid Res.* **3**, 197 (1965).
958. Wood, P. D. S., and Sodhi, H. S., *Proc. Soc. Exp. Biol. Med.* **118**, 590 (1965).
959. Wood, R., Baumann, W. J., Snyder, F., and Mangold, H. K., *J. Lipid Res.* **10**, 128 (1969).
960. Wood, R., and Harlow, R. D., *Arch. Biochem. Biophys.* **131**, 495 (1969).
961. Wood, R., and Harlow, R. D., *Lipids* **5**, 776 (1970).
962. Wood, R., Harlow, R. D., and Lambremont, E. N., *Lipids* **4**, 159 (1969).
963. Wood, R., and Snyder, F., *J. Amer. Oil Chem. Soc.* **43**, 53 (1966).
964. Wood, R., and Snyder, F., *Arch. Biochem. Biophys.* **131**, 478 (1969).
965. Wood, R. D., Raju, P. K., and Reiser, R., *J. Amer. Oil Chem. Soc.* **42**, 161 (1965).
966. Wyatt, C. J., Pereira, R. L., and Day, E. A., *J. Dairy Sci.* **50**, 1760 (1967).
967. Yaguchi, M., Tarassuk, N. P., and Abe, N., *J. Dairy Sci.* **47**, 1167 (1964).
968. Yakubov, M. K., *Maslob.-Zhir. Prom.* **21**, No. 1, 14 (1956).
969. Young, T. E., and Maggs, R. J., *Anal. Chim. Acta* **38**, 105 (1967).
970. Youngs, C. G., *J. Amer. Oil Chem. Soc.* **36**, 664 (1959).

971. Youngs, C. G., *J. Amer. Oil Chem. Soc.* **38,** 62 (1961).
972. Youngs, C. G., and Baker, C. D., personal communication (1964).
973. Youngs, C. G., and Sallans, H. R., *J. Amer. Oil Chem. Soc.* **35,** 388 (1958).
974. Youngs, C. G., and Subbaram, M. R., *J. Amer. Oil Chem. Soc.* **41,** 218 (1964).
975. Yurkowski, M., and Brockerhoff, H., *Biochim. Biophys. Acta* **125,** 55 (1966).
976. Yurkowski, M., and Walker, B. L., *Biochim. Biophys. Acta* **231,** 145 (1971).
976a. Zhelvakova, E. G., Magnashevskii, V. A., Ermakova, L. I., Shvets, V. I., and Preobrazhenskii, N. A., *Zh. Org. Khim.* **6,** 1987 (1970) (available in English translation).
976b. Zhelvakova, E. G., Smirnova, G. V., Shvets, V. I., and Preobrazhenskii, N. A., *Zh. Org. Khim.* **6,** 1992 (1970) (available in English translation).
977. Zhukov, A. V., and Vereshchagin, A. G., *J. Chromatogr.* **51,** 155 (1970).
978. Zöllner, N., and Kirsch, K., *Z. Gesamte Exp. Med.* **135,** 545 (1962).

# AUTHOR INDEX

Numbers in parentheses are reference numbers. Numbers in italics show the page on which the complete reference is listed.

## A

Aasen, A. J., 208(1, 558), 209(1, 558), 210(558), *282, 298*
Abe, N., 183(967), *309*
Abel, E. W., 70(2), 108(2), *282*
Achaya, K. T., 26(746), 40(4), 54(746), 84(3, 715), 101(3, 715), 102(3), 177(3), *282, 302, 303*
Ackman, R. G., 25(9), 31(5, 6), 33(5), 34(5), 36(5), 39(7), 121(8), 122(8), 129(575), 151(575), *282, 298*
Addison, R. F., 25(9), *282*
Agranoff, B. W., 262(10), *282*
Ahmad, S., 28(412a), *294*
Ahrens, E. H., Jr., 20(380), 21(380), 24(380), 32(693), 36(392), 44(259), *289, 293, 301*
Ainsworth, A. N., 162(872), *307*
Aïvazis, G. A. M., 19(294), *290*
Akehurst, E. E., 54(69), 216(69), 217(69), *284*
Åkesson, B., 16(12, 15), 32(12), 51(12), 66(12, 15), 69(13), 72(13), 103(13), 154(14), 169(12), 189(12), 192(12), 197(12), 199(12, 15), 257(12), 263(867), 264(11, 15), *282, 306*
Alaupovic, P., 21(200), *287*
Albrink, M. J., 28(16), *282*
Alexander, D. E., 198(212), 235(212), 241(212, 945), 242(212), 254(945), *288, 309*
Alford, J. A., 185(17, 18), 186(17), 187(17), *282*
Amat, F., 54(19), *282*
Amberger, C., 139(20, 21), *282*
Amenta, J. S., 28(22), *283*
Anderson, R. E., 32(26), 179(25), 196(25), *283*
Andrews, P., 183(231), *288*
Aneja, R., 211(667), *301*
Anker, L., 69(27), 73(27), 76(27), 77(27), *283*
Antila, M., 183(589), *299*
Antonis, A., 26(30a), 27(643), *283, 300*
Aparicio, M., 67(476), 69(31), *283, 295*
Appelqvist, L.-A., 235(32), 236(32), *283*
Appelwhite, T. H., 43(291), *290*
Archibald, F. M., 26(33), *283*
Ard, J. S., 187(525), *297*
Arunga, R. O., 257(33a), *283*

# AUTHOR INDEX

Arvidson, G., 16(15), 66(15), 154(14), 199(15), 264(15), *282*
Atherton, H. V., 151(236), *288*
Atkinson, S. M., 27(64a), *284*
Aue, W. A., 70(34, 349a), 79(34), 108(34, 349, 349a), *283, 292*
Auffret, M., 51(710), *302*
Augl, J. M., 108(349), *292*
Augustin, M. P., 108(35), 165(35), 166(35), *283*

## B

Baer, E., 224(36), 225(38), 226(36), *283*
Bagby, M. O., 226(39), 227(39), 229(39), 243(39), *283*
Bailey, A. E., 214(41), 221(40), *283*
Bailey, R. A., 176(882), *307*
Baker, C. D., 157(272), *310*
Balatre, P., 40(42), *283*
Balls, A. K., 173(43), 178(44), *283*
Bamford, P. C., 68(696), 69(695, 696), 70(695, 696), 72(696), 73(696), 96(696), 97(696), *302*
Bandi, Z. L., 64(45), *283*
Bandyopadhyay, C., 26(46), 54(46), 60(46), 74(140), 76(140), 77(140), *283, 286*
Barber, M., 207(48), 208(47, 48), 209(48, 851), 211(47), 213(47), *283, 306*
Barbier, F., 27(703), *302*
Barclay, M., 22(842), *306*
Barford, R. A., 32(49, 586), 84(51a), 174(587), 175(587), 177(587), 179(51, 587), 254(50), *283, 299*
Barker, C., 140(52), 146(52), *283*
Baron, D. N., 27(64a), *284*
Barr, J. K., 115(53), *283*
Barrall, E. M., 223(54), *284*
Barratt, M. D., 211(667), *301*
Barrett, C. B., 10(55), 14(55), 49(55), 52(55), 53(56), 57(55, 56), 63(55), *284*
Barron, E. J., 174(343), 198(343), 263(57), *284, 292*
Barrowman, J., 174(657), *300*
Baskys, B., 175(58), *284*

Bauer, F. J., 20(223), *288*
Bauman, R. A., 78(894), 79(894), 80(894), *307*
Baumann, W. J., 23(59), 137(959), *284, 309*
Beall, J. R., 29(948), *309*
Beck, L. W., 10(616), 14(616), 167(615, 616), 173(615, 616, 617), 175(615), 176(616), *299*
Bederke, K., 137(524), *297*
Beeson, J. H., 24(60), *284*
Begemann, P. H., 40(61), 41(61), 48(908), *284, 308*
Beldowicz, M., 69(321), 72(321), *291*
Belfrage, P., 27(62), *284*
Bell, J., 12(63, 64), *284*
Bell, J. L., 27(64a), *284*
Belleville, J., 177(179), *287*
Benedict, J. H., 173(617), *299*
Bennett, C. E., 104(610), *299*
Bentz, A. P., 214(65), *284*
Benzonana, G., 175(66), 179(67), 184(829, 830, 831), *284, 305*
Bergelson, L. D., 23(68), 108(918), *284, 308*
Berger, K. G., 54(69), 216(69), 217(69), *284*
Berger, R. I., 176(514), *296*
Bergmann, L., 230(70), *284*
Bergström, S., 177(71), *284*
Berner, D. L., 187(72), *284*
Bernhart, F. W., 16(889), 107(889), *307*
Bernstein, H. J., 223(391), *293*
Bernstein, I. M., 253(73), *284*
Beroza, M., 110(74), *284*
Berry, D. A., 113(839), *306*
Berthelot, M., 12(75), *284*
Bertin, P., 40(42), *283*
Bertsch, R. J., 84(51a), *283*
Beveridge, J. M. R., 104(634), 116(542), 125(541), 165(634), *297, 300*
Bezard, J., 123(125), 135(76, 125), 174(183), *284, 285, 287*
Bhattacharya, R., 40(77), 252(77), *284*
Bhattacharyya, D., 52(142), 68(143, 144), 69(141), 72(143), 73(143), 74(140), 76(140), 77(140), *286*
Bhide, B. V., 44(645), 149(645), *300*
Bhutey, P. G., 68(838), 77(838), *306*

Bickford, W. G., 158(697), *302*
Biernoth, G., 28(78), *284*
Bird, P. R., 199(79), *284*
Biserte, G., 151(833, 834), 153(833, 834), *305*
Björnstad, P., 26(413), *294*
Black, B. C., 78(80), *284*
Black, H. C., 45(81), *284*
Blank, M. L., 26(85), 28(84), 32(85), 35(734), 42(733), 43(732, 733), 54(85), 57(85), 75(85), 151(82), 152(82), 156(733), 157(732, 733, 735), 158(735), 254(736), 255(83), 257(736), 258(736), *284, 303, 308*
Blankenhorn, D. H., 26(161), *286*
Bligh, E. G., 17(86), *285*
Block, W. D., 27(86a), *285*
Blyth, A. W., 12(87), *285*
Bobbitt, J. M., 22(88), *285*
Boekenoogen, H. A., 40(61), 41(61), *284*
Boelhouwer, C., 232(883), *307*
Boerma, H., 14(185), 70(185), *287*
Bombaugh, K. J., 29(89), 163(89, 90, 91), *285*
Bömer, A., 139(92, 93), *285*
Bonar, A. R., 70(857), 72(857), 76(857), 78(858), 79(94, 858), 80(94), 81(858), 93(858), 95(858), *285, 306*
Bonell, E. J., 105(635), *300*
Boni, K. A., 113(839), *306*
Bonsen, P. P. M., 194(95), *285*
Borgström, B., 173(96, 98, 99), 174(99, 657), 175(97, 252), 176(99), 177(71), 184(680a), 187(680), *284, 285, 289, 301*
Bottino, N. R., 32(26), 50(100), 51(101), 52(100), 59(100), 60(100), 177(102), 179(25), 196(25), *283, 285*
Boucrot, P., 16(103), 254(180), *285, 287*
Boudet, F., 12(707), *302*
Bougault, J., 172(104), *285*
Bourdon, R., 27(704), *302*
Bowie, J. H., 140(105), *285*
Bowland, J. P., 254(861), *306*
Bowman, M. C., 110(74), *284*
Braae, B., 44(106), *285*
Braconnot, H., 10(107), *285*
Bradford, R. H., 21(200), *287*

Breckenridge, W. C., 26(538), 53(537, 544), 54(538), 55(535), 66(537), 104(109), 105(535), 108(535), 109(535), 111(535), 112(535), 113(535), 114(535), 116(534), 123(109, 535), 130(534), 131(534), 134(534), 135(109, 389a), 136(109), 151(536), 247(108), *285, 297*
Breidenbach, B. G., 214(65), *284*
Brekke, O. L., 93(239), 253(239), *289*
Brockerhoff, H., 10(111), 15(111), 32(110), 160(975), 167(975), 168(975), 169(119), 170(975), 171(975), 172(117, 975), 174(118, 343), 175(975), 176(122), 177(110, 116, 975), 179(110, 111, 121, 122), 182(122, 975), 188(111), 189(110, 111, 114, 119, 120, 975), 190(111, 114, 115), 191(119), 192(114), 195(114, 120, 122), 196(111, 114, 119, 975), 198(343), 199(119), 201(113), 235(122, 236), 241(122), 245(119, 120), 246(120, 121), 247(112, 117, 120, 121), 257(121), 259(110), *285, 292, 310*
Bromig, K., 139(21), *282*
Brown, C. A., 39(123), *285*
Brown, H. C., 39(123), *285*
Brown, J. L., 16(124), *285*
Brunner, J. R., 140(886), *307*
Bryson, M. J., 262(755), *303*
Budwig, J., 162(477), *295*
Bugaut, M., 123(125), 135(76, 125), *284, 285*
Buide, N., 154(267, 268), 160(267), *290*
Burchfield, H. P., 105(126), 121(126), *285*
Burgher, R. D., 39(7), *282*
Burns, D. T., 52(127), *286*
Burton, T. H., 175(309), *291*
Buteau, G. H., Jr., 198(128), *286*
Butterfield, R. O., 84(816), 86(129), 93(816), *286, 305*
Buziassy, C., 172(130, 676), *286, 301*

**C**

Calderon, M., 54(935), *309*
Calhoun, R. R., Jr., 175(588), *299*
Callery, I. M., 107(131), *286*
Cameron, D. W., 140(105), *285*

Campbell, F. M., 19(677), 27(677), *301*
Camurati, F., 54(261), *289*
Cannon, J. A., 10(238), 14(238), 67(238), 84(238), 90(238), 93(238), *288*
Capella, P., 16(132), 220(132), *286*
Carlson, L. A., 24(133), 26(133), *286*
Carpenter, D. L., 9(419), 24(417), 25(417), 178(785), 185(427), 186(427), *294, 304*
Carr, M. J., 262(755), *303*
Carracedo, C. F., 104(134), 114(134), 115(134), 137(729), *286, 303*
Carreau, J.-P., 55(135), *286*
Carroll, K. K., 19(136, 137), 20(136, 137), *286*
Casparrini, G., 211(394), 212(394), 213(394), *293*
Cattaneo, P., 45(206), *288*
Cavina, G., 29(138), *286*
Ceglowska, K., 54(319), 72(319), *291*
Chacko, G. K., 254(139), *286*
Chakrabarty, M. M., 52(142), 68(143, 144), 69(141), 72(143), 73(143), 74(140), 76(140), 77(140), 78(862), 255(862), *286, 306*
Chandan, R. C., 177(422), 183(145, 422, 425), 184(422, 425), *286, 294*
Chapman, D., 16(154), 207(151), 214(147), 218(150), 220(146, 150, 154), 221(147, 149, 150, 153, 154), 223(148, 152, 155), *286*
Chapman, D. G., 20(777), 153(777), *304*
Chapman, J. R., 208(47), 211(47), 213(47), *283*
Chen, P. C., 141(156), *286*
Cherayil, G. D., 19(157), *286*
Chernick, S. S., 27(158), *286*
Chevreul, M. E., 9(159, 160), 10(159), 11(159), 139(159), *286*
Chin, H. P., 26(161), *286*
Chino, H., 19(162), 175(162), *286*
Chobanov, D., 40(163, 165), 41(164), 172(163), *287*
Christian, B. C., 40(166), *287*
Christie, W. W., 26(176), 32(169), 54(176), 61(167), 75(176), 160(170), 168(170), 170(170), 177(171), 189(170), 192(170), 196(170), 199(170,

172), 243(168), 247(171, 174), 248(174), 254(172, 175), 257(172, 173), 260(172), *287*
Christopherson, S. W., 32(177), *287*
Cirimele, M., 16(132), 220(132), *286*
Claesson, S., 162(178), *287*
Clark, D. A., 20(564), *298*
Clement, G., 16(183), 135(76), 174(182), 175(181), 177(179), *284, 287*
Clement, J., 16(103, 183), 174(182, 184), 175(181), 254(180), *285, 287*
Coenen, J. W. E., 14(185), 70(185), *287*
Coleman, M. H., 2(186), 173(189), 174(187), 176(187), 180(187, 188), 182(187), 227(190), 228(190), 249(187), 250(191), 251(186), 252(186), 253(186), 274 (190), 280(190), *287*
Collin, G., 40(193), 252(192), *287*
Conacher, H. B. S., 243(194), *287*
Constantin, M. J., 175(216), 179(251), 180(251), 197(251), *288, 289*
Cope, D. G., 137(875), 154(875), 176(878), 177(878), 243(875, 878), 255(875), *307*
Cornelissen, P. J. H. C., 27(752a), *303*
Cornwell, D. G., 26(387), 43(748), *293, 303*
Cosyns, L., 26(521a), *297*
Cotgreave, T., 29(195, 196), *287*
Craig, B. M., 32(197), 42(900), 84(3), 101(3), 102(3), 177(3), 254(230), *282, 287, 288, 307*
Cramp, D. G., 19(198), 24(198), 27(198), *287*
Crawford, R. V., 146(199), *287*
Crider, Q. E., 21(200), *287*
Crossley, A., 16(154), 172(201), 220(154), 221(154), *286, 287*
Crowe, P. F., 21(396), *293*
Crump, G. B., 104(290), *290*
Cuero, J. M., 51(202), 53(202), 64(202), *288*
Culp, T. W., 51(203), 269(203), *288*

**D**

Dailey, R. E., 174(870), 175(870), *306*
Dalgliesh, C. E., 32(204), *288*

Dallas, M. S. J., 10(55), 14(55), 49(55), 51(205), 52(55, 127), 53(56), 57(55, 56), 63(55), *284, 286, 288*
Dalton, C., 27(205a, 602), *288, 299*
Dark, W. A., 29(89), 163(89, 90, 91), *285*
Das, B., 67(478, 485, 498), 68(478, 485), 70(478, 485), 71(485), 72(478, 485), 73(478), 76(478, 485, 498), 77(485), 87(478), 91(478, 485), 92(478), 100(478), *295, 296*
Dasso, I., 45(206), *288*
Daubert, B. F., 8(207), 84(389), 85(389), 141(226), 253(207, 226), *288, 293*
Davies, A. C., 16(154), 220(154), 221(154), *286*
Day, A. J., 21(208), *288*
Day, E. A., *309*
Dean, M. P., 231(404), *293*
de Bruyne, P., 164(906), 165(906), *308*
Decoteau, A. E., 46(254), 163(254), *289*
de Haas, G. H., 175(929), 192(209, 210), 194(95), 198(209, 903), 199 (79, 903), *284, 285, 288, 308*
de Kruyff, B., 264(211), *288*
de la Roche, I. A., 198(212), 235(212), 241(212), 242(212), 254(945), *288, 309*
deMan, J. M., 19(505), 45(505), 46 (505), 141(156), 160(505), 254 (861), *286, 296, 306*
den Boer, F. C., 58(213), 59(213), *288*
Desnuelle, P., 10(796), 14(796), 148 (215), 160(217), 162(797), 167(795, 796), 173(795, 796), 174(788, 790), 175(214, 216, 788, 789, 929), 176 (250, 796), 178(788), 179(67, 250, 251), 180(250, 251), 184(829, 830, 831), 187(217), 197(251), *284, 288, 289, 304, 305, 308*
DeStefano, J. J., 70(508), 108(508), *296*
de Vries, B., 10(218), 14(185, 218), 28(441), 29(220), 49(218), 52(221, 222), 54(441, 442), 55(219, 220), 56(220), 57(220, 221), 62(220, 221, 442), 64(221), 70(185), *287, 288, 294*
Diamond, M. J., 43(291), *290*
Dieckert, J. W., 158(697), *302*

Dils, R., 104(943), 108(943), 112(943), 119(943), 120(943), 122(943), 128 (943), 130(943), 131(943), 137 (944), *308*
Dimick, P. S., 48(887), *307*
Distler, E., 20(223), *288*
Dittmer, J. C., 21(224), *288*
Dixon, C. W., 111(225), 135(225), *288*
Doerschuk, A. P., 141(226), 253(226), *288*
Dolendo, A. L., 209(227), *288*
Dolev, A., 55(228, 229), 56(228), 60(228, 229), *288*
Dollear, F. G., 39(698), *302*
Dowdell, R. J., 235(32), 236(32), *283*
Downey, R. K., 254(230), *288*
Downey, W. K., 25(232), 183(231), *288*
Downing, D. T., 28(233), *288*
Duffy, P., 139(234), *288*
Dugan, L. R., Jr., 175(311), *291*
Duncan, W. R. H., 254(235), *288*
Dunphy, P. J., 74(955), 76(955), 77(955), *309*
DuPre, E. F., 221(692), *301*
Duthie, A. H., 151(236), 183(415), *288, 294*
Dutton, H. J., 2(240), 10(238), 14(238), 67(238), 84(237, 238, 241, 258, 813, 814, 816, 817), 85(813, 814), 86(129, 258), 93(813, 814, 816, 817), 161(753), 286(237), 290(238), 293(238, 239, 240, 241), 253(239), *286, 288, 289, 303, 305*
Dyer, W. J., 17(86), *285*

**E**

Earle, F. R., 87(655), 129(510), 154(511, 512, 653, 875), 160(512), 176(512), 177(509), 206(511, 512), 220(512), 223(512), 226(512), 227 (511, 512), 229(512), 243(512, 875), 255(875), *296, 300, 307*
Eaton, C. A., 129(575), 151(175), *298*
Ebing, W., 107(242), *289*
Eckey, E. W., 254(243), *289*
Eggstein, M., 27(244), *289*
Eibner, A., 46(245), 148(246), 149 (245), *289*

Ellingboe, J., 78(247), 79(247), 80(247), 289
El-Meguid, S. S. A., 26(161), 286
Elovson, J., 16(15), 66(15), 154(14), 199(15), 264(15), 282
Engstrom, R. G., 27(888), 307
Entenman, C., 17(248, 249), 18(249), 19(248), 26(841), 289, 306
Entressangles, B., 176(250), 179(67, 250, 251), 180(250, 251), 197(251), 284, 289
Erlanson, C., 175(252), 289
Ermakova, L. I., 229(976a), 310
Eshelman, L. R., 40(253), 41(253), 46(254), 163(254), 289
Ettre, L. S., 105(255), 289
Evans, C. D., 84(258), 86(258), 155(256), 237(257), 253(256), 289
Evans, R. S., 29(335), 292

**F**

Fairbairn, D., 198(128), 286
Fales, H. M., 211(258a), 289
Fallon, H. J., 262(554), 298
Farquhar, J. W., 44(259), 289
Farquhar, M., 24(576), 52(576), 53(576), 54(576), 298
Fedeli, E., 16(132), 54(261), 151(260), 154(262), 220(132), 286, 289
Fehl, A. J., 221(594), 299
Feuge, R. O., 214(263), 221(692), 289, 301
Fidge, N. H., 21(208), 288
Field, H., Jr., 174(870), 175(870), 306
Filer, L. J., Jr., 16(264), 289
Filipek-Wender, H., 174(657), 300
Fillerup, D. L., 20(265), 289
Finch, R. W., 108(265a), 289
Findley, T. W., 43(266), 289
Fioriti, J. A., 108(269), 131(269), 154(267, 268), 160(267), 290
Fischer, G. A., 29(270), 290
Fischer, H. O. L., 224(36), 225(849), 226(36), 229(849), 283, 306
Fischer, L., 163(436), 294
Fischer, R., 231(271, 272), 290
Fletcher, M. J., 24(273), 27(273), 290
Fodor, P. J., 174(274, 275), 290
Fogerty, A. C., 64(429), 294

Folch, J., 17(276), 290
Fomon, S. J., 16(264), 289
Fontell, K., 35(665), 293(276a), 290, 301
Forster, T. L., 183(656), 300
Forster, T. L., 183(416, 420), 184(420), 294
Fosslien, E., 19(277), 290
Fox, P. F., 183(278), 290
Frankel, E. N., 183(279, 280, 281), 290
Franzke, C., 26(282), 155(283), 290
Frazer, A. C., 175(284), 290
Freeman, C. P., 19(407), 84(847), 180(407), 293, 306
Freeman, I. P., 172(201), 287
Freeman, N. K., 27(285), 219(285, 286), 290
Friedrich, J. P., 141(287), 290
Fritz, J. S., 44(288), 290
Fritz, P. J., 175(289), 290
Fryer, F. H., 104(290), 290
Fu, H.-C., 16(756), 179(345), 255(345), 292, 303
Fuller, G., 43(291), 290
Fulton, W. C., 250(191), 287

**G**

Gaffney, P. J., Jr., 183(292, 293), 290
Galanos, D. S., 16(295), 19(294), 290
Galletti, F., 19(296), 27(296), 290
Galoppini, C., 254(297), 290
Gander, G. W., 183(298, 299, 415, 416), 184(298, 299), 290, 294
Gardant, D., 220(711), 302
Garner, C. W., 175(300), 177(301), 290
Garton, G. A., 254(235), 288
Gauglitz, E. J., Jr., 152(318), 291
Gayen, A. K., 74(140), 76(140), 77(140), 286
Geddes, W. F., 84(671), 94(671), 96(671), 301
Gellerman, J. L., 32(799), 74(800), 75(800), 76(800), 304
Gellhorn, A., 24(607), 299
Gerike, U., 19(305), 27(305), 291
Gessmann, G. W., ii(302), 290
Getz, H. R., 229(561), 298
Ghirardi, P., 28(612), 299
Gidez, L. I., 175(303), 291

AUTHOR INDEX 317

Gilbert, L. I., 19(162), 175(162), *286*
Glass, R. L., 32(177), 84(671), 94(671), 96(671), 151(304), *287, 291, 301*
Gödicke, W., 19(305), 27(305), *291*
Golborn, P., 216(306), *291*
Gold, M., 52(307, 308), *291*
Goldman, 175(309), *291*
Goldrick, R. B., 27(583), *298*
Goldwater, W. H., 36(392), *293*
Golikova, V. S., 220(310), *291*
Good, E. I., 19(522), 24(522), *297*
Goodman, L. P., 175(311), *291*
Gorbach, G., 28(312), *291*
Gordis, E., 16(313, 314), 52(313), 54(313), 254(314), *291*
Gornall, D. A., 53(544), 78(545), 108(545), 112(545), *297*
Gould, I. A., 183(292, 293, 825, 826), *290, 305*
Gouw, T. H., 231(315, 316, 317), 232(315, 316, 317), *291*
Graff, G., 208(558), 209(558), 210(558), *298*
Gries, F. A., 26(936), *309*
Grier, O. T., 19(954), *309*
Gruger, E. H., Jr., 152(318), *291*
Grynberg, H., 54(319), 69(321), 72(319, 321), 235(320), 236(320), *291*
Guffy, J. C., 223(54), *284*
Gunde, B. G., 46(322), 149(322), *291*
Gunstone, F. D., 26(330), 52(328, 330), 53(325), 54(332), 55(325), 56(325), 57(328, 331), 59(328), 60(328, 332), 61(327, 328), 62(331), 65(328), 66(328), 139(325, 330), 140(325, 330), 146(325, 326, 328, 329, 330, 331), 147(325), 154(332), 158(328), 177(332), 234(324), 237(323, 324, 325), 238(328), 243(194, 324), 253(323), 254(328), 255(328, 331), 256(328), *287, 291*
Gupta, A. J., 68(143), 72(143), 73(143), *286*
Gupta, S. S., 146(333), *291*

**H**

Haab, W., 84(334), 85(334), 86(334), *291*
Haahti, E., 29(335), 163(675), *291, 301*

Habgood, H. W., 105(347), *292*
Haeberer, E. T., 43(596), *299*
Hagemann, J. W., 32(717), 35(717), 64(336), 154(717, 875), 243(717, 875), 255(875), *291, 302, 307*
Hagony, P. L., 176(337), *291*
Haighton, A. J., 41(338), 78(338), 79(338), 81(338), 100(338), 101(338), 220(338), *291*
Hajra, A. K., 262(10), *282*
Hall, S. W., 158(664), 159(664), 177(664), *300*
Hamilton, J. G., 25(723), *302*
Hamilton, R. J., 53(325), 55(325), 56(325), 139(325), 140(325), 146(325), 147(325), 162(339), 234(324), 237(324, 325), 243(324), *291*
Hammond, E. G., 15(340), 40(253), 41(253), 46(254), 50(666), 61(666), 78(80), 163(254), 164(435), 165(435, 597), 187(72), 252(341), 271(340), 272(340), 273(340), 276(340), 281(340), *284, 289, 292, 294, 299, 301*
Hammonds, T. W., 35(342), *292*
Hanahan, D. J., 174(343), 198(343), *292*
Hannewijk, J., 41(338), 78(338), 79(338), 81(338), 100(338), 101(338), 220(338), *291*
Hanson, A. V., 251(708), 252(708), *302*
Harding, U., 27(343a), *292*
Harlow, R. D., 51(203), 104(346, 578), 105(578), 107(577, 578, 579), 108(577, 578, 961, 962), 109(578, 579), 110(578), 111(346, 578, 579), 112(579), 113(577, 578, 579), 114(344, 577, 578, 579), 115(578), 117(578), 120(346, 578), 121(578), 122(578), 123(346), 124(578), 125(346, 577, 578, 579, 580), 126(346, 579), 127(577, 578), 128(577, 579), 129(577, 579), 130(579), 179(345), 255(345, 580), 257(960), 269(203), *288, 292, 298, 309*
Harper, W. J., 183(292, 293, 825, 826), *290, 304*
Harris, J., 177(876), *307*
Harris, W. E., 105(347), *292*
Hart, P., 194(555), *298*
Hashi, K., 148(348), *292*

Hashim, S. A., 27(523), 31(523), 219(523), *297*
Hastings, C. R., 70(34, 349a), 79(34), 108(34, 349, 349a), *283, 292*
Haux, P., 26(350), *292*
Hayashi, J. A., 27(722), *302*
Heemskerk, C. H. T., 199(79), *284*
Heims, K.-O., 26(282), *290*
Heinzel, G., 27(343a), *292*
Heise, R., 12(351, 352), 139(351, 352), *292*
Helder, R. A. P., 27(752a), *303*
Henly, R. S., 107(868), 109(527), *297, 306*
Hennig, H. J., 67(479), 103(479), *295*
Herb, S. F., 32(49, 586), 36(353), 43(596), 174(587), 175(587), 177(587), 179(587), 254(50), *283, 292, 299*
Herrick, A. B., 78(894), 79(894), 80(894), *307*
Hicks, M. A., 81(815), 84(815), 85(815), 93(815), *305*
Hilditch, T. P., 2(365), 10(355), 12(355, 365), 13(355, 365), 40(4, 77, 166, 193, 355, 358, 366), 46(322), 139(364), 140(52, 367), 143(367), 144(360, 367), 145(360), 146(52, 326, 333, 359), 148(354, 355, 362), 149(322), 215(356, 361), 252(77, 192, 363), 253(357, 363), 254(368), *282, 283, 284, 287, 291, 292*
Hill, E. E., 16(370), 53(369), 264(370), *292*
Hillsberry, J., 21(200), *287*
Hinnekamp, E. R., 113(839), *837*
Hirayama, O., 45(371), 70(679), 72(373), 73(679), 75(373, 374), 77(373, 374), 100(679), 102(375), 158(371), 160(371), 180(372), 181(372, 376), 263(377), *292, 301*
Hirsch, J., 20(380), 21(380), 24(380), 29(379), 78(378, 379, 518), 79(378, 379), 80(378, 379), 81(379), 86(379), 87(379), 103(379), *293, 297*
Hirschmann, H., 3(381), *293*
Hirvisalo, E. L., 60(767), *303*
Hites, R. A., 15(382), 209(382), 211(382), *293*
Hladik, J., 155(725), *302*

Ho, R. J., 29(383, 384), *293*
Hoefnagel, M. A., 220(385), *293*
Hoffmann, R. L., 155(256), 253(256), *289*
Hofmann, A. F., 180(386), *293*
Holla, K. S., 26(387), *293*
Hollenberg, C. H., 254(388), 263(773), *293, 304*
Hollingsworth, C. A., 84(389), 85(389), *293*
Holman, R. T., 26(933), 35(665), 84(598), 158(598), 162(339), 207(866), 208(1, 558), 209(1, 430, 558, 851), 210(558), 293(276a), *282, 290, 291, 294, 298, 299, 301, 306, 308*
Holub, B. J., 65(390), 135(389a), 137(546), *293, 297*
Hopkins, C. Y., 223(391), *293*
Hornby, G. M., 243(194), *287*
Horner, J., 231(272), *290*
Horning, E. C., 20(395), 21(395), 24(395), 32(204), 36(392), 107(393), 109(393), 211(394), 212(394), 213(394), *288, 293*
Horning, M. G., 20(395), 21(395), 24(395), 32(204), 211(394), 212(394), 213(394), *288, 293*
Hornstein, I., 21(396), *293*
Horrocks, L. A., 26(387), *293*
Horvath, W. L., 197(397), *293*
Hoyle, R. J., 169(119), 174(118), 179(121), 189(119, 120), 190( '9), 191(119), 195(120), 196(119), 199(119), 245(119, 120), 246(120, 121), 247(120, 121), 257(121), *285*
Huber, C. O., 28(786), *304*
Hubscher, G., 262(398), *293*
Hudson, B. J. F., 172(201), *287*
Huebner, V. R., 10(402), 14(402), 104(399, 400, 401, 402), 113(402), 130(399), 131(401), 137(399), *293*
Hughes, E. B., 162(872), *307*
Hujii, K., 72(373), 75(373), 77(373), 263(373), *292*
Husbands, D. R., 53(369, 403), *292, 293*
Hustad, G. O., 231(404), *293*
Hwang, P. C., 169(119), 189(119, 120), 190(119), 191(119), 195(120), 196(119), 199(119), 245(119, 120), 246(120), 247(120), *285*

## I

Ikekawa, N., 104(793), 109(793, 794), 111(793), 112(793), *304*
Inkpen, J. A., 161(405), *293*
Inouye, Y., 75(374), 77(374), 102(374), *292*
Insull, W., Jr., 44(259), *289*
Iskhakov, N. I., 75(928), 98(928), *308*
Ismail, I. A., 61(327), *291*

## J

Jaakonmäki, I., 29(335), *291*
Jacini, G., 16(132), 154(262), 220(132), *286, 289*
Jack, E. L., 19(402), 84(334, 847), 85(334), 86(334), 180(407), *291, 293, 306*
Jack, R. C. M., 218(408), *293*
Jackson, F. L., 214(595), 215(595), 221(595), *299*
Jackson, J. E., 45(409), *293*
Jagannathan, S. N., 19(410), *293*
Jaillard, J., 151(833, 834), 153(833, 834), *305*
James, A. T., 14(411), 29(948), *293, 309*
Jamieson, G. R., 31(412), 34(412), *294*
Jarrett, K. J., Jr., 27(86a), *285*
Jeejeebhoy, K. N., 28(412a), *294*
Jellum, E., 26(413), *294*
Jenness, R., 151(304), *291*
Jensen, R. G., 9(419), 16(782), 24(417), 25(417), 29(521), 32(784), 174(421, 783), 175(414), 176(414, 418), 177(422, 423, 780), 178(424, 785), 180(421), 183(415, 416, 420, 422, 425), 183(298, 299), 184(298, 299, 420, 422, 425), 185(427, 608, 781), 186(427, 608, 781), 187(608, 694, 781, 782), 197(781), 199(781), 203(426), 204(426, 781), 205(781, 782), 247(724), 255(782), *290, 294, 297, 299, 302, 304*
Jezyk, P. F., 16(428), *294*
Jones, Q., 154(653), *300*
Johnson, A. R., 64(429), *294*
Johnson, C. B., 209(430), *294*
Johnson, L. F., 223(434), *294*
Johnson, R. M., 25(563), 161(563), *298*
Johnston, J. M., 16(124), 262(431), 263(431, 432, 824), 264(433), *285, 294, 305*
Johnston, P. V., 211(709), *302*
Jones, E. C., 148(354), *292*
Jones, E. P., 84(816), 93(816), *305*
Jones, G. V., 164(435), 165(435), 252(341), *292, 294*
Jones, Q., 156(513), *296*
Joustra, M., 163(436), *294*
Jover, A., 27(437), *294*
Jurriens, G., 16(443), 28(441), 52(221, 222, 438, 444), 54(441, 442), 57(221), 59(438, 443), 62(221, 442), 64(221), 123(443), 157(439), 174(440), 176(440, 443), 179(440), 255(443, 444), 256(443, 444), *288, 294*

## K

Kabara, J. J., 29(270), *290*
Kaimal, T. N. B., 42(445), 176(446), *295*
Kanuk, M. J., 108(269), 131(269), *290*
Kapoulas, V. M., 16(295), 19(294), *290*
Karlsson, K.-A., 16(448), 151(447, 448), *295*
Karmen, A., 29(449), *295*
Karnovsky, M. L., 176(450), *295*
Kartha, A. R. S., 7(451), 40(451), 41(451, 455, 460, 463, 464, 465, 466, 467, 468, 469, 470), 148(451, 455, 460, 461, 462, 463, 470), 172(431, 455, 458, 459), 173(458, 459), 252(452, 453, 454, 456), 454(457), *295*
Kashket, S., 27(470a), *295*
Kates, M., 32(472), 175(471), 176(882), *283, 295, 307*
Katz, I., 27(473), *295*
Kaufmann, H. P., 2(494), 16(496, 497), 29(489, 490), 39(487), 44(474, 480), 45(487), 54(488, 495, 496), 59(495), 61(497), 67(475, 476, 478, 479, 480, 481, 482, 483, 484, 485, 486, 487, 490, 491, 492, 493, 494, 495, 496, 497, 498, 499), 68(478, 480, 484, 485, 495), 69(483, 484, 485), 70(478, 482, 484, 485, 494), 71(484, 485), 72(478,

482, 483, 484, 485, 495), 73(499), 74(484), 75(490), 76(478, 485, 498, 499), 77(480, 482, 483, 485, 490), 86(484), 87(478, 494), 90(499), 91(478, 485, 494), 92(478, 499), 99(487), 100(478), 101(487), 102 (483), 103(479), 121(491), 161(500), 162(477), 177(486), *295, 296*
Kaverina, A. V., 73(683), 96(683), *301*
Kay, H. D., 183(501), *296*
Keeney, M., 27(473), *295*
Keeney, P. G., 220(502), *296*
Kelley, T. F., 21(503), *296*
Kelley, W., 207(48), 208(48), 209(48), *283*
Kelly, L. A., 27(888), *307*
Kennedy, E. P., 262(504, 946, 947), *296, 309*
Kennett, B. H., 64(429), *294*
Keogh, M. K., 25(232), *288*
Keppler, J. G., 40(61), 41(61), *284*
Kerkhoven, E., 19(505), 45(505), 46(505), 160(505), *296*
Khan, R. P., 28(702), *302*
Khoe, T. H., 39(487), 44(480), 45(487), 67(480, 481, 486, 487), 68(480), 77(480, 486), 99(487), 101(487), *295, 296*
Kidson, C., 24(607), *299*
King, R. N., 29(89), 163(89), *285*
Kircher, H. W., 64(506), *296*
Kirkland, J. J., 70(508), 79(507), 108(508), *296*
Kirsch, K., 28(978), *310*
Kiyasu, J. Y., 262(947), *309*
Kleiman, R., 129(510), 131(728), 137 (877, 878), 154(511, 512), 156(513), 160(512), 176(512, 877, 878), 177 (509, 878), 206(511, 512), 220(512), 223(512), 226(512), 227(512), 229 (512), 243(512, 878), *296, 302, 307*
Klein, E., 175(58), 176(514), *284, 296*
Klein, R. A., 209(515), *296*
Klimont, J., 139(516, 517), *297*
Klopfenstein, W. E., 48(690), 137(690), *301*
Knittle, J. L., 78(518), *297*
Knox, K. L., 32(204), *288*
Ko, H., 19(775), 27(518a, 775), *297, 304*
Koch, R., ii(519), *297*

Kolesova, N. P., 108(918), *308*
Koley, S. N., 45(769), 46(769), *304*
Kolloff, R. H., 107(520), *297*
Komarek, R. J., 29(521), *297*
Korn, E. D., 263(632), *300*
Kozak, G., 28(412a), *294*
Kraml, M., 26(521a), *297*
Krehl, W. A., 19(522), 24(522), *297*
Krell, K., 27(523), 31(523), 219(523), *297*
Kresze, G., 137(524), *297*
Kretzschmann, F., 155(283), *290*
Kreutz, F. H., 27(244), *289*
Krewson, C. F., 174(783), 187(525), *297, 304*
Kroesen, A. C. J., 16(443), 59(443), 123(443), 176(443), 255(443), 256 (443), *294*
Kruidenier, A. J., 232(883), *307*
Krukovsky, V. N., 183(526), *297*
Kruppa, R. F., 107(868), 109(527), *297, 306*
Kuemmel, D. F., 109(528), *297*
Kuiken, K. A., 262(755), *303*
Kuksis, A., 26(538), 31(532), 48(531), 53(537, 544), 54(538), 55(535), 65 (390, 543), 66(537, 543), 78(545), 87 (633), 104(109, 530, 535, 539, 540, 634), 105(531, 533, 535), 108(530, 535, 545), 109(529, 530, 531, 535, 540), 111(530, 531, 533, 535, 539, 540), 112(531, 535, 545), 113(530, 531, 535, 539), 114(530, 531, 535), 115(539), 116(531, 534, 539, 542), 122(531), 123(109, 535), 125(531, 541), 129(531), 130(534), 131(534), 134(531, 534, 539), 135(109, 389a), 136(109), 137(531, 537, 546), 138 (531a, 533, 533a), 151(536), 165 (634), 247(108), *285, 293, 297, 300*
Kummerow, F. A., 158(835), *305*
Kunnes, R. S., 197(719), *302*
Kurtz, F. E., 107(880), *307*
Kwapniewski, Z., 68(547), 72(547), 75(845), 93(547), *297, 306*
Kwolek, W. F., 39(654), *300*

**L**

Laboureur, P., 184(548, 549), *297*
Labrousse, M., 184(548, 549), *297*

Lakshminarayana, G., 2(550), 51(605), 41(552, 553), 42(445), 148(552), 162(551, 552), 176(446), *295, 297, 298, 299*
Lamb, R. G., 262(554), *298*
Lambertsen, G., 293(276a), *290*
Lambremont, E. N., 108(962), *309*
Lamp, B. G., 77(604), *299*
Lancaster, C. R., 93(239), 253(239), *289*
Lands, W. E. M., 16(370), 53(369), 188(556), 189(556), 192(556), 193(556), 194(555, 556, 843), 195(556), 197(556), 198(556), 203(557), 204(843), 257(843), 264(370), *292, 298, 306*
Larsen, F. N., 70(349a), 108(349a), *292*
Laudat, P., 16(557a), *298*
Lauer, W. M., 208(1, 558), 209(1, 558), 210(558), *282, 298*
Laurell, S., 16(559), 27(559a), 151(559), *298*
Lavery, H., 213(560), *298*
Lawrence, J. G., 29(827), *305*
Lawson, D. D., 229(561), *298*
Lea, C. H., 10(355), 12(355), 13(355), 40(193, 355, 562), 73(672), 148(355), *287, 292, 298, 301*
Lederkremer, J. M., 25(563), 161(563), *298*
Leeder, L. G., 20(564), *298*
Leegwater, D. C., 116(564a), *298*
Lees, M., 17(276), *290*
Lefort, D., 108(565), 135(565), *298*
Legari, J. J., 20(779), 137(779), *304*
Levitskii, A. P., 176(566), *298*
Levangie, R. F., 163(90, 91), *285*
Lever, W. F., 175(58), *284*
Levoue, G., 16(183), *287*
Lie Ken Jie, M., 61(327), *291*
Lindgren, F. T., 219(286), *290*
Linsen, B. G., 14(185), 70(185), *287*
Lipsky, S. R., 36(392), *293*
List, G. R., 237(257), *289*
Litchfield, C., 24(576), 26(567), 51(203), 52(576), 53(576), 54(576), 68(567), 69(567), 70(567, 569), 72(567), 74(567), 75(567), 76(567), 86(567), 87(567), 104(346, 578), 105(578), 107(577, 578, 579), 108(577), 109(578, 579), 110(578), 111(346, 578, 579), 112(579), 113(577, 578, 579), 114(577, 578, 579), 115(578), 117(578), 120(346, 578), 121(578), 122(578), 123(346), 124(578), 125(346, 577, 578, 579, 580), 126(346, 579), 127(577, 578), 128(577, 579), 129(575, 577, 579), 130(579), 151(575), 176(569), 179(345, 569), 189(120), 195(120), 235(573), 236(571, 573), 237(573), 239(573), 240(573), 241(570, 572), 243(568), 244(568), 245(120, 574), 246(120, 574), 247(120, 574), 254(580), 255(345, 580), 269(203, 567), *285, 288, 292, 298*
Littlewood, A. B., 105(582), *298*
Lloyd, M. R., 27(583), *298*
Lofland, H. B., Jr., 26(583a), *298*
Lohse, L. W., 151(304), *291*
Longenecker, H. E., 251(584), *298*
Lopez-S., 19(522), 24(522), *297*
Loriette, C., 177(179), 254(180), *287*
Loschiavo, S. R., 16(879, 879a), *307*
Lotti, G., 254(297), *290*
Lovegren, N. V., 214(263), *289*
Lovern, J. A., 2(585), *298*
Lowe, P. A., 264(433), *294*
Luddy, F. E., 32(49, 586), 174(587), 175(587, 588), 177(587), 179(51, 587), 254(50), *283, 299*
Ludwig, J., 104(539), 111(539), 115(539), 116(539), 134(539), *297*
Luhtala, A., 183(589), *299*
Lundberg, W. O., 45(409), 231(806, 807, 808), *293, 305*
Lundquist, A., 27(62), *284*
Lutton, E. S., 213(593), 214(592, 595), 215(590, 595), 221(590, 591, 594, 595), 234(618), 247(628), *299, 300*
Lyman, R. B., Jr., 176(514), *296*
Lynes, A., 29(196), *287*

**M**

Maddison, L., 215(356), *292*
Madelmont, C., 108(565), 135(565), *298*

Marker, G., 43(596), *299*
Mager, M., 27(850), *306*
Maggs, R. J., 29(969), *309*
Magidman, P., 32(49, 586), 174(587), 175(587), 177(587), 179(51, 587), 254(50), *283, 299*
Magnashevskii, V. A., 229(976a), *310*
Magnusson, J. R., 165(597), *299*
Mahadevan, V., 225(38), *283*
Maier, R., 84(598), 158(598), 209(851), *299, 306*
Malangeau, P., 27(599, 704), *299, 302*
Malins, D. C., 35(600), 152(318), *291, 299*
Malkin, T., 214(601), 221(601), *299*
Mallon, J. P., 27(205a, 602), *288, 299*
Mallya, M. V., 148(931), *308*
Mal'tseva, V. I., 73(684), 96(684), 98(684), *301*
Mangold, H. K., 23(59, 805), 35(600), 44(603), 51(202), 53(202), 64(45, 202), 69(603), 70(603), 72(603), 74(800), 75(800), 76(800), 77(604), 100(603), 103(603), 137(959), 231(806, 807, 808), *283, 288, 299, 304, 305, 309*
Mangold, W. K., 24(59), *284*
Mani, V. V. S., 51(605), *299*
Mank, A. P. J., 164(906), 165(906), *308*
Mankel, G., 67(491), 121(491), *296*
Mann, A. N., 19(954), *309*
Manning, J. A., 27(791), *304*
Manzo, E. Y., 46(254), 163(254), *289*
Marai, L., 26(538), 53(537, 544), 54(538), 65(543), 66(537, 543), 78(545), 108(545), 112(545), 137(537), *297*
Marchis-Mouren, G., 174(790), 175(789), 179(67), *284, 304*
Marcus, S. J., 46(254), 163(254), *289*
Margolis, S., 263(856), *306*
Marinetti, G. V., 153(606), 154(606), *299*
Mark, T. A., 24(417), 25(417), 185(427), 186(427), *294*
Marks, P. A., 24(607), *299*
Marks, T. A., 185(608), 186(608), 187(608), *299*

Markus, Z., 39(487), 45(487), 67(482, 483, 484, 485, 486, 487), 68(484, 485), 69(483, 484, 485), 70(482, 484, 485), 71(484, 485), 72(482, 483, 484, 485), 74(484), 76(485), 77(482, 483, 485, 486), 86(484), 91(485), 99(487), 101(487), 102(483), *395, 396*
Marquinez, E., 54(19), *282*
Marsh, J. B., 28(609), *299*
Martin, A. J., 104(610), *299*
Martin, A. J. P., 14(411), *293*
Martin, D., 54(19), *282*
Martin, J. B., 173(617), *299*
Martin, V. G., 36(353), *292*
Martinek, R. G., 26(611), *299*
Martinez, F. W., Jr., 104(610), *299*
Marzo, A., 28(612), *299*
Masoro, E. J., 254(613), *299*
Mastruk, S., 41(901), *308*
Mathieu, A., 214(712), 216(713), *302*
Matlack, M. B., 173(43), 178(44), *283*
Matsui, M., 104(793), 109(793, 794), 111(793), 112(793), *304*
Mattil, K. F., 47(681), 251(614), *299, 301*
Mattson, F. H., 10(616), 14(616), 16(264, 621, 624), 36(392), 167(615, 616), 173(615, 616, 617, 623), 174(626, 627), 175(615, 625), 176(616, 619, 620, 627), 179(622), 180(622), 234(618, 619, 623), 235(619, 623), 236(619), 237(623), 238(619, 623), 247(628), 262(624), 263(621), *289, 293, 299, 300*
Mayer, B. J., 16(889), 107(889), *307*
Maza, M. P., 54(935), 154(934), *309*
Mazliak, P., 263(629, 630), *300*
Mazuelas-Vela, F., 16(631), *300*
McBride, O. W., 263(632), *300*
McCarthy, M. J., 87(633), 104(540, 634), 109(540), 111(540), 116(542), 125(541), 165(634), *297, 300*
McConnell, D. G., 84(258), 86(258), 155(256), 237(257), 253(256), *289*
McGinnis, A. J., 16(879, 879a), *307*
McKinley, W. P., 20(779), *304*
McLellan, G. H., 27(634a), *300*
McNair, H. M., 105(635), *300*
Mead, J. F., 20(265), 31(855), 36(392), *293, 289, 306*

# AUTHOR INDEX 323

Means, J. C., 209(227), *288*
Meara, M. L., 84(641), 86(641), 146(641), 215(636, 637, 638, 639, 640), 253(357), *292, 300*
Mehlenbacher, V. C., 44(642), *300*
Melius, P., 175(289, 822, 823), *290, 305*
Mendelsohn, D., 27(643), *300*
Meng, H. C., 29(384), *293*
Menna, A. J., 175(588), *299*
Meroni, G., 28(612), *299*
Merren, T. O., 207(48), 208(48), 209(48), *283*
Mestan, J., 68(646), 69(646), 72(646), *300*
Metcalfe, L. D., 32(644), *300*
Mhaskar, V. V., 44(645), 149(645), *300*
Michalec, C., 68(646), 69(646), 72(646), *300*
Michalski, E., 28(647), *300*
Michel, G., 162(648), *300*
Mickle, J. B., 19(407), 180(407), *293*
Mikolajczak, K. L., 87(655), 108(651), 155(649, 652), 158(650), *300*
Miller, E., 125(580), 255(580), *298*
Miller, O. N., 25(723), *302*
Miller, R. C., 254(840), *306*
Miller, R. W., 154(511, 512, 653), 160(512), 176(512), 206(511, 512), 220(512), 223(512), 226(512), 227(511, 512), 229(512), 243(512), *296, 300*
Mills, M. R., 161(940, 941), *309*
Milne, G. W. A., 211(258a), *289*
Mitrofanova, T. K., 220(310), *291*
Miwa, T. K., 39(654), 87(655), *300*
Mollica, A., 29(138), *286*
Mondall, B., 68(144), *286*
Montenegro, L., 137(729), *303*
Montgomery, M. W., 183(656), *300*
Moore, J. H., 26(176), 54(176), 75(176), 160(170), 168(170), 170(170), 177(171), 189(170), 192(170), 196(170), 199(170, 172), 247(174), 248(174), 254(172, 175), 257(172, 173), 260(172), *287*
Moretta, L., 29(138), *286*
Moretti, G., 29(138), *286*
Morgan, M. E., 183(415), *294*
Morgan, R. G. H., 174(657), *300*
Morris, L. J., 35(663, 665), 50(663, 666), 51(658, 659, 666), 53(663), 55(660), 61(666), 158(664), 159(664), 177(664), 226(661), 227(661, 662), 228(661, 662), 229(661, 662), 255(662), *300, 301*
Morrison, A., 211(667), *301*
Morrison, W. R., 32(668), 257(33a), *283, 301*
Moscatelli, E. A., 107(393), 109(393), *293*
Mounts, T. L., 84(241), 93(241), *289*
Mukherjee, K. D., 29(489), 54(488), *296*
Mulder, H., 183(852), *306*
Muldrey, J. E., 51(669), *301*
Murphy, R. F., 25(232), *288*
Murray, K. E., 64(429), *294*
Murty, N. L., 32(197), *287*
Musil, F., 19(277), *290*

## N

Nakae, T., 181(376), *292*
Nakajima, H., 223(670), *301*
Narasingarao, M., 44(919), 45(919), 148(919), *308*
Narayanan, R., 41(464, 465, 466, 467), 148(463), *295*
Natelson, S., 26(350), *292*
Naudet, M., 148(215), 175(216), *288*
Nawar, W. W., 172(130, 676), *286, 301*
Nelson, J. H., 84(671), 94(671), 96(671), *301*
Neudoerffer, T. S., 73(672), *301*
Ng, Y. C., 219(286), *290*
Nichols, A. V., 219(286), *290*
Nickell, E. C., 42(739), 43(738, 739), 57(737), 78(673), 79(673), 80(673), 81(673), 87(673), 88(673), 92(673), 100(673), 141(674), *301, 303*
Nickless, G., 70(2), 108(2), *282*
Nicolaides, N., 43(854), *306*
Nicolosi, R. J., 29(674a), *301*
Nikkari, T., 163(675), *301*
Nilsson, K., 151(447), *295*
Noble, A. C., 172(676), *301*
Noble, R. D., 26(176), 54(176), 75(176), *287*
Noble, R. P., 19(677), 27(677), *301*

Noda, M., 45(678), 69(678), 70(678, 679), 73(679), 75(678), 77(678), 100(678, 679), *301*
Noma, A., 184(680a), 187(680), *301*
Norrby, A., 16(448), 151(448), *295*
Norris, F. A., 47(681), 251(614), *299, 301*
Notarnicola, L., 27(682), 219(682), *301*
Novitskaya, G. V., 70(926), 72(926), 73(683, 684, 926), 75(926), 90(926), 96(683, 684, 926), 98(684), *301, 308*
Nowakowska, J., 84(817), 93(817), *305*
Nutter, L. J., 28(688), 47(740), 53(740), 57(687), 78(686, 687), 81(686), 176(740), 199(685), *301, 303*
Nyström, E., 78(247), 79(247), 80(247), 163(689), *289, 301*

**O**

O'Brien, J. F., 48(690), 137(690), *301*
O'Connor, R. T., 214(691), 221(691, 692), *301*
Oette, K., 32(693), *301*
Ohama, S., 263(377), *292*
Olcott, H. S., 55(228, 229), 56(228), 60(228, 229), *288*
Olney, C. E., 187(694), *302*
Ord, W. O., 68(696), 69(695, 696), 70(695, 696), 72(696), 73(696), 96(696), 97(696), *302*
Ormand, W. L., 104(290), *290*
Ory, R. L., 158(697), *302*
Overley, C. A., 45(81), *284*

**P**

Pack, F. C., 39(698), *302*
Padley, F. B., 2(700), 10(55), 14(55), 26(330), 52(328, 330), 29(701), 49(55), 52(55), 53(56), 57(55, 56, 328), 59(328), 60(328), 61(328), 63(55), 64(699), 65(328), 66(328), 139(330), 140(330), 146(328, 329, 330), 158(328), 234(324), 237(324), 238(328), 243(194, 324), 254(328), 255(328), 256(328), *284, 287, 291, 302*
Pallansch, M. J., 107(880), *307*

Pande, S. V., 28(702), *302*
Paquot, C., 214(712), 216(713), *302*
Parijs, J., 27(703), *302*
Parry, R. M., Jr., 174(421, 783), 177(422), 180(421), 183(420, 422), 184(420, 422), *294, 304*
Parsons, J. L., 45(769), 46(769), *304*
Pascher, I., 151(447), *295*
Pasero, L., 179(67), *284*
Patton, S., 48(887), *307*
Paultauf, F., 163(432), *294*
Pays, M., 27(599, 704), *299, 302*
Pearson, J. A., 64(429), *294*
Pecsar, R. E., 24(60), *284*
Peisker, K. V., 32(705), *302*
Pelick, N., 104(706), *302*
Pelouze, J., 12(707), *302*
Pennock, J. F., 74(955), 76(955), 77(955), *309*
Pereira, R. L., 177(423), 178(424), 183(425), 184(425), *294, 309*
Perichon, J., 220(711), *302*
Perkins, E. G., 209(227), 211(709), 251(708), 252(708), 254(139), *286, 288, 302*
Perron, R., 51(710), 108(565), 135(565), 214(712), 216(713), 220(711), *298, 302*
Persmark, U., 60(714), *302*
Peters, H., 155(256), 253(256), *289*
Peterson, L., 176(514), *296*
Petit, J., 108(565), 135(565), *298*
Peurifoy, P. V., 29(873), *307*
Pfeffer, P. E., 223(714a), 224(714a), *302*
Phalnikar, N. L., 44(645), 149(645), *300*
Philip, K. J., 84(715), 101(715), *302*
Phillips, B. E., 32(717), 35(717), 47(716), 154(717), 241(718), 243(717), *302*
Piantadosi, C., 175(848), 179(848), *306*
Pickett, B. W., 29(521), *297*
Pierce, D. A., 185(17), 186(17), 187(17), *282*
Pierce, J. H., 172(201), *287*
Pieringer, R. A., 188(556), 189(556), 192(556), 193(556), 194(556), 195(556), 197(397, 556, 719), 198(566), *293, 298, 302*

Piguelevsky, G. V., 43(720, 721), 149 (720, 721), *302*
Pinter, J. K., 27(722), *302*
Pinter, K. G., 25(723), *302*
Pitas, R. E., 32(784), 176(418), 178(785), 247(724), *294, 302, 304*
Planck, R. W., 39(698), *302*
Platt, D. S., 26(30a), *283*
Pohanka, D. S., 48(887), *307*
Pokorny, J., 155(725), 161(726), *302*
Pollard, F. H., 70(2), 108(2), *282*
Pollerberg, J., 223(884), *307*
Popkova, G. A., 23(68), *284*
Popov, A., 40(165), *287*
Poukka, R., 39(727), *302*
Pourchez, A., 108(565), 135(565), *298*
Powell, R. G., 131(728), *302*
Prada, D., 137(729), *303*
Preobrazhenskii, N. A., 72(792), 220(310), 229(976a, 976b), *291, 304, 310*
Prieto, A., 104(134), 114(134), 115(134), 137(729), *286, 303*
Priori, O., 10(730), 14(730), 67(730), *303*
Privett, O. S., 26(85), 28(84, 688), 32(85), 35(731, 734), 42(733, 739), 43(732, 733, 738, 739), 47(740), 53(740, 772), 54(85), 57(85, 687, 737), 75(85), 78(673, 686, 687), 79(673), 80(673), 81(673, 686), 87(673), 88(673), 92(673), 100(673), 141(674), 151(82), 152(82), 156(733), 157(732, 733, 735), 158(735), 176(740), 199(685), 254(736), 255(83), 257(736), 258(736), *284, 301, 303, 304, 308*
Probst, A. H., 254(953), *309*
Prochazkova, O., 23(68), 161(726), *302*
Prokazova, N. V., 24(68), *284*
Puleo, L. E., 262(741), *303*
Purdy, S. J., 29(742), *303*

## Q

Quackenbush, F. W., 46(952), 161(405), 254(952), *293, 309*
Quinlin, P., 20(743), *303*
Quinn, J. G., 9(419), 24(417), 25(417), 176(418), 178(785), 185(427, 608), 186(427, 608), 187(608, 694), 203(426), 204(426), *294, 299, 302, 304*
Qureshi, M. I., 26(330), 52(330), 53(325), 54(332), 55(325), 56(325), 57(331), 60(332), 62(331), 139(325, 330), 140(325, 330), 146(325, 330, 331), 147(325), 154(332), 177(332), 237(324, 325), 243(324), 255(331), *291*

## R

Radin, N. S., 17(744), 19(745), *303*
Rainey, N., 107(880), *307*
Rajagopal, N. S., 50(951), 52(951), 53(951), 57(951), 61(951), 62(951), 63(951), 70(951), 72(951), 73(951), 76(951), 77(951), 87(951), 93(951), 268(951), *309*
Rajiah, A., 26(746), 54(746), *303*
Raju, P. K., 32(747), 48(965), *303, 309*
Ramachandran, S., 43(748), 175(749), *303*
Randrup, A., 27(750), *303*
Rao, G. A., 262(741, 751, 752), 264(433), *294, 303*
Rao, P. V., 43(748), *303*
Raulin, J., 55(135), 177(179), 254(180), *286, 287*
Rayman, M. M., 175(309), *291*
Rebello, D., 41(552, 553), 148(552), 162(551, 552), *298*
Regouw, B. J. M., 27(752a), *303*
Reinbold, C. L., 161(753), *303*
Reiser, R., 2(754), 16(756), 24(576), 32(26, 747), 48(965), 51(203), 52(576), 53(576), 54(576), 104(346, 578), 105(578), 107(578, 579), 108(578), 109(578, 579), 110(578), 111(346, 578, 579), 112(579), 113(578, 579), 114(578, 579), 115(578), 117(578), 120(346, 578), 121(578), 122(578), 123(346), 124(578), 125(346, 578, 579, 580), 126(346, 579), 127(578), 128(579), 129(579, 130(579), 177(102), 179(25), 196(25), 254(581, 757), 255(581), 262(741, 751, 752, 755), 269(203), *283, 285, 288, 292, 298, 303, 309*

Renkonen, O., 51(763), 53(760), 60(767), 63(768), 65(761, 762, 763, 766), 66(768), 137(764), 153(759), 155(759), 161(758), 176(759), *303, 304*
Renkonen, O.-V., 60(767), *303*
Rheineck, A. E., 45(769), 46(769), *304*
Richards, R. E., 223(155), *286*
Richardson, T., 231(404), *293*
Riemenschneider, R. W., 32(49), 174(587), 175(587), 177(587), 179(587), 187(525), 254(50), *283, 297, 299*
Rigollot, B., 174(184), *287*
Riiner, Ü., 214(770), *304*
Rikkinen, L., 63(768), 66(768), *304*
Riley, J. P., 146(326), *291*
Roberts, R. N., 26(771), *304*
Robertson, G., 19(198), 24(198), 27(198), *287*
Robertson, G. H., 12(87), *285*
Roehm, J. N., 53(772), *304*
Rohwedder, W. K., 211(772a), *304*
Romanus, O., 35(734), *303*
Roncari, D. A. K., 263(773), *304*
Rose, A., 104(706), *302*
Rosen, P., 44(259), *289*
Rothbart, H. L., 84(51a), 223(714a), 224(714a), *283, 302*
Rowe, C. E., 219(774), *304*
Royer, M. E., 19(775), 27(518a, 775), *297, 304*
Ruck, J. B., 21(396), *293*
Rüstow, B., 155(283), *290*
Rugenstein, H., 155(283), *290*
Ryhage, R., 207(776), 209(776), 211(776), *304*
Rzhekhin, V. P., 84(837), *306*

## S

Sahasrabudhe, M. R., 20(777, 779), 137(778), 153(777), *304*
Saletore, S. A., 40(358), *292*
Sallans, H. R., 140(973), 142(973), 143(973), *310*
Sampugna, J., 9(419), 16(782), 24(417), 25(417), 32(784), 174(421, 783), 176(418), 177(422, 42ä, 780), 178(424, 785), 180(421), 183(299, 416, 422, 422, 425), 184(299, 420, 422, 425), 185(427, 608, 781), 186(427, 608, 781), 187(608, 694, 781, 782), 197(781), 199(781), 203(426), 204(426, 781), 205(781, 782), 247(724), 255(782), *290, 294, 299, 302, 304*
Samuelsson, B., 16(448), 151(448), *295*
Sand, J. R., 28(786), *304*
Santerre, R. F., 29(674a), *301*
Saprokhina, A. E., 43(720), 149(720), *302*
Saran, B., 148(787), *304*
Sarda, L., 174(788, 790), 175(788, 789, 929), 178(788), 179(67), *284, 304, 308*
Sardesai, V. M., 27(791), *304*
Sardini, D., 28(612), *299*
Sari, H., 176(250), 179(250), 180(250), *289*
Sarycheva, I. K., 72(792), *304*
Sato, K., 104(793), 109(793, 794), 111(793), 112(793), *304*
Savary, P., 10(796), 14(796), 160(217), 162(797), 167(795, 796), 173(795, 796), 176(796), 179(251), 180(251), 187(217), 197(251), *288, 289, 304*
Sawyer, D. T., 115(53), *283*
Scanlan, J. T., 43(266), *289*
Scaria, K. S., 19(157), *286*
Schäuffelhut, F., 137(524), *297*
Scherpereel, P., 151(833, 834), 153(833, 834), *305*
Schild, E., 148(246), *289*
Schiller, C. M., 263(432), *294*
Schlenk, H., 32(799), 74(800), 75(800), 76(800), 77(604), 141(798), *299, 304*
Schlenk, W., Jr., 214(803), 215(803), 221(801, 802, 803), 222(801, 802, 803), 225(803), 226(803), 229(803), 230(801, 802, 803), 255(803), *304, 305*
Schlierf, G., 29(804), *305*
Schmid, H. H. O., 23(59, 805), 231(806, 807, 808), *284, 305*
Schmidinger, K., 46(245), 149(245), *289*
Schmidt, C. W., 162(477), *295*
Schmidt, F. H., 27(809), *305*

Schmit, J. A., 28(84), 111(225), 119(810), 135(225), 157(735), 158(735), *284, 288, 303, 305*
Schmitz, A. A., 32(644), *300*
Schnurbusch, H., 29(490), 67(490), 75(490), 77(490), *296*
Schogt, J. C. M., 48(908), *308*
Scholfield, C. R., 2(240), 81(815), 82(812), 83(811), 84(241, 258, 811, 812, 813, 814, 815, 816, 817), 85(812, 813, 814, 815), 86(258, 813), 93(240, 241, 811, 812, 813, 814, 815, 816, 817), 237(257), *289, 305*
Schønheyder, F., 173(820, 821), 175(819), 178(818), 180(821), *305*
Schoor, W. P., 175(822, 823), *305*
Schouten, L., 28(441), 52(444), 54(441, 442), 62(442), 255(444), 256(444), *294*
Schultz, F. M., 263(824), *305*
Schultz, L. D., 263(432), *294*
Schuster, G., 172(104), *285*
Schwartz, D. P., 183(825, 826), *305*
Scott, R. P. W., 29(827), *305*
Seavell, A. J., 140(359), 146(359), *292*
Sedlak, M., 39(828), *305*
Seher, A., 67(491), 121(491), *296*
Selvaraj, Y., 41(468, 469), *295*
Semeriva, M., 184(829, 830, 831), *305*
Serck-Hanssen, K., 155(832), *305*
Serdarevich, B., 19(137), 20(137), *286*
Sethi, S. C., 39(123), *285*
Sezille, G., 151(833, 834), 153(833, 834), *305*
Sgoutas, D., 158(835), *305*
Shahani, K. M., 177(421, 422), 180(421), 183(145, 422, 425), 184(422, 425), *286, 294*
Sharp, P. F., 183(526), *297*
Sheath, J., 27(836), *305*
Shenstone, F. S., 64(429), *294*
Shepherd, G. F., 52(127), *286*
Shkuropatova, Z. I., 84(837), *306*
Shone, G., 35(342), *292*
Shoolery, J. N., 223(434), *294*
Shrivastava, R. K., 68(838), 77(838), 144(360), 145(360), *292, 306*
Shvets, V. I., 220(310), 229(976a, 976b), *291, 310*
Sims, R. J., 108(269), 131(269), 154(267, 268), 160(267), *290*
Sinclair, R. G., 113(839), *306*
Singh, B. K., 148(787), *304*
Siniscalchi, P., 29(138), *286*
Sink, J. D., 254(840), *306*
Sipos, J. C., 121(8), 122(8), 129(575), 151(575), *282, 298*
Sitnikova, M. L., 108(918), *308*
Sjövall, J., 78(247), 79(247), 80(247), 163(689), *289, 301*
Skidmore, W. D., 26(841), *306*
Skipski, V. P., 22(842), 26(33), *283, 306*
Skvortsova, S. V., 67(927), 71(927), 73(927), 75(928), 98(928), *308*
Slakey, P. M., 16(370), 188(556), 189(556), 192(556), 193(556), 194(556, 843), 195(556), 197(556), 198(556), 203(557), 204(843), 257(843), 264(370), *292, 298, 306*
Slawson, V., 31(855), *306*
Sliwiok, J., 16(844), 68(547), 72(547), 75(845), 93(547), *297, 306*
Smead, D. L., 109(527), *297*
Smirnova, G. V., 229(976b), *310*
Smith, C. R., Jr., 32(717), 35(717), 47(716), 108(651), 131(728), 154(717), 155(649, 652), 156(513), 158(650), 226(39), 227(39), 229(39), 241(718), 243(39, 717), *283, 290, 296, 300, 302*
Smith, E. D., 107(846), *306*
Smith, J. L., 185(18), *282*
Smith, L. C., 175(300), 177(301), *290*
Smith, L. M., 19(407), 32(668), 84(334, 847), 85(334), 86(334), 180(407), *291, 293, 301*
Smith, R. R., Jr., 226(39), 227(39), 229(39), 241(39), *283*
Smith, S. C., 29(674a), *301*
Snyder, F., 50(963), 137(959), 169(964), 175(848), 179(848), 189(964), 192(964), *306, 309*
Sodhi, H. S., 29(958), *309*
Söderqvist, B., 163(436), *294*
Sokolova, A. E., 84(837), *306*
Soloni, F. G., 27(848a), *306*

Sonanini, D., 69(27), 73(27), 76(27), 77(27), *283*
Sorrells, K. E., 107(846), *306*
Sorrels, M. F., 254(757), 262(751, 752), *303*
Sowden, J. G., 225(849), 229(849), *306*
Spencer, G. F., 177(876), *307*
Spijkers, J. B. F., 27(752a), *303*
Spinella, C. J., 27(850), *306*
Spirito, J. A., 27(888), *307*
Sprecher, H. W., 209(851), *306*
Stachnyk, O., 26(538), 53(537, 544), 54(538), 66(537), 137(537, 546), *297*
Stadhouder, J., 183(852), *306*
Stahl, E., 22(853), *306*
Stainsby, W. J., 148(362), 215(361), *292*
Stanley, G. H. S., 17(276), *290*
Stapor, M., 28(647), *300*
Starostina, T. A., 43(721), 149(721), *302*
Stein, R. A., 31(855), 43(854), *306*
Steinberg, D., 263(856), *306*
Steiner, E. H., 70(857), 72(857), 76(857), 78(858), 79(858), 81(858), 84(641), 86(641), 93(858), 95(858), 146(641), *300, 306*
Stenhagen, E., 207(776, 860), 209(776), 211(776), 214(859), *304, 306*
Stine, C. M., 140(886), *307*
Stinson, C. G., 254(861), *306*
Stoffel, W., 44(259), *289*
Storrs, E. E., 105(126), 121(126), *285*
Stretton, R. J., 52(127), *286*
Stumpf, P. K., 263(57), *284*
Subbaram, M. R., 16(863), 26(746), 29(863), 41(974), 42(974), 54(746), 55(863), 56(863), 78(862, 974), 108(974), 132(974), 133(974), 255 (862, 864, 865, 974), *303, 306, 310*
Suggs, F. G., 185(17), 186(17), 187(17), *282*
Sulc, M., 68(646), 69(646), 72(646), *300*
Sun, K. K., 207(866), *306*
Sundler, R., 263(867), *306*
Supina, W. R., 104(706), 107(868), *302, 306*
Suzuki, B., 148(869), *306*
Sweeley, C. C., 107(393), 109(393), *293*
Swell, L., 135(869a), 174(870), 175 (870), *306*
Swern, D., 43(266), *289*
Sylvester, N. D., 162(871, 872), *307*
Szakasits, J. J., 29(873), *307*
Szczepanska, H., 54(319), 69(321), 72(319, 321), 235(320), 236(320), *291*

## T

Taber, J. J., 84(389), 85(389), *293*
Takahashi, Y., 42(874), 157(874), *307*
Tallent, W. H., 137(877, 878), 154 (875), 176(877, 878), 177(509, 876, 878), 241(718), 243(875, 878), 255 (875), *296, 302, 307*
Tamaki, Y., 16(879, 879a), *307*
Tamsma, A., 107(880), *307*
Tarassuk, N. P., 183(278, 279, 280, 281, 967), *290, 309*
Tarenghi, A., 154(262), *289*
Tattrie, N. H., 176(882), 198(881), *307*
Tels, M., 232(883), *307*
Teupel, M., 223(884), *307*
Therriault, D. G., 25(885), *307*
Thompson, M. P., 140(886), *307*
Thorp, J. M., 26(30a), *283*
Tibbling, G., 27(559a), *298*
Tillotson, J. A., 74(800), 75(800), 76(800), *304*
Timmen, H., 48(887), *307*
Timms, A. R., 27(888), *307*
Tjarks, L. W., 108(651), *300*
Tobias, J., 209(227), *288*
Töregard, B., 60(714), *302*
Tomarelli, R. M., 16(889), 107(889), *307*
Tove, S. B., 254(889a), *307*
Traisnel, M., 40(42), *283*
Trappe, W., 161(890), *307*
Treadwell, C. R., 174(870, 902), 175(870), *306, 308*
Trechant, J., 105(891), *307*
Trevelyan, W. E., 18(892), *307*
Trippel, A. I., 77(893), *307*
Trowbridge, J. R., 78(894), 79(894), 80(894), *307*
Truter, E. V., 29(742), *303*

Tryding N., 177(71), 178(895), *284, 307*
Tsuda, S., 2(898), 249(896), 253(896, 897), *307*
Tucker, I. W., 178(44), *283*
Tucknott, O. G., 107(899), *307*
Tulloch, A. P., 32(197), 42(900), *287, 307*
Turner, D. A., 36(392), *293*
Turpeinen, O., 39(727), *302*
Tyutyunnikov, B. N., 41(901), *308*

## U

Uden, P. C., 70(2), 108(2), *282*
Ulshöfer, H. W., 23(59), *284*
Upadhyay, G. S., 41(470), 148(470), *295*
Ushakov, A. N., 23(68), 108(918), *284, 308*
Utkina, O. V., 72(792), *304*
Utrilla, R. M., 54(19), *282*

## V

Vagelos, P. R., 262(904), *308*
Vahouny, G. V., 174(902), *308*
van Beers, G. J., 41(338), 78(338), 79(338), 81(338), 100(338), 101(338), 220(338), *291*
van Deenen, L. L. M., 62(916), 66(915, 916), 176(916), 192(209, 210), 198(209, 903), 199(79, 903), 264(211), *284, 288, 308*
van den Bosch, H., 262(904), *308*
Vandenheuvel, F. A., 39(905), *308*
van den Tempel, M., 164(906), 165(906), *308*
Vanderburg, G. A., 177(102), *285*
van der Ven, B., 48(907, 908), *308*
Vander Wal, R. J., 2(909), 29(912), 42(912, 913), 53(912), 54(912), 65(912, 913, 914), 250(910), 257(911), *308*
van Gend, H. W., 116(564a), *298*
van Golde, L. M. G., 62(916), 66(915, 916), 176(916), 264(211), *288, 308*
van Handel, E., 24(917), 26(917), *308*
van Veen, A., 220(385), *293*
Vargaftik, M. N., 72(792), *304*
Vasenius, L., 39(727), *302*
Vaughan, M., 263(856), *306*
Vaver, V. A., 23(68), 108(918), *284, 308*
Venkatarao, C., 44(919), 45(919), 148(919), *308*
Venkatarao, P., 84(715), 101(715), *302*
Venkateswarlu, A., 44(919), 45(919), 148(919), *308*
Venkitasubramanian, T. A., 28(702), *302*
Verdino, B., 26(85), 32(85), 54(85), 57(85), 75(85), 254(736), 257(736), 258(736), *284, 303, 308*
Vereshchagin, A. G., 26(977), 44(921), 67(924, 927), 69(925), 70(924, 926), 71(924, 927), 72(921, 926), 73(683, 924, 926, 927), 75(921, 926, 928), 77(921), 86(923), 87(922, 923), 90(926), 96(683, 924, 926), 98(921, 928), 99(921), *301, 308, 310*
Verger, R., 175(929), *308*
Verkade, P. E., 220(385), *293*
Vermeire, P., 27(703), *302*
Vidyarthi, N. L., 44(930), 148(931), *308*
Vihko, R., 29(335), *291*
Vinkenborg, C., 75(932), *308*
Vioque, E., 26(933), 54(935), 154(934), *308, 309*
Viswanathan, C. V., 67(492, 493, 499), 72(499), 73(499), 74(499), 76(499), 90(499), 92(499), *296*
Vlugter, J. C., 231(315, 316, 317), 232(315, 316, 317), *291*
Vogelberg, K. H., 26(936), *309*
Vollgraf, I., 26(282), *290*
Volpenhein, R. A., 16(221, 224), 173(623), 174(626, 627), 175(625), 176(619, 620, 627), 179(622), 180(622), 234(619, 623), 235(619, 623), 236(619), 237(623), 238(619, 623), 247(628), 262(624), 263(621), *299, 300*
Volqvartz, K., 173(820, 821), 175(819), 178(818), 180(821), *305*
von Dahl, K., 27(809), *305*

von Rudloff, E., 41(937), 227(937), *309*
Vorob'ev, N. V., 99(938), *309*
Voudouris, E. C., 16(295), *290*

## W

Wagle, S. R., 175(749), *303*
Walker, B. L., 176(976), *310*
Walker, F. T., 161(939, 940, 941), *309*
Walsh, V. G., 175(284), *290*
Waterman, H. I., 232(883), *307*
Waters, W. A., 41(942), *309*
Watkins, J. L., 254(840), *306*
Watson, J. A., 27(722), *302*
Watts, R., 104(943), 108(943), 112(943), 119(943), 120(943), 122(943), 128(943), 130(943), 131(943), 137(944), *309*
Weaber, J. R., 16(889), 107(889), *307*
Weber, E. J., 198(212), 235(212), 241(212, 945), 242(212), 254(945), *288, 309*
Weeber, Y. M. M., 27(752a), *303*
Weersing, S., 174(902), *308*
Weinstein, D. B., 28(609), *299*
Weiser, H. J., 20(743), *303*
Weiss, S. R., 262(946, 947), *309*
Wells, M. A., 21(224), *288*
Werthessen, N. T., 29(948), *309*
Wessels, H., 2(494), 16(496, 497), 50(951), 52(951), 53(951), 54(495, 496), 59(495), 61(497, 951), 62(951), 63(951), 67(494, 495, 496, 497, 498, 499), 68(495), 70(494, 951), 72(495, 499, 951), 73(499, 951), 74(499, 950), 75(949), 76(498, 499, 950, 951), 77(950, 951), 87(494, 950, 951), 89(950), 90(499), 91(494), 92(499), 93(951), 268(951), *296, 309*
Westöö, G., 177(71), *284*
Wharry, D. M., 50(666), 61(666), *301*
White, H. B., Jr., 46(952), 254(953), *309*
Whitner, V. S., 19(954), *309*
Whittle, K. J., 74(955), 76(955), 77(955), *309*
Widenmeyer, L., 148(246), *289*
Wiebe, T., 27(62), *284*

Williams, A. A., 107(899), *307*
Williams, E. A., 24(395), *293*
Williams, M. C., 254(757), *303*
Williams, P. N., 2(365), 12(365), 13(365), 40(366), 139(364), 140(367), 143(367), 144(367), 252(363), 253(363), *292*
Wills, E. D., 175(957), 178(957), 187(957), *309*
Winder, W. C., 231(404), *293*
Witter, R. F., 19(954), *309*
Wolf, L.-M., 16(557a), *298*
Wolf, W., 161(500), *296*
Wolff, D., 176(450), *295*
Wolff, I. A., 39(654), 87(655), 129(510), 154(511, 512, 653, 875), 155(652), 160(512), 176(512), 177(509, 876), 206(511, 512), 220(512), 223(512), 226(512), 227(511, 512), 229(512), 243(512, 875), 255(875), *296, 300, 307*
Wolmark, N., 179(121), 246(121), 247(121), 257(121), *285*
Wolstenholme, W. A., 208(47), 211(47), 213(47), *283*
Wood, G. E., 44(288), *290*
Wood, P., 29(804), *305*
Wood, P. D. S., 29(958), *309*
Wood, R., 48(965), 50(963), 108(961, 962), 137(959), 169(964), 189(964), 192(964), 257(960), *309*
Woods, L. A., 29(873), *307*
Wyatt, C. J., *309*
Wynne, R. B., 119(810), *305*

## Y

Yaguchi, M., 183(967), *309*
Yakubov, M. K., 41(968), *309*
Yarger, K., 32(204), *288*
Yates, S. G., 156(513), *296*
Yen, C., 21(200), *287*
Yip, Y. K., 175(749), *303*
Yokoyama, Y., 148(869), *306*
Yorke, R. W., 223(155), *286*
Young, T. E., 29(969), *309*
Youngs, C. G., 16(863), 29(863), 41(971, 974), 42(971, 974), 55(863),

56(863), 78(862, 971, 974), 84(3, 971), 100(971), 101(3), 102(3), 108(974), 132(974), 133(974), 140(973), 142(973), 143(973), 157(972), 177(3), 253(970), 255(862, 844, 865, 974), *282, 306, 309, 310*
Yurkowski, M., 160(975), 167(975), 168(975), 170(975), 171(975), 172(975), 175(975), 176(122, 976), 177(975), 179(122), 182(122, 975), 189(975), 195(122), 196(975), 235(122), 236(122), 241(122), *285, 310*

## Z

Zaky, Y. A. H., 254(368), *292*
Zeman, I., 155(725), *302*
Zhelvakova, E. G., 229(976a, 976b), *310*
Zhukov, A. V., 26(977), *310*
Ziegler, J. H., 254(840), *306*
Zlatkis, A., 105(255), *289*
Zöllner, N., 28(978), *310*
Zschocke, A., 188(556), 189(556), 192(556), 193(556), 194(556), 195(556), 197(556), 198(556), *298*
Zubov, P. I., 220(310), *291*

# SUBJECT INDEX

*Note:* Individual triglycerides are indexed by the types of fatty acids they contain rather than under specific compound names. Thus "butyrodipalmitin" is indexed under both "Triglycerides, of $n$-$C_4$–$n$-$C_{11}$ acids" and under "Triglycerides, of $n$-$C_{12}$–$n$-$C_{18}$ acids." Similarly, "triricinolein" is listed under "Triglycerides, hydroxy." This will help the reader find appropriate analytical procedures for the specific type of triglyceride he is studying, even though the exact compounds he is working with may not have been analyzed previously.

## A

Acetic acid, *see also* Triglycerides, of acetic acid
  positional distribution in natural triglycerides, 243
Acetodiglycerides, *see* Triglycerides, of acetic acid
Acetylation reaction, 37, 47–48, *see also* Triglycerides, of acetic acid
  aids chromatographic separation of derived diglycerides, 65, 103, 137, 272
  aids stereospecific analysis of hydroxy-triglycerides, 200
Acetylenic acids, *see* Triglycerides, of acetylenic acids
Acyl migration in partial glycerides
  caused by acid, 175
  by heat, 172
  during deacylation reactions, 168–172, 175, 179–181, 183, 185
  gas–liquid chromatography, 137

  liquid–liquid partition chromatography, 103
  mass spectrometry, 213
  silver ion adsorption chromatography, 65–66
  stereospecific analysis, 197
  prevention with boric acid, 160, 169, 175, 185
  on silicic acid, 160, 169, 175, 185
*Allanblackia stuhlmannii* seed triglycerides, fractional crystallization, 12
Allenic acids, *see* Triglycerides, of allenic acids
Aluminum oxide, *see* Chromatography, aluminum oxide adsorption
Animal triglycerides
  biosynthesis, 261–264
  complexity, 9
  positional distribution of fatty acids, 243–248
  triglyceride composition patterns, 257–264

# SUBJECT INDEX

Antioxidants, stabilize polyunsaturates, 18, 53, 200
Arachidic acid, see also Triglycerides, of $n$-$C_{19}$–$n$-$C_{24}$ acids
   positional distribution in natural triglycerides, 235
Arachidonic acid, see Triglycerides, of $n$-$C_{19}$–$n$-$C_{24}$ acids and Triglycerides, separation by unsaturation
Argentation chromatography, see Chromatography, silver ion adsorption
Azelaic acid, see Triglycerides, oxidized

## B

Beef tallow triglycerides
   deacylation with pancreatic lipase, 181
   fractional crystallization, 12
   nonhomogeneous origin, 254
Behenic acid, see also Triglycerides, of $n$-$C_{19}$–$n$-$C_{24}$ acids
   positional distribution in natural triglycerides, 235
Beluga whale head triglycerides
   gas–liquid chromatography, 129
   silicic acid adsorption chromatography, 151
Biosynthesis of triglycerides, 261–264
   acyl transferase specificity, 263–264
   comparison with fatty acid distribution hypotheses, 253, 263–264
   dihydroxyacetone phosphate pathway, 261–262, 264
   glycerol phosphate pathway, 16, 261–264
   monoglyceride pathway, 16, 261–264
   turnover in vivo, 254, 264
Bird triglycerides, positional distribution of fatty acids, 246
Bromination, 44–45, see also Triglycerides, brominated
   debromination of products, 45
   directly on TLC plates, 45
Butterfat triglycerides
   complexity, 9
   deacylation with pancreatic lipase, 178
   distillation, 165
   fractional crystallization, 12

   gas–liquid chromatography, 104, 125, 134
   hydroxy triglycerides, 48
   keto triglycerides, 48
   liquid–liquid partition chromatography, 84
   positional distribution of fatty acids, 247
   silicic acid adsorption chromatography, 151–152
Butyric acid, see also Triglycerides, of $n$-$C_4$–$n$-$C_{11}$ acids
   positional distribution in natural triglycerides, 247

## C

Candlenut oil triglycerides
   comparison of experimental and calculated triglyceride compositions, 255–256
   deacylation with pancreatic lipase, 256
   silver ion adsorption chromatography, 255–256
Capric acid, see Triglycerides, of $n$-$C_4$–$n$-$C_{11}$ acids
Caproic acid, see Triglycerides, of $n$-$C_4$–$n$-$C_{11}$ acids
Caprylic acid, see Triglycerides, of $n$-$C_4$–$n$-$C_{11}$ acids
Carbon number
   definition, 105
   separation by, see Triglycerides, separation by molecular weight
Caryocar villosum seed triglycerides, X-ray diffraction, 221
Cashew nut seed triglycerides
   deacylation with pancreatic lipase, 238
   positional distribution of fatty acids, 238
Castor oil triglycerides
   countercurrent distribution, 84, 101–102
   liquid–liquid partition chromatography, 84, 101–102
   paper chromatography, 162
   silicic acid adsorption chromatography, 154–155
Cat fat triglycerides, deacylation with pancreatic lipase, 179

Charcoal adsorption chromatography, see Chromatography, charcoal adsorption
Chromatography
　aluminum oxide adsorption, 161–162
　　acetodiglycerides, 161
　　column chromatography, 24–25, 161–162
　　hydrolysis during, 161
　　isolation of triglycerides, 24–25
　　mercurated triglycerides, 161–162
　　oxidized triglycerides, 162
　　separation by molecular weight, 161
　　　by unsaturation, 161–162
　　thin-layer chromatography, 25, 161
　argentation, see Chromatography, silver ion adsorption
　charcoal adsorption, 162
　　column chromatography, 162
　　separation by molecular weight, 162
　　　by unsaturation, 162
　column, see listings under specific type of chromatography, e.g., Chromatography, liquid–liquid partition, column chromatography and Chromatography, silicic acid adsorption, column chromatography
　Florisil adsorption, 19–20, 160–161
　　column chromatography, 19–20, 24
　　isolation of triglycerides, 19–20, 24
　　mercurated triglycerides, 160–161
　　oxidized triglycerides, 160
　　oxygenated triglycerides, 20
　　slight separation by unsaturation and molecular weight, 20
　gas–liquid, 104–138, see also Fatty acid analysis by gas–liquid chromatography
　　acetodiglycerides, 104, 130–131, 137
　　applications, 126–138, 266–269
　　carbon number, definition, 105
　　carrier gas, 114–115
　　　flow rate, 107, 110–111, 115, 117–118
　　　type, 114–115, 118
　　column, 107–114
　　　capillary, 107
　　　conditioning, 113–114, 118
　　　liquid phase, 108–109, 118, 128
　　　preparation, 112–114
　　　single vs. dual, 112
　　　size, 111–112, 118
　　　solid support, 107–108, 118
　　　temperature, 105, 115–119
　　　tubing material, 105, 109–111, 117–118
　　combined with
　　　deacylation with pancreatic lipase, 176
　　　liquid–liquid partition chromatography, 75, 81, 266–269
　　　mass spectrometry, 211, 213, 266–269
　　　silver ion adsorption chromatography, 54, 57, 266–269
　　detector, 117
　　　gas flow rates, 117
　　　linear response, 119–120
　　　temperature, 118
　　　type, 107
　　diglycerides, 137–138, 176
　　epoxy triglycerides, 131–132
　　flash heater, see Chromatography, gas–liquid, sample injection
　　gas chromatograph, 105–107
　　historical development, 10, 14, 104
　　hydrogenated triglycerides, 123, 126, 128
　　hydroxy triglycerides, 131
　　isomer separations, 129–131
　　　acyl group chain length isomers, 131
　　　acyl group positional isomers, 130
　　　alicyclic acyl groups, 129–130
　　　branched-chain acyl groups, 129
　　measurement of total triglycerides, 26, 30, 54, 57, 75, 81
　　methods, 105–126
　　operating conditions, 114–118
　　oxidized triglycerides, 132–133
　　peak identification, 118–119
　　preparative separations, 133–135
　　quantitation, 118–126
　　　accuracy, 124–126
　　　calibration, 120–124
　　　fatty acid carbon recovery, 125–126
　　　linearity of detector response, 119–120

Chromatography (*continued*)
    response factors, 111–112, 121–124
    radioisotope detection, 135–137
    resolution, 109–111, 115, 117
    sample injection
        flash heater design, 106
        on-column injection, 105–106, 113
        septum, 105–107, 114
        syringe, 105, 113–114
        temperature, 107, 114, 118
    sample size, 114, 118
    separation by carbon number, 126–128, 266–269
    by unsaturation, 128–129
  ion-exchange, 162–163
    column chromatography, 162–163
    mercaptoacetic acid addition products, 163
    oxidized triglycerides, 162–163
  liquid–liquid partition, 67–103
    acetodiglycerides, 103
    $Ag^+$/olefin $\pi$-complexing, 71–72, 96
    applications, 86–103, 266–269, 271–280
    brominated triglycerides, 98–100
    calculating component triglycerides in fractions separated, 94
    choice of method, 68
    column chromatography, 78–81, 92–93
        applications, 88, 92–95, 100–101, 103
        quantitation, 80–81, 281
        separation procedure, 79–80
        solid support, 78
        solvent system, 78–79
    combined with
        deacylation reactions, 271–280
        gas–liquid chromatography, 75, 81, 266–269
        mass spectrometry, 266–269
        silver ion adsorption chromatography, 266–269, 271–280
        stereospecific analysis, 271–280
    consecutive separation before and after bromination, 98–100
    consecutive separation in normal and $Ag^+$-containing solvent systems, 96–98

  countercurrent distribution, 81–86, 93–96
    apparatus, 81–83
    applications, 93–96, 100–102
    principle of separation, 82–83
    quantitation, 86
    separation procedure, 85–86
    solvent system, 81–82, 84–85
  critical partners
    definition, 87
    separation of, 91–93, 96–100
  diglycerides, 102–103, 271–280
  elution order, 86–90
  epoxy triglycerides, 84
  historical development, 10, 14, 67
  hydrogenated triglycerides, 101–102
  hydroxy triglycerides, 84, 101–102
  isolation of total triglycerides, 25
  isomer separations
    double bond geometrical isomers, 89, 100–101
    double bond positional isomers, 89, 279
  mercurated triglycerides, 100
  methods, 68–86
  need for automated method, 281
  oxidized triglycerides, 78, 84, 100, 279
  paper chromatography, 68–77, 90–92
    applications, 90–92, 98–100, 103
    locating bands, 74, 76–77
    preparation of paper, 69
    preparative separations, 73–75, 92
    quantitation, 74–75
    separation procedure, 72–74
    solvent system, 69–72
  partition number, 86–90
    definition, 86–87
    equivalent, 87–90
    integral, 86–90
  separation by carbon number, 86, 276
    by partition number, 86–96, 266–269, 274–278
    by unsaturation, 95–100
  solvent systems
    $AgNO_3$-containing, 71–72, 84
    organic solvents, 69–72, 78–85
    oxidizing solvents, 100

Chromatography (*continued*)
  thin-layer chromatography, 68–77, 90–93
    applications, 90–93, 97, 100–103, 268–269
    locating bands, 74, 76–77
    preparation of plates, 68–69
    preparative separations, 74–75, 91–93
    quantitation, 74–75
    separation procedure, 72–74
    solvent system, 69–73
    *trans* unsaturation, 89, 100–101
  paper, 162, *see also* Chromatography, liquid–liquid partition, paper chromatography
    cellulose adsorption, 162
    hydroxy triglycerides, 162
  partition, *see* Chromatography, liquid–liquid partition
  permeation, 163
    column chromatography, 163
    isolation of triglycerides, 25
    separation by molecular weight, 163
  reversed-phase, *see* Chromatography, liquid–liquid partition
  silicic acid adsorption, 150–160
    acetodiglycerides, 151–155
    brominated triglycerides, 158
    column chromatography, 20–22, 24, 150–155, 159–160
    diglycerides, 159–160, 169, 175, 183, 185, 202, 204, 273–279
    epoxy triglycerides, 154, 156
    estolide triglycerides, 158–159
    hydroxy triglycerides, 154–155
    isolation of triglycerides, 20–25
    isomer separations, acyl group positional isomers, 153–155
    keto triglycerides, 155
    mercurated triglycerides, 158–160
    oxidized triglycerides, 22, 157
    ozonized triglycerides, 156–158
    separation by molecular weight, 150–152
      by number of ester groups, 157–159
      by unsaturation, 153, 158–159
    thin-layer chromatography, 22–23, 150–160
  silver ion adsorption, 49–66
    acetodiglycerides, 65–66
    acetylenic triglycerides, 64
    $Ag^+$/olefin $\pi$-complex, 49, 57, 62–64
    $Ag(NH_3)_2^+$/olefin $\pi$-complex, 50
    applications, 57–66, 266–269, 271–280
    calculating component triglycerides in fractions separated, 60–61
    choice of method, 50
    column chromatography, 55–57, 60–62
      adsorbent, 55
      applications, 60–62
      quantitation, 56–57, 281
      separation procedure, 55–66
      solvent, 55–56
    combined with
      deacylation reactions, 227–228, 256, 258, 271–280
      differential cooling curves, 217
      fractional crystallization, 147
      gas–liquid chromatography, 54, 57, 266–269
      liquid–liquid partition chromatography, 266–269, 271–280
      mass spectrometry, 266–269
      rotation of polarized light, 227–228
      stereospecific analysis, 202–203, 260, 271–280
    diglycerides, 65–66, 227–228, 271–280
    elution order, 57–58, 65–66
    historical development, 10, 14, 49
    isomer separations
      acyl group chain length isomers, 63
      acyl group positional isomers, 58, 63, 66
      double bond geometrical isomers, 62–64
      double bond positional isomers, 57–64, 66
      enantiomers, 63
    methods, 50–57
    need for automated method, 281
    optimum level of $AgNO_3$ in adsorbent, 50
    oxidized triglycerides, 64–65

SUBJECT INDEX 337

Chromatography (continued)
  poor resolution when a fatty acid contains > 2 double bonds, 59–60, 276–278, 281
  separation by numbers of cis double bonds, 57–61, 266–269
  thin-layer chromatography, 50–54, 58–60
    adsorbent, 50–51
    applications, 57–66, 266–269, 271–280
    locating bands, 52–53
    quantitation, 53–54
    sample recovery, 53–54
    separation procedure, 52–53
    solvent, 51–52
    trans unsaturation, 62–64
  thermal-gradient, 164–165
    apparatus, 164
    column chromatography, 164–165
    separation by molecular weight, 165
    by unsaturation, 165
  thin-layer, see listings under specific type of chromatography, e.g., Chromatography, liquid–liquid partition, thin-layer chromatography and Chromatography, silver ion adsorption, thin-layer chromatography
  zeolite adsorption, 24
Cis–trans isomerization reaction, 46, see also Triglycerides, cis–trans isomerized
Cocoa butter triglycerides
  confectionary chocolate, 16
  deacylation with Geotrichum candidum lipase, 187
  with pancreatic lipase, 176, 179
  enantiomorphic triglycerides, 227, 255, 257
  fractional crystallization, 12
  gas–liquid chromatography, 125, 132–133
  infrared spectroscopy, 220
  liquid–liquid partition chromatography, 79, 93, 95, 101–102
  mass spectrometry, 210–211
  rotation of polarized light by derived diglycerides, 227
  stereospecific analysis, 235

  thermal gradient chromatography, 165
  X-ray diffraction, 221–222
Coconut oil triglycerides
  deacylation with Grignard reagent, 271, 274, 276, 280
  with pancreatic lipase, 181, 274, 276
  distillation, 165–166
  gas–liquid chromatography, 115–116
  liquid–liquid partition chromatography, 271, 274, 276, 280
  positional distribution of fatty acids, 241
  scheme for complete analysis, 274, 276, 280
  silver ion adsorption chromatography, 274, 276, 280
  stereospecific analysis, 274, 276, 280
Cod liver oil triglycerides
  deacylation with Grignard reagents, 169
  with pancreatic lipase, 177, 179
  fatty acid composition, 33
  silver ion adsorption chromatography, 60
Cod liver phospholipids, fatty acid composition, 36
Column chromatography, see listings under specific type of chromatography, e.g., Chromatography, liquid–liquid partition, column chromatography and Chromatography, silicic acid adsorption, column chromatography
Combining triglyceride analysis techniques, 265–281, see also other listings under individual analytical techniques
  detailed examples, 267–269, 275–276, 278–280
  maximum analysis, 272–280
  positional analysis techniques, 270–271
  separation techniques, 266–269
  use of derived diglycerides, 271–280
Complexity of triglyceride mixtures, 8–9
  calculating all possible triglyceride species from $n$ fatty acids, 8–9
Conjugated polyunsaturated acids, see Triglycerides, of conjugated polyunsaturated acids

Corn oil triglycerides
  deacylation with pancreatic lipase, 179
  nonhomogeneous origin, 254
  positional distribution of fatty acids, 179, 192, 200, 235, 241–242
  stereospecific analysis, 192, 200, 235, 241–242
Cottonseed oil triglycerides
  comparison of experimental and predicted triglyceride compositions, 255–256
  deacylation with pancreatic lipase, 238, 256, 271
  gas–liquid chromatography, 104
  liquid–liquid partition chromatography, 84, 91, 99
  positional distribution of fatty acids, 238
  silver ion adsorption chromatography, 255–256, 271
Countercurrent distribution, see also Chromatography, liquid–liquid partition
  apparatus, 81–83
  applications, 93–96, 100–102
  isolation of total triglycerides, 25
  principle of separation, 82–83
  quantitation, 86
  separation procedure, 85–86
  solvent system, 81–82, 84–85
Crambe abyssinica seed triglycerides, gas–liquid chromatography, 126–127
Critical solution temperature, variation with triglyceride composition, 231–232
Cruciferae seed triglycerides, positional distribution of fatty acids, 237–241
Crystallization, fractional, 139–149
  from $AgNO_3$ solution, 140, 146–147
  applications, 142–149
  brominated triglycerides, 148–149
  calculating component triglycerides in fractions separated, 145–146
  cis–trans isomerized triglycerides, 149
  combined with silver ion adsorption chromatography, 147
  crystallization sequence, 143–145
  epoxy triglycerides, 149
  equipment, 141–142
  historical development, 9–14, 139–140
  hydrogenated triglycerides, 148
  methods, 140–142
  oxidized triglycerides, 147–148
  procedure, 141–142
  separation by number of saturated acyl groups, 142–146
  by unsaturation, 146–147
  solvents
    $AgNO_3$-containing, 140
    organic, 140
  triglyceride solubilities, 142–143
Cyclopentene acids, see Triglycerides, of alicylic acids
Cyclopropane and cyclopropene acids, see Triglycerides, of alicyclic acids

D

Deacylation reactions, 167–187, 188–196, 270–271
  castor bean acid lipase, 187
  chemical methods, 168–173
  enzymatic methods, 173–187
  Geotrichum candidum lipase, 185–187, see also Lipase, Geotrichum candidum
  Grignard reagents, 168–172, see also Grignard reagents, deacylation with
  hydroxylamine, 172
  lithium aluminum hydride, 172
  microbial lipases, 187
  milk lipase, 183–184, see also Lipase, milk
  pancreatic lipase, 173–183, see also Lipase, pancreatic
  potassium carbonate, 172–173
  "representative" diglycerides and monoglycerides, definition of, 167–168
  Rhizopus arrhizus lipase, 184–185, 187
  sodium hydroxide, 172
  sodium methoxide, 172
  thermal hydrolysis, 172
  uses of, 167
  Vernonia anthelmintica lipase, 187
Decanoic acid, see also Triglycerides, of $n$-$C_4$–$n$-$C_{11}$ acids
  positional distribution in seed triglycerides, 241, 243

Density
  polymorphism, 214
  variation with triglyceride composition, 231–232
Derivative formation prior to triglyceride analysis, 37–48
  acetylation, 37, 47–48
  bromination, 44–45
  cis–trans isomerization, 46
  dioxolane ring formation from epoxy group, 131–132
  elaidinization, 46
  epoxidation, 43–44
  estolide ester cleavage, 47
  hydrazone formation, 48
  hydrogenation, 37–39
  interesterification, 47
  mercaptoacetic acid addition, 37, 46
  mercuration, 45–46
  ozonization, 37, 42–43
  permanganate oxidation, 39–42
  reactions at double bonds, 38–46
    at ester linkages, 46–47
    of hydroxy, epoxy, and keto groups, 47–48, 131–132
  trifluoroacetate esters, 137
  trimethylsilyl ethers, 48
Diacyl glyceryl ethers, separation from triglycerides, 23
Dielectric constant, variation with triglyceride composition, 231–232
Differential cooling curves, 216–218
  combined with silver ion adsorption chromatography, 217
Differential scanning calorimetry, 216–218
Differential thermal analysis, 216–218
Diglyceride kinase, use in stereospecific analysis, 193, 197–198
Diglycerides
  of $n$-acids
    of acetic acid, 160
    acetylation, 47–48, 65, 103, 137, 272
    acyl migration, see Acyl migration in partial glycerides
    deacylation with pancreatic lipase, 176, 178, 205
    derived diglycerides used for analysis of unresolvable triglyceride mixtures, 271–280
    dihydroxyacetone derivatives of 1,3-diglycerides, 103
    gas–liquid chromatography, 137–138, 176
    liquid–liquid partition chromatography, 102–103, 271–280
    mass spectrometry, 211–213
    permanganate oxidation, 227–229
    products of triglyceride deacylation by
      Geotrichum candidum lipase, 186–187, 195–196
      Grignard reagents, 169–172, 195–196
      milk lipase, 177, 184
      pancreatic lipase, 177, 179, 195
      "representative," 167–172, 179, 184, 187, 195–196
    rotation of polarized light by asymmetric diglycerides, 227–229
    separation by molecular weight, 66, 102, 137, 160, 271–280
      of double bond positional isomers, 66
      by partition number, 102–103, 271–280
      of $sn$-1,2(2,3)- and $sn$-1,3-isomers, 65, 103, 137–138, 159–160, 204, 271–280
      of $\beta$-StOAc and $\beta$-OStAc, 66
      by unsaturation, 65–66, 138, 271–280
    silicic acid adsorption chromatography, 159–160, 169, 175, 183, 185, 202, 204, 273–279
    silver ion adsorption chromatography, 65–66, 227–228, 271–280
    stereospecific analysis, 200, 203–205, 271–280
    trifluoroacetate derivative, 137
    trimethylsilyl ether derivative, 48, 137–138, 211–213, 227–229
  epoxy
    gas–liquid chromatography, 137
    silicic acid adsorption chromatography, 160
  hydroxy
    gas–liquid chromatography, 137
    stereospecific analysis, 200

oxidized, rotation of polarized light, 227–229
Diol lipids, neutral, separation from triglycerides, 23
Distillation, separation by molecular weight, 165–166
Distribution of fatty acids in natural triglyceride mixtures, 233–264
  acetic acid, 243
  animal triglycerides, 9, 243–248, 257–264
  arachidic acid, 235
  behenic acid, 235
  Bernstein's hypothesis, 253
  biosynthesis of triglycerides, 253, 261–264
  butyric acid, 247
  comparison of experimental and calculated positional distributions of fatty acids, 235–236, 238–240, 242, 244–248
    of experimental and predicted triglyceride compositions, 255–261
  computer programs for distribution hypotheses, 251–252
  criteria for applying distribution hypotheses, 255
  decanoic acid, 241, 243
  docosahexaenoic acid, 243–245
  docosapentaenoic acid, 244
  docosenoic acid, 235–236, 245–247
  eicosapentaenoic acid, 245–246
  eicosenoic acid, 236
  erucic acid, 235–236
  Evans hypothesis for seed triglycerides, 237–238, 240–241
  even distribution hypothesis, 252–253
  Gunstone-Mattson hypothesis for seed triglycerides, 237, 240
  hexanoic acid, 247
  lauric acid, 241, 243
  lignoceric acid, 235
  linoleic acid, 236–242, 246–248
  linolenic acid, 236–241
  Litchfield correlation formulas for Cruciferae seed triglycerides, 237, 239–241
  monoacid hypothesis, 253
  myristic acid, 241, 243, 246
  nonhomogeneous origin of many natural fats, 253–255
  1-nonselective hypothesis, 253
  oleic acid, 236–242, 247–248
  ordered distribution hypothesis, 253
  palmitic acid, 234–235, 242, 247–248
  partial random hypothesis, 253
  plant triglycerides, 9, 234–243, 255–257, 261, 263
  positional distribution patterns, 234–248
  1-random–2-random–3-random hypothesis, 249–250, 255, 257, 259–261, 263–264
  1-random–2,3-random hypothesis, 253
  1,3-random–2-random hypothesis, 250–251, 254–259, 261, 264
  1,2,3-random hypothesis, 251–252
  restricted random distribution hypothesis, 252
  saturated acids, 234–235, 241–243, 246–248
  stearic acid, 235, 242, 247–248
  tetracosenoic acid, 236
  triglyceride composition patterns, 248–261
  unsaturated acids, 235–248
  validity of distribution hypotheses, 253–261
  vernolic acid, 243
  widest distribution hypothesis, 252–253
  Young's hypothesis, 253
Docosahexaenoic acid, see also Triglycerides, or $n$-$C_{19}$–$n$-$C_{24}$ acids and Triglycerides, separation by unsaturation
  oxidation during analysis, 195, 200
  positional distribution in natural triglycerides, 243–245
  resistance to lipolysis, 169, 177
Docosapentaenoic acid, see also Triglycerides, of $n$-$C_{19}$–$n$-$C_{24}$ acids and Triglycerides, separation by unsaturation
  oxidation during analysis, 195, 200
  positional distribution in natural triglycerides, 244
Docosenoic acid, see also Erucic acid; Triglycerides, of $n$-$C_{19}$–$n$-$C_{24}$ acids and Triglycerides, separation by unsaturation

## SUBJECT INDEX 341

Docosenoic acid (*continued*)
  positional distribution in natural triglycerides, 235–236, 245–247
Dog fat triglycerides, deacylation with pancreatic lipase, 179

### E

Egg yolk (chicken) lecithin, derived diglycerides, silicic acid adsorption chromatography, 155
Egg yolk (chicken) triglycerides
  comparison of experimental and calculated triglyceride compositions, 257
  silicic acid adsorption chromatography, 153–154, 157–158
  stereospecific analysis, 257
Eicosapentaenoic acid, *see also* Triglycerides, of $n$-$C_{19}$–$n$-$C_{24}$ acids *and* Triglycerides, separation by unsaturation
  oxidation during analysis, 195, 200
  positional distribution in natural triglycerides, 245–246
  resistance to lipolysis, 169, 177
Eicosenoic acid, *see also* Triglycerides, of $n$-$C_{19}$–$n$-$C_{24}$ acids *and* Triglycerides, separation by unsaturation
  positional distribution in natural triglycerides, 236
Elaidic acid, *see* Triglycerides, double bond positional isomers *and* Triglycerides, of *trans* acids
Eliadinization reaction, 46
*Ephedra nevadensis* seed triglycerides
  gas–liquid chromatography, 269
  liquid–liquid partition chromatography, 269
Epoxidation reaction, 43–44, *see also* Triglycerides, epoxy
  directly on TLC plates, 44
Ergot oil triglycerides, silicic acid adsorption chromatography, 158–159
Erucic acid, *see also* Triglycerides, of $n$-$C_{19}$–$n$-$C_{24}$ acids *and* Triglycerides, separation by unsaturation
  positional distribution in natural triglycerides, 235–236

Estolide triglycerides, 47, *see also* Triglycerides, estolide
  estolide ester cleavage, 47
*Euonymus verrucosus* seed triglycerides
  nuclear magnetic resonance, 223
  positional distribution of fatty acids, 243
  rotation of polarized light, 206, 227
*Euphorbia lagascae* seed triglycerides, silicic acid adsorption chromatography, 154, 156
Extraction of lipids, 17–19
  precautions for preventing contamination and chemical alteration, 18
  procedures, 17–19
  solvents, 17–19

### F

Fatty acid analysis by gas–liquid chromatography, 31–36
  accuracy, 36
  column, 32–33
  identification of peaks, 33–35
    capillary column, 35
    different liquid phase, 33, 35
    hydrogenation, 35
    isolation–oxidation, 35
    known standards, 34
    predicting retention times, 34
    TLC fractionation, 35
  measurement of total triglycerides, 25–26, 30, 54, 75, 81
  methyl ester preparation, 31–32
  quantitation, 35–36
Fish triglycerides
  deacylation with Grignard reagent, 169
    with pancreatic lipase, 169, 177, 179
  gas–liquid chromatography, 125–126, 128
  liquid–liquid partition chromatography, 98, 278, 281
  need for better separation techniques, 280–281
  positional distribution of fatty acids, 243–247
  silver ion adsorption chromatography, 59–60, 278, 281
  stereospecific analysis, 195

Florisil, see Chromatography, Florisil adsorption
Fractional crystallization, see Crystallization, fractional
Frog fat triglycerides, deacylation with pancreatic lipase, 179

## G

Gas–liquid chromatography, see Chromatography, gas–liquid
Gel permeation chromatography, see Chromatography, permeation
*Geotrichum candidum* lipase, see Lipase, *Geotrichum candidum*
*Glomerella cingulata* conidial triglycerides, infrared spectroscopy, 218
Glycerol
  analysis of, 26–28, 31, 54
  stereochemistry, 2–4
*Gmelina asiatica* seed triglycerides, silver ion adsorption chromatography, 62
Grignard reagents, deacylation with, 168–172
  acyl migration, 169–172
  determining positional distribution of fatty acids, 170–172, 270
  diglyceride products, 169–172, 195–196
  ethyl magnesium bromide, 168
  methyl magnesium bromide, 168, 171
  methyl magnesium iodide, 168
  monoglyceride products, 171–172
  reaction conditions, 168–169
  specificity, 168–169
  used in combination with
    liquid–liquid partition chromatography, 271–278
    rotation of polarized light by derived diglycerides, 227–229
    silver ion adsorption chromatography, 227–228, 271–278
    stereospecific analysis, 189–196, 202, 271–278

## H

Hexanoic acid, see also Triglycerides, of $n$-$C_1$–$n$-$C_{11}$ acids
  positional distribution in natural triglycerides, 247

History of triglyceride analysis, 9–15
  1915–1955, 9–14
  1956–1972, 14–15
  discovery of mixed acid triglycerides, 12
  fractional crystallization, 9–14, 139–140
  gas–liquid chromatography, 10, 14, 104
  liquid–liquid partition chromatography, 10, 14, 67
  mass spectrometry, 15
  pancreatic lipase hydrolysis, 10, 14, 173
  permanganate oxidation, 10, 12–14, 39–40
  silver ion adsorption chromatography, 10, 14, 49
  stereospecific analysis, 10, 15, 188
Horse fat triglycerides
  analysis of derived diglycerides, 276–278, 280
  deacylation with Grignard reagent, 276–278, 280
    with pancreatic lipase, 179, 276–278
  liquid–liquid partition chromatography, 276–278, 280
  scheme for complete analysis, 274, 276–278, 280
  silver ion adsorption chromatography, 276–278, 280
  stereospecific analysis, 276–278, 280
Human triglycerides
  absorption in the intestine, 16, 261–262
  adipose tissue, 259
  extraction from serum, 19
  plasma, 25, 254
  serum, 15, 19, 31
  stereospecific analysis, 259
*Hydnocarpus wightiana* seed triglycerides, gas–liquid chromatography, 129–130
Hydrazone derivatives of keto triglycerides, 48
Hydrogenation, 38–39, see also Triglycerides, hydrogenated
  directly on thin-layer and paper chromatograms, 39, 101
  NaBH$_4$ titration method, 39
  platinum oxide catalyst procedure, 38–39

Hydroxy triglycerides, see Triglycerides, hydroxy

## I

Illipe butter triglycerides
  deacylation with pancreatic lipase, 173–174
  positional distribution of fatty acids, 174, 249–250
*Impatiens edgeworthii* seed fat
  positional distribution of fatty acids, 243
  rotation of polarized light, 227
Infrared spectroscopy, 218–221
  acetodiglycerides, 219–220
  band assignments, 218
  identification of acyl group chain length, 220
  of acyl group positional isomers, 220
  measurement of total triglycerides, 27, 54, 218–219
  polymorphism, 214, 220
Interesterification reaction, 47
Invertebrate triglycerides, positional distribution of fatty acids, 243–247
Ion-exchange chromatography, see Chromatography, ion-exchange
Isano oil triglycerides, silicic acid adsorption chromatography, 155
Isolation of triglycerides, 19–25
  column chromatography on
    aluminum oxide, 24–25
    Florisil, 19–20, 24
    silicic acid, 20–22, 24
    zeolite, 24
  countercurrent distribution, 25
  glass fiber impregnated with silicic acid, 25
  permeation chromatography, 25
  single solvent fractionation, 24–25
  thin-layer chromatography on aluminum oxide, 25
  on silicic acid, 22–23
Isomeric triglycerides, see Triglycerides, acyl group chain length isomers; Triglycerides, acyl group positional isomers; Triglycerides, of alicyclic acids; Triglycerides, of branched-chain acids; Triglycerides, double-bond geometrical isomers; *and* Triglycerides, double-bond positional isomers
Isovaleric acid, see Triglycerides, of branched-chain acids
Ixonanthaceae seed triglycerides, positional distribution of fatty acids, 241

## J

*Jatropha curcas* seed triglycerides
  fractional crystallization, 146–147
  silver ion adsorption chromatography, 147

## K

Keto triglycerides, see Triglycerides, keto
Kokum butter triglycerides
  deacylation with $K_2CO_3$, 173
  with pancreatic lipase, 173
  fractional crystallization, 12
  X-ray diffraction, 221

## L

Lard triglycerides
  comparison of experimental and calculated triglyceride compositions, 257, 260
  deacylation with Grignard reagent, 170–171
  with pancreatic lipase, 171, 176, 179–181
  enantiomorphic triglycerides, 227
  fractional crystallization, 9, 11
  graininess in shortenings and margarines, 16
  infrared spectroscopy, 220
  nonhomogeneous origin, 254
  permanganate oxidation, 65
  positional distribution of fatty acids, 247–248
  rotation of polarized light by derived diglycerides, 227
  silver ion adsorption chromatography, 65, 260
  stereospecific analysis, 247–248, 260
Lauraceae seed triglycerides, positional distribution of fatty acids, 241

Lauric acid, see also Triglycerides, of $n$-$C_{12}$–$n$-$C_{18}$ acids
  positional distribution in seed triglycerides, 241, 243
Lesquerella seed triglycerides, silicic acid adsorption chromatography, 154
Lignoceric acid, see also Triglycerides, of $n$-$C_{19}$–$n$-$C_{24}$ acids
  positional distribution in natural triglycerides, 235
Lindera praecox seed triglycerides, gas–liquid chromatography, 125–127
Linoleic acid, see also Triglycerides, of $n$-$C_{12}$–$n$-$C_{18}$ acids and Triglycerides, separation by unsaturation
  positional distribution in natural triglycerides, 236–242, 246–248
Linolenic acid, see also Triglycerides, of $n$-$C_{12}$–$n$-$C_{18}$ acids and Triglycerides, separation by unsaturation
  positional distribution in natural triglycerides, 236–241
Linseed oil triglycerides
  aluminum oxide adsorption chromatography, 161
  countercurrent distribution, 95
  deacylation with pancreatic lipase, 238
  liquid–liquid partition chromatography, 84, 90–92, 95
  nonhomogeneous origin, 254
  paper chromatography, 162
  positional distribution of fatty acids, 238
  silver ion adsorption chromatography, 60
  stereospecific analysis, 235
Lipase
  castor bean, 187
  Geotrichum candidum, 185–187
    acyl migration, 185
    diglyceride products, 186–187, 195–196
    monoglyceride products, 187
    preparation, 185
    reaction conditions, 185
    specificity for cis-9-unsaturation, 185–186
    used in combination with stereospecific analysis, 195–196, 204
  microbial, 187
  milk, 183–184
    acyl migration, 183
    determining positional distribution of fatty acids, 184
    diglyceride products, 177, 184
    free fatty acid products, 177
    monoglyceride products, 177, 184
    preparation and purification, 183
    reaction conditions, 183
    specificity
      fatty acid, 184
      positional, 183–184
      triglyceride, 184
    used in combination with stereospecific analysis, 195
  pancreatic, 173–183
    acetodiglycerides, 176
    acyl migration, 175, 179–181
    applications, 174, 177, 178, 181, 234–235, 237–239, 241, 243, 256, 258, 270–280
    contamination with ester hydrolases, 174–175
    deacylation of diglycerides, 176, 178, 205
    determining positional distribution of fatty acids, 179–180
    diglyceride products, 177, 179, 195
    free fatty acid products, 177, 182–183
    historical development, 10, 14, 173
    hydrolysis of solid SSS, 179
    inhibitors, 176
    monoglyceride products, 171, 177, 179–182
    preparation and purification, 174–175
    reaction conditions, 175–176
    source
      dog, 174
      hog (pancreatin, steapsin), 174
      human, 174
      rat, 174
      skate, 174
    specificity
      diglyceride, 178
      fatty acid, 176–177
      positional, 173, 176
      triglyceride, 178

Lipase (*continued*)
    used in combination with
        gas–liquid chromatography, 176
        liquid–liquid partition chromatography, 271–280
        rotation of polarized light by derived diglycerides, 227–229
        silver ion adsorption chromatography, 227–228, 256, 258, 271–280
        stereospecific analysis, 189–196, 271–280
    *Rhizopus arrhizus,* 184–185, 187
    *Vernonia anthelmintica,* 187

## M

Magnesium silicate, *see* Chromatography, Florisil adsorption
Maize seed triglycerides, *see* Corn oil triglycerides
Malabar tallow triglycerides
    deacylation with $K_2CO_3$, 173
        with pancreatic lipase, 173
    enantiomorphic triglycerides, 227, 257
    rotation of polarized light by derived diglycerides, 227
Margarine oil triglycerides
    consistency, 15
    silver ion adsorption chromatography, 60
Mass spectrometry, 206–213
    acetodiglycerides, 208–209
    combined with
        gas–liquid chromatography, 211, 213, 266–269
        liquid–liquid partition chromatography, 266–269
        silver ion adsorption chromatography, 266–269
    diglycerides, 211–213
    estolide triglycerides, 209
    isomer identification
        acyl group chain length isomers, 208
        acyl group positional isomers, 208
        double bond positional isomers, 209–210
    major triglyceride fragments, 207–209
    natural triglyceride mixtures, 209–211

    pure triglycerides, 207–209
    separation by molecular weight, 209–211, 266–269
        by unsaturation, 209–211, 266–269
Measurement of total triglycerides, 25–31
    accuracy, 31
    acetylacetone method, 27, 31
    *o*-aminophenol method, 27
    automatic column chromatography monitor, 29, 31, 56, 80
        GLC detector, 29, 80
        refractive index, 29, 80, 86
    chromotropic acid method, 26, 31, 54, 75
    copper soap method, 27
    dichromate/$H_2SO_4$ oxidation method, 28
    enzymatic methods, 27, 31
    gas–liquid chromatography of
        glycerol, 26, 54
        methyl esters, 25–26, 30, 54, 75, 81
        triglycerides, 26, 30, 54, 57, 75, 81
    glycerol kinase/glycerol dehydrogenase method, 27, 31
    glycerol kinase/pyruvate kinase/lactate dehydrogenase method, 27, 31
    hydroxamic acid method, 26, 31, 54
    infrared spectroscopy, 27, 54, 218–219
    3-methylbenzothiazoline-2-one method, 27
    nuclear magnetic resonance, 223
    paper chromatography, *see* Measurement of total triglycerides, thin-layer chromatography
    *p*-phenazobenzoyl chloride method, 27
    phenylhydrazine method, 27, 31
    radioactivity, 29
    semiautomated methods, 26–27, 31
    sensitivity, 26–29, 31
    spectrophotometry, 26–28
    sulfophosphovanillin method, 28
    thin-layer chromatography
        densitometry, 28, 54, 57, 75
        fluorimetry, 29
        spot area, 29, 54, 75
        volatilize for GLC detector, 29
    titration with
        cerium perchlorate, 28
        KOH, 28

NaOH, 28, 81
periodate, 28, 54
weight, 29, 54, 56, 75, 81, 86
Melting point, 213–215
enantiomorphic triglycerides, 215
identification of acyl group positional isomers, 215
polymorphism, 213–214
techniques, 214
Mercaptoacetic acid addition to double bonds, 37, 46
Mercuration reaction, 45–46, see also Triglycerides, mercurated
regeneration of original compound, 46
Milk lipase, see Lipase, milk
*Monarda fistulosa* seed triglycerides, liquid–liquid partition chromatography, 97–98
*Monnina emarginata* seed triglycerides, estolide ester cleavage, 47
Monoglycerides of *n*-acids
acyl migration, see Acyl migration in partial glycerides
products of triglyceride deacylation by *Geotrichum candidum* lipase, 187
Grignard reagents, 171–172
milk lipase, 177, 184
pancreatic lipase, 171, 177, 179–182
"representative," 167–168, 171–172, 180–182, 184, 187
Mullet body triglycerides, gas–liquid chromatography, 128
Mutton tallow, see Sheep body fat
*Mycobacterium marianum* triglycerides, aluminum oxide adsorption chromatography, 162
Myristicaceae seed triglycerides, positional distribution of fatty acids, 241
Myristic acid, see also Triglycerides, of $n$-$C_{12}$–$n$-$C_{18}$ acids
positional distribution in natural triglycerides, 241, 243, 246

## N

Nervonic acid, see Tetracosenoic acid
Nomenclature
of fatty acids, abbreviations, 6

of triglycerides, 2–8
abbreviations, 4–7
acid–alcohol system, 5
alcohol–acid system, 4–5
diacid, 6
estolide, 8
glyceride type, 7
mixed-acid, 7
monoacid, 4
prefixes indicating acyl group positional isomers, 4, 7
simplified system, 4–5
*sn*-positional numbering convention, 3–4
symmetrical, 7
triacid, 7
unsymmetrical, 7
Nuclear magnetic resonance, 222–224
acetodiglycerides, 223
chemical shift reagents, 223–224
identification of acyl group positional isomers, 223–224
measurement of total triglycerides, 223
polymorphism, 214

## O

Octanoic acid, see Triglycerides, of $n$-$C_4$–$n$-$C_{11}$ acids
Oiticica oil triglycerides, silicic acid adsorption chromatography, 155
Oleic acid, see also Triglycerides, of $n$-$C_{12}$–$n$-$C_{18}$ acids *and* Triglycerides, separation by unsaturation
positional distribution in natural triglycerides, 236–242, 247–248
Olive oil triglycerides
aluminum oxide adsorption chromatography, 162
deacylation with pancreatic lipase, 179, 181
detecting adulteration, 16
gas–liquid chromatography, 104, 132–133
infrared spectroscopy, 220
liquid–liquid partition chromatography, 91, 93
silver ion adsorption chromatography, 61
stereospecific analysis, 235

Omphaloskepsis, 172–173
Optical rotatory dispersion, see Rotation of polarized light
Oxidation with $KMnO_4$, see Permanganate oxidation
Oxidized triglycerides, see Triglycerides, oxidized
Ozonization reaction, 42–43, see also Triglycerides, ozonized

## P

Palm oil triglycerides (fruit coat fat of *Elaeis guineensis*)
  differential cooling curves, 217
  enantiomorphic triglycerides, 227
  liquid–liquid partition chromatography, 101
  possible nonhomogeneous origin, 255
  rotation of polarized light by derived diglycerides, 227
  silver ion adsorption chromatography, 62
Palmae seed triglycerides
  positional distribution of fatty acids, 241
  possible nonhomogeneous origin, 255
Palmitic acid, see also Triglycerides, of $n$-$C_{12}$–$n$-$C_{18}$ acids
  positional distribution in natural triglycerides, 234–235, 242, 247–248
Palmitoleic acid, see Triglycerides, of $n$-$C_{12}$–$n$-$C_{18}$ acids *and* Triglycerides, separation by unsaturation
Pancreatic lipase, see Lipase, pancreatic
Paper chromatography, see Chromatography, paper
Partition chromatography, see Chromatography, liquid–liquid partition
Partition number
  definition, 86–87
  equivalent, 87–90
  integral, 86–90
  separation by, 86–95, 266–269
Peanut oil triglycerides
  deacylation with pancreatic lipase, 179
  liquid–liquid partition chromatography, 268–269
  nonhomogeneous origin, 254
  silver ion adsorption chromatography, 268–269
  stereospecific analysis, 235
Peccary fat triglycerides, positional distribution of fatty acids, 247
Permanganate oxidation, 39–42, see also Triglycerides, oxidized
  acetodiglycerides, 227–229
  allyl esters of product glycerides, 42, 64–65
  diglycerides, 227–229
  Hilditch method using $KMnO_4$ in acetone, 12–14, 40–42
  historical development, 10, 12–14, 39–40
  Kartha method using $KMnO_4$ in acetone/acetic acid, 40–42
  methyl esters of product glycerides, 42
  reaction by-products, 40–42
  von Rudloff method using $KMnO_4/KIO_4$ in *tert*-butanol, 40–42, 227–228
Permeation chromatography, see Chromatography, permeation
Petroselinic acid, see Triglycerides, of $n$-$C_{12}$–$n$-$C_{18}$ acids; Triglycerides, double bond positional isomers; *and* Triglycerides, separation by unsaturation
Phenyl dichlorophosphate, 189–190, 192, 196
Phospholipase A
  reaction conditions, 199
  sources, 198–199
  specificity, 198–199
  use in stereospecific analysis, 15, 189–192, 198–199, 203, 205
Phytanic acid, see Triglycerides, of branched-chain acids
Piezoelectric effect, 230
Pig triglycerides
  adipose tissue, see Lard triglycerides
  blood, 257, 260
  comparison of experimental and calculated triglyceride compositions, 257, 260

liver, 257, 260
milk, 257, 260
silver ion adsorption chromatography, 260
stereospecific analysis, 260
Piquia seed triglycerides, *see Caryocar villosum* seed triglycerides
Plant triglycerides
  biosynthesis, 263
  complexity, 9
  positional distribution of fatty acids, 234–243
  triglyceride composition patterns, 255–257, 261
Polar bear blubber triglycerides
  positional distribution of fatty acids, 191, 243–247
  stereospecific analysis, 191, 200
Polymorphism of triglycerides
  density, 214
  infrared spectroscopy, 214, 220
  melting point, 213–214
  nuclear magnetic resonance, 214
  X-ray diffraction, 214, 221–222
Poppyseed triglycerides, liquid–liquid partition chromatography, 99
Positional distribution of fatty acids in natural triglyceride mixtures, *see* Distribution of fatty acids in natural triglyceride mixtures *or* name of specific natural triglyceride mixture

## R

Radioactive triglycerides, *see* Triglycerides, radioactive
Rapeseed oil triglycerides
  deacylation with pancreatic lipase, 238
  gas–liquid chromatography, 125
  nonhomogeneous origin, 254
  positional distribution of fatty acids, 235–241
  stereospecific analysis, 235–236
Rat liver lecithin, silver ion adsorption chromatography of derived diglyceride acetates, 66
Rat triglycerides
  adipose tissue, 125–127, 257–258
  biosynthesis, 254, 257, 261–264
  comparison of experimental and calculated triglyceride compositions, 257–258
  deacylation with pancreatic lipase, 258
  gas–liquid chromatography, 125–127
  influence of diet on composition, 254
  intestine, 136, 261–262
  kidney, 257–258
  liver, 194, 200, 254, 257–258, 262
  lymph, 262
  nonhomogeneous origin, 254
  plasma, 257–258
  silver ion adsorption chromatography, 257–258
  stereospecific analysis, 194, 200, 257
Refractive index, variation with triglyceride composition, 231–232
Refsum's disease, correlated with phytanate triglycerides, 16
Reversed-phase chromatography, *see* Chromatography, liquid–liquid partition
*Rhizopus arrhizus* lipase, *see* Lipase, *Rhizopus arrhizus*
Ricinoleic acid, *see* Triglycerides, hydroxy
Rose seed triglycerides
  deacylation with pancreatic lipase, 238
  positional distribution of fatty acids, 238
  silver ion adsorption chromatography, 61
Rotation of polarized light, 224–229
  acetodiglycerides, 206, 226–229
  combined with
    deacylation reactions, 227–229
    permanganate oxidation, 227–229
    silver ion adsorption chromatography, 227–228
  cryptoactive triglycerides, 225–226
  diglycerides, 227–229
  enantiomorphic triglycerides, 206, 224–229
  identification of acyl group positional isomers, 206, 224–229
  optical rotatory dispersion data, 225–226

## S

Safflower seed triglycerides
  countercurrent distribution, 95
  deacylation with pancreatic lipase, 270
  stereospecific analysis, 270
Salvadoraceae seed triglycerides, positional distribution of fatty acids, 241
*Sapium sebiferum* triglycerides
  fruit coat, 158–160, 221
  mass spectrometry, 209
  positional distribution of fatty acids, 243
  seed, 209, 243
  silicic acid adsorption chromatography, 158–160
  X-ray diffraction, 221
Saturated fatty acids, *see also* listings of individual acids
  positional distribution in natural triglycerides, 234–235, 241–243, 246–248
Seal blubber triglycerides
  deacylation with Grignard reagent, 182
    with pancreatic lipase, 177, 179, 182
  positional distribution of fatty acids, 243–247
Seed triglycerides
  biosynthesis, 263
  complexity, 9
  dicotyledons, 241
  monocotyledons, 241
  positional distribution of fatty acids, 234–243
  triglyceride composition patterns, 255–257, 261
Sesame seed triglycerides
  deacylation with pancreatic lipase, 238
  positional distribution of fatty acids, 238
Sheep body fat
  fractional crystallization, 144–146
  nonhomogeneous origin, 254
Shortening triglycerides
  consistency, 15–16
  silver ion adsorption chromatography, 60
Silicic acid, chromatography, *see* Chromatography, silicic acid adsorption
Silver ion adsorption chromatography, *see* Chromatography, silver ion adsorption
Sound velocity, variation with triglyceride composition, 231–232
Soybean oil triglycerides
  aluminum oxide adsorption chromatography, 161
  countercurrent distribution, 95
  deacylation with pancreatic lipase, 181, 238
  liquid–liquid partition chromatography, 91, 93, 95, 99
  nonhomogeneous origin, 254
  positional distribution of fatty acids, 181, 235, 238
  silver ion adsorption chromatography, 59
  stereospecific analysis, 235
Stearic acid, *see also* Triglycerides, of $n$-$C_{12}$–$n$-$C_{18}$ acids
  positional distribution in natural triglycerides, 235, 242, 247–248
Stereospecific analysis, 188–205
  accuracy, 189, 192, 194–195, 200
  applications, 199–205, 235–236, 241–248, 259–260, 270–280
  Brockerhoff $sn$-1,2(2,3)-diglyceride method, 189–192, 194–195, 203
  Brockerhoff $sn$-1,3-diglyceride method, 190, 192, 195
  choice of method, 194–195
  deacylation of triglycerides to representative diglycerides, 195–196
  determining composition of derived diglycerides, 203–205, 271–280
    of triglyceride mixtures, 200–205, 278–279
  diglyceride kinase, 193, 197–198
  diglycerides, 200, 203–205, 271–280
  historical development, 10, 15, 188
  hydrolysis of phospholipid with phospholipase A, 198–199
  Lands method, 192–195
  need for more rapid method, 281
  phosphorylation of diglycerides, 196–198
    chemical synthesis, 196–197
    enzymatic synthesis, 197–198

used in combination with
  deacylation with *Geotrichum candidum* lipase, 195–196, 204
  with Grignard reagents, 189–196, 202, 271–278
  with milk lipase, 195
  with pancreatic lipase, 189–196, 202, 271–280
  liquid–liquid partition chromatography, 271–280
  silver ion adsorption chromatography, 202–203, 260, 271–280
Stillingia tallow triglycerides, *see Sapium sebiferum* triglycerides, fruit coat
Sunflower seed oil triglycerides
  comparison of experimental and calculated triglyceride compositions, 255–256
  deacylation with pancreatic lipase, 256, 271
  silver ion adsorption chromatography, 255–256, 271

### T

Tetracosenoic acid, *see also* Triglycerides, of $n$-$C_{19}$–$n$-$C_{24}$ acids *and* Triglycerides, separation by unsaturation
  positional distribution in natural triglycerides, 236
Thermal-gradient chromatography, *see* Chromatography, thermal-gradient
Thin-layer chromatography, *see* listings under specific type of chromatography, e.g., Chromatography, liquid-liquid partition, thin-layer chromatography *and* Chromatography, silver ion adsorption, thin-layer chromatography
*Trans* fatty acids, *see* Triglycerides, of *trans* acids
Triglycerides
  of acetic acid
    aluminum oxide adsorption chromatography, 161
    critical solution temperature, 231
    deacylation with pancreatic lipase, 176, 178
    density, 231
    dielectric constant, 231
    gas–liquid chromatography, 104, 130–131, 137
    infrared spectroscopy, 219–220
    liquid–liquid partition chromatography, 103
    mass spectrometry, 208–209
    nuclear magnetic resonance, 223
    permanganate oxidation, 227–229
    positional distribution in natural triglycerides, 243
    preparation from diglycerides, 47–48
    refractive index, 231
    rotation of polarized light, 206, 226–229
    separation by molecular weight, 130, 137, 151–155
      of acyl group positional isomers, 66, 130, 153–155
      by unsaturation, 65–66
    silicic acid adsorption chromatography, 151–155
    silver ion adsorption chromatography, 65–66
    sound velocity, 231
  of acetylenic acids, silver ion adsorption chromatography, 64
  acyl group chain length isomers (such as PPP vs. MPSt)
    gas–liquid chromatography, 131
    mass spectrometry, 208
    silver ion adsorption chromatography, 63
  acyl group positional isomers ($sn$-1- vs. $sn$-2- vs. $sn$-3-)
    acetodiglycerides, 66, 130, 153–155
    analytical scheme for triacid triglycerides, 201–205, 278–279
    critical solution temperature, 231
    deacylation with *Geotrichum candidum* lipase, 204
      with Grignard reagents, 169–172, 270
      with milk lipase, 184
      with pancreatic lipase, 179–182, 270–271, 279
    density, 231
    dielectric constant, 231
    gas–liquid chromatography, 130–131
    infrared spectroscopy, 220

Triglycerides (*continued*)
  liquid–liquid partition chromatography, 89, 279
  mass spectrometry, 208
  melting point, 215
  nomenclature, 4–5, 7
  nuclear magnetic resonance, 223–224
  piezoelectric effect, 230
  refractive index, 231
  rotation of polarized light, 206, 224–229
  silicic acid adsorption chromatography, 153–155
  silver ion adsorption chromatography, 58, 63, 66
  sound velocity, 231
  stereospecific analysis, 199–205, 270–271
  X-ray diffraction, 221–222
  of alicyclic acids
    cyclopentene acids, 64, 129–130
    cyclopropane acids, 129
    cyclopropene acids, 64, 176
    deacylation with pancreatic lipase, 176
    gas–liquid chromatography, 129–130
    silver ion adsorption chromatography, 64
  of allenic acids
    positional distribution of allenic acids in natural triglycerides, 241
    silver ion adsorption chromatography, 64
  asymmetric, *see* Triglycerides, enantiomorphic
  of branched-chain acids
    2-alkyl and 3-alkyl acids inhibit deacylation with pancreatic lipase, 177
    gas–liquid chromatography, 129
    isovaleric acid, 129, 151, 226
    phytanic acid, 16, 151, 153
    pivalic acid, 226
    Refsum's disease correlated with phytanoyl triglycerides, 16
    rotation of polarized light, 226
    silicic acid adsorption chromatography, 151, 153
  brominated
    debromination, 45
    fractional crystallization, 148–149
    liquid–liquid partition chromatography, 98–100
    preparation, 44–45
    silicic acid adsorption chromatography, 158
  of $n$-$C_4$–$n$-$C_{11}$ acids
    aluminum oxide adsorption chromatography, 161
    charcoal adsorption chromatography, 162
    countercurrent distribution, 84
    critical solution temperature, 231
    deacylation with Grignard reagents, 172, 271–272, 274, 276, 280
      with milk lipase, 184
      with pancreatic lipase, 176–181, 271–272, 274, 276
    density, 231–232
    dielectric consant, 231–232
    distillation, 165–166
    fractional crystallization, 12
    gas–liquid chromatography, 104, 115–116, 121–122, 124–127, 130–131, 134–135
    liquid–liquid partition chromatography, 84, 271, 274, 276, 280
    mass spectrometry, 208–209, 211
    permeation chromatography, 163
    positional distribution in natural triglycerides, 241, 243, 247
    refractive index, 231–232
    rotation of polarized light, 225–227
    silicic acid adsorption chromatography, 151–152, 154
    silver ion adsorption chromatography, 57–58, 271, 274, 276, 280
    sound velocity, 231–232
    stereospecific analysis, 200, 272, 274, 276, 280
  of $n$-$C_{12}$–$n$-$C_{18}$ acids
    aluminum oxide adsorption chromatography, 161–162
    charcoal adsorption chromatography, 162
    countercurrent distribution, 81–86, 93–96
    critical solution temperature, 231–232

Triglycerides (*continued*)
  deacylation with *Geotrichum candidum* lipase, 185–187
    with Grignard reagents, 168–172, 270–280
    with milk lipase, 183–184
    with pancreatic lipase, 173–183, 270–280
  density, 231–232
  dielectric constant, 231–232
  differential cooling curves, 216–218
  Florisil adsorption chromatography, 19–20, 160–161
  fractional crystallization, 139–149
  gas–liquid chromatography, 104–137, 266–269, 271–280
  infrared spectroscopy, 218–221
  ion exchange chromatography, 162–163
  liquid–liquid partition chromatography, 67–102, 266–269, 271–280
  mass spectrometry, 206–213, 266–269
  melting point, 213–215
  nuclear magnetic resonance, 222–224
  paper chromatography, 162
  permeation chromatography, 163
  piezoelectric effect, 230
  positional distribution in natural triglycerides, 234–248
  refractive index, 231–232
  rotation of polarized light, 224–229
  silicic acid adsorption chromatography, 150–160
  silver ion adsorption chromatography, 49–66, 266–269, 271–280
  sound velocity, 231–232
  stereospecific analysis, 188–205, 270–280
  thermal gradient chromatography, 164–165
  X-ray diffraction, 221–222
  zeolite adsorption chromatography, 24
  of $n$-$C_{19}$–$n$-$C_{24}$ acids
    deacylation with Grignard reagents, 169
    with pancreatic lipase, 177, 238
    gas–liquid chromatography, 112, 114, 120, 122–128
  liquid–liquid partition chromatography, 72, 268
  mass spectrometry, 207
  permeation chromatography, 163
  positional distribution in natural triglycerides, 191, 235–238, 243–247
  silicic acid adsorption chromatography, 151
  silver ion adsorption chromatography, 57–59, 62, 268
  stereospecific analysis, 191, 195, 235–236, 243–247
  *cis–trans* isomerized
    fractional crystallization, 149
    preparation, 46
  of conjugated polyunsaturated acids
    rotation of polarized light, 226, 229
    silver ion adsorption chromatography, 64
  of dicarboxylic acids, *see* Triglycerides, oxidized
  double bond geometrical isomers (*cis* vs. *trans*)
    critical solution temperature, 231
    density, 231
    dielectric constant, 231
    elaidic acid, 62–64, 100–101, 177, 186, 231
    liquid–liquid partition chromatography, 89, 100–101
    refractive index, 231
    silver ion adsorption chromatography, 62–64
    sound velocity, 231
  double bond positional isomers (such as 18:1ω9 vs. 18:1ω7 or **0**02 vs. **0**11)
    analytical scheme for triglyceride mixtures, 278–279
    critical solution temperature, 231
    deacylation with *Geotrichum candidum* lipase, 186
      with pancreatic lipase, 177, 279
    liquid–liquid partition chromatography, 89, 279
    mass spectrometry, 209–210
    permanganate oxidation, 278–279
    petroselinic acid, 61, 63, 186, 231, 278

Triglycerides (*continued*)
  silver ion adsorption chromatography, 57–64, 66
  stereospecific analysis, 279
  vaccinic acid, 62, 186, 278–279
  enantiomorphic
    melting point, 215
    piezoelectric effect, 230
    rotation of polarized light, 206, 224–229
    seed triglycerides, 227, 255, 257
    silver ion adsorption chromatography, 63
    stereospecific analysis, 199–205
    X-ray diffraction, 221–222
  epoxy
    countercurrent distribution, 84
    deacylation with pancreatic lipase, 176–177, 243
    dioxolane ring formation, 131–132
    fractional crystallization, 149
    gas–liquid chromatography, 131–132
    liquid–liquid partition chromatography, 84
    positional distribution in natural triglycerides, 243
    preparation from nonepoxy triglycerides, 43–44
    silicic acid adsorption chromatography, 154, 156
    stereospecific analysis, 200
  estolide
    definition, 8
    estolide ester cleavage, 47
    mass spectrometry, 209
    positional distribution in natural triglycerides, 241
    silicic acid adsorption chromatography, 158–159
  hydrogenated
    fractional crystallization, 148
    gas–liquid chromatography, 123, 126, 128
    liquid–liquid partition chromatography, 101–102
    preparation, 38–39
    silver ion adsorption chromatography, 63
  hydroxy
    acetylation, 47–48
    countercurrent distribution, 101–102
    deacylation with pancreatic lipase, 176–177
    estolide ester cleavage, 47
    gas–liquid chromatography, 131
    hydrazone ester derivative reaction, 48
    liquid–liquid partition chromatography, 84, 101–102
    paper chromatography, 162
    silicic acid adsorption chromatography, 154–155
    stereospecific analysis, 200
    trimethylsilylation, 48
  isolation of, *see* Isolation of triglycerides
  isomeric, *see* Triglycerides, acyl group chain length isomers; Triglycerides, acyl group positional isomers; Triglycerides, of alicyclic acids; Triglycerides, of branched-chain acids; Triglycerides, double bond geometrical isomers; Triglycerides, double bond positional isomers; *and* Triglycerides, of *trans* acids
  keto
    deacylation with pancreatic lipase, 176–177
    hydrazone derivative reaction, 48
    silicic acid adsorption chromatography, 155
    stereospecific analysis, 200
  measurement of total, *see* Measurement of total triglycerides
  mercaptoacetic acid addition products
    ion-exchange chromatography, 163
    preparation, 37, 46
  mercurated
    aluminum oxide adsorption chromatography, 161–162
    Florisil adsorption chromatography, 160–161
    liquid–liquid partition chromatography, 100
    preparation, 45–46
    regeneration of original compound, 46
    silicic acid adsorption chromatography, 158–160

Triglycerides (*continued*)
  of odd chain length acids
    gas–liquid chromatography, 126, 128
    liquid–liquid partition chromatography, 88
  oxidized
    allyl esters of, 42, 64–65
    aluminum oxide adsorption chromatography, 162
    deacylation with $K_2CO_3$, 172–173
      with pancreatic lipase, 279
    Florisil adsorption chromatography, 160
    fractional crystallization, 147–148
    gas–liquid chromatography, 132–133
    ion-exchange chromatography, 162–163
    liquid–liquid partition chromatography, 78, 84, 100, 279
    methyl esters of, 42
    preparation, 39–42
    rotation of polarized light by derived diglycerides, 227–229
    silicic acid adsorption chromatography, 22, 157, 279
    silver ion adsorption chromatography, 64–65
  ozonized
    preparation, 42–43
    reaction by-products, 42–43
    reduction to "aldehyde cores," 43
    silicic acid adsorption chromatography, 156–158
  physical properties of pure compounds
    critical solution temperature, 231–232
    density, 231–232
    dielectric constant, 231–232
    differential cooling curves, 216–218
    infrared spectroscopy, 218–221
    mass spectrometry, 206–209
    melting point, 213–215
    nuclear magnetic resonance, 222–224
    piezoelectric effect, 230
    polymorphism, 213–214, 220–222
    refractive index, 231–232
    rotation of polarized light, 224–229
    sound velocity, 231–232
    X-ray diffraction, 214, 221–222
  radioactive
    gas–liquid chromatography, 135–137
    stereospecific analysis, 195
  separation by molecular weight
    acetodiglycerides, 130, 137, 151–152
    aluminum oxide adsorption chromatography, 161
    charcoal adsorption chromatography, 162
    distillation, 165–166
    Florisil adsorption chromatography, 20
    gas–liquid chromatography, 126–128, 266–269
    liquid–liquid partition chromatography, 86, 276
    mass spectrometry, 209–211, 266–269
    permeation chromatography, 163
    silicic acid adsorption chromatography, 150–153
    thermal gradient chromatography, 165
  separation by number of ester groups, silicic acid adsorption chromatography, 157–159
  separation by number of saturated acyl groups, fractional crystallization, 142–146
  separation by partition number, liquid–liquid partition chromatography, 86–96, 266–269
  separation by unsaturation
    acetodiglycerides, 65–66
    aluminum oxide adsorption chromatography, 161–162
    charcoal adsorption chromatography, 162
    Florisil adsorption chromatography, 20, 160–161
    fractional crystallization, 146–147
    gas–liquid chromatography, 128–129
    liquid–liquid partition chromatography, 95–100, 266–269
    mass spectrometry, 209–211, 266–269
    silicic acid adsorption chromatography, 153, 158–160
    silver ion adsorption chromatography, 57–66, 266–269

Triglycerides (*continued*)
  thermal gradient chromatography, 165
  total, *see* Measurement of total triglycerides
  of *trans* acids
    critical solution temperature, 231
    deacylation with *Geotrichum candidum* lipase, 186
      with pancreatic lipase, 177
    density, 231
    dielectric constant, 231
    elaidic acid, 62–64, 100–101, 177, 186, 231
    liquid–liquid partition chromatography, 89, 100–101
    refractive index, 231
    silver ion adsorption chromatography, 62–64
    sound velocity, 231
*Trilobium confusum* beetle, triglycerides produce aggregation reaction, 16
Trimethylsilyl ether derivatives, *see also* Diglycerides, of *n*-acids; *and* Triglycerides, hydroxy
  preparation, 48
Tuna muscle triglycerides, gas–liquid chromatography, 125, 128
Turkey fat triglycerides, deacylation with pancreatic lipase, 179
Turtle triglycerides, positional distribution of fatty acids, 243–247

## U

Ucuhuba seed triglycerides
  gas–liquid chromatography, 269
  silver ion adsorption chromatography, 269
Unsaturated fatty acids, *see also* listings under individual acids
  positional distribution in natural triglycerides, 235–248
Unsolved problems of triglyceride analysis, 280–281

## V

Vaccinic acid, *see* Triglycerides, of $n$-$C_{12}$–$n$-$C_{18}$ acids; Triglycerides, double bond positional isomers; *and* Triglycerides, separation by unsaturation
Vernolic acid, *see also* Triglycerides, epoxy
  positional distribution in natural triglycerides, 243
*Virola surinamensis* seed triglycerides, *see* Ucuhuba seed triglycerides

## W

Watercress seed triglycerides, gas–liquid chromatography, 125
Whale triglycerides
  deacylation with Grignard reagent, 169
    with pancreatic lipase, 177
  gas–liquid chromatography, 129
  liquid–liquid partition chromatography, 98, 278, 281
  need for better separation techniques, 280–281
  positional distribution of fatty acids, 243–247
  silicic acid adsorption chromatography, 151
  silver ion adsorption chromatography, 59, 278, 281
  stereospecific analysis, 195
Wheat seed triglycerides
  countercurrent distribution, 94–96
  liquid–liquid partition chromatography, 94–96
  positional distribution of fatty acids, 257
  stereospecific analysis, 257

## X

X-ray diffraction, 221–222
  enantiomorphic triglycerides, 221–222
  identification of acyl group positional isomers, 221–222
  polymorphism, 214, 221–222

## Y

Yeast triglycerides, isolation by silicic acid adsorption chromatography, 22 23

THE LIBRARY